Concepts in Syngas Manufacture

CATALYTIC SCIENCE SERIES

Series Editor: Graham J. Hutchings *(Cardiff University)*

Published

Vol. 1   Environmental Catalysis
         edited by F. J. J. G. Janssen and R. A. van Santen

Vol. 2   Catalysis by Ceria and Related Materials
         edited by A. Trovarelli

Vol. 3   Zeolites for Cleaner Technologies
         edited by M. Guisnet and J.-P. Gilson

Vol. 4   Isotopes in Heterogeneous Catalysis
         edited by Justin S. J. Hargreaves, S. D. Jackson and G. Webb

Vol. 5   Supported Metals in Catalysis
         edited by J. A. Anderson and M. F. García

Vol. 6   Catalysis by Gold
         edited by G. C. Bond, C. Louis and D. T. Thompson

Vol. 7   Combinatorial Development of Solid Catalytic Materials:
         Design of High-Throughput Experiments, Data Analysis,
         Data Mining
         edited by M. Baerns and M. Holeňa

Vol. 8   Petrochemical Economics: Technology Selection in a Carbon Constrained World
         by D. Seddon

Vol. 9   Deactivation and Regeneration of Zeolite Catalysts
         edited by M. Guisnet and F. R. Ribeiro

Vol. 10 Concepts in Syngas Manufacture
         by J. Rostrup-Nielsen and L. J. Christiansen

CATALYTIC SCIENCE SERIES — VOL. 10
Series Editor: Graham J. Hutchings

Concepts in Syngas Manufacture

Jens Rostrup-Nielsen
Lars J. Christiansen
Haldor Topsoe A/S, Denmark

Imperial College Press

Published by

Imperial College Press
57 Shelton Street
Covent Garden
London WC2H 9HE

Distributed by

World Scientific Publishing Co. Pte. Ltd.
5 Toh Tuck Link, Singapore 596224
USA office: 27 Warren Street, Suite 401-402, Hackensack, NJ 07601
UK office: 57 Shelton Street, Covent Garden, London WC2H 9HE

British Library Cataloguing-in-Publication Data
A catalogue record for this book is available from the British Library.

CONCEPTS IN SYNGAS MANUFACTURE
Catalytic Science Series — Vol. 10

Copyright © 2011 by Imperial College Press

All rights reserved. This book, or parts thereof, may not be reproduced in any form or by any means, electronic or mechanical, including photocopying, recording or any information storage and retrieval system now known or to be invented, without written permission from the Publisher.

For photocopying of material in this volume, please pay a copying fee through the Copyright Clearance Center, Inc., 222 Rosewood Drive, Danvers, MA 01923, USA. In this case permission to photocopy is not required from the publisher.

ISBN-13 978-1-84816-567-0
ISBN-10 1-84816-567-6

Printed in Singapore.

Preface

"...a thin film of carbonaceous matter destroys the igniting power of platinum, and a slight coating of sulphuret deprives palladium of this property..."

Davy Humphrey (1817) [141]

The aim of this book is twofold as reflected by the two parts of the book. Part One deals with a broad introduction to the routes to synthesis gas and to its applications. As the synthesis gas technologies involve big exchanges of energy, this part includes a description of thermodynamic methods necessary for process analysis and optimisation. Part Two deals with an in-depth analysis of the steam reforming process which is the mostly applied method for the manufacture of synthesis gas. The reforming process is a complex coupling of catalysis and heat transfer. Therefore this part describes the modelling of the process and an analysis of the secondary phenomena dominating the catalytic reactions. The book ends with a summary of the science of the steam reforming reaction. In this way the procedure has been "peeling the onion", starting with the general aspects of the subject and approaching the scientific understanding.

The development of the synthesis gas technologies is not a result of a linear process starting with fundamental research. It is an example of parallel efforts in industry and academia characterised by fruitful interaction. It is the hope that the book may serve as a link between the scientist in the research laboratory and the operator of the industrial plant, as did a previous treatise, "Catalytic Steam Reforming", published in 1984 [389].

The authors have worked on the subject for a number of decades. We have had the privilege to work with a multiple approach. In parallel to the development work in the laboratory, at pilot units and industrial plants we have been engaged in fundamental and explorative research (JRN) and reactor modelling (LJC).

Professor Graham Hutchings encouraged us to contribute this book to Catalysis Science Series published by Imperial College Press. JRN acknowledges his contact with British catalysis groups as having started by reading Griffith and Marsh, *Contact Catalysis* [215], continued by discussions with the group at Solihull in the 1960s, and carried further by the activities at Imperial College via his friend David Trimm. This gave JRN the pleasure to become acquainted with "competitors" from ICI.

We thank the Haldor Topsøe company for permission to publish the book and we are thankful for the support from our colleagues. In particular we give thanks for valuable input from Kim Aasberg-Petersen, Jens Sehested, John Bøgild Hansen and Mette Stenseng. Professor Jens Nørskov (DTU, Lyngby) has advised on the scientific aspects and Professor Bryan Haynes (Univ. of Sydney) on heat transfer problems in microreactors. Ms Susanne Mainz, Ms Birthe Bruun Nielsen, Ms Sussie Nygaard, and Ms Birgit Rossil showed great patience in preparing the manuscript.

Jens Rostrup-Nielsen and Lars J. Christiansen
Lyngby, Denmark, October 2010

Contents

Preface v

Part I Basic Principles

1 Routes to Syngas 3
   1.1 General trends 3
      1.1.1 Towards focus and sustainability 3
      1.1.2 Direct or indirect conversion 10
   1.2 Manufacture by steam reforming of hydrocarbons 14
      1.2.1 Reactions and thermodynamics 14
      1.2.2 Product gas composition 26
      1.2.3 Thermodynamics of higher hydrocarbons 30
      1.2.4 The tubular reformer 31
      1.2.5 Carbon formation. Higher hydrocarbons 34
      1.2.6 Non-tubular reforming 36
   1.3 Other manufacture routes 38
      1.3.1 Partial oxidation 38
      1.3.2 Autothermal reforming 41
      1.3.3 Catalytic partial oxidation 43
      1.3.4 Air-blown technologies and membranes 48
      1.3.5 Choice of technology 49
   1.4 Other feedstocks 51
      1.4.1 Alcohols, oxygenates 51
      1.4.2 Coal, gasification 55
      1.4.3 Biomass 63
   1.5 Gas treatment 64
      1.5.1 Purification 64
      1.5.2 Water gas shift 67
      1.5.3 Acid gas removal 70

2 Syngas Applications 73
   2.1 Thermodynamic framework for syngas processes 73
      2.1.1 Syngas properties 74
      2.1.2 Synthesis process properties 79
      2.1.3 Process analysis 80
   2.2 Hydrogen 85
      2.2.1 Routes to hydrogen 85
      2.2.2 Hydrogen by steam reforming of hydrocarbons 87

        2.2.3 The steam export problem 92
        2.2.4 Membrane reforming 94
        2.2.5 Hydrogen via catalytic partial oxidation (CPO) 95
    2.3 Fuel cells 96
        2.3.1 Fuel processing system 96
        2.3.2 Internal reforming 99
        2.3.3 Process schemes for SOFC 104
    2.4 CO rich gases 105
        2.4.1 Town gas 105
        2.4.2 Oxogas 106
        2.4.3 Reducing gas 110
    2.5 Ammonia 112
    2.6 Methanol and synfuels 117
        2.6.1 Methanol as intermediate 117
        2.6.2 Methanol plant 118
        2.6.3 Methanol via gasification 123
        2.6.4 Combined syntheses and co-production 124
        2.6.5 Fischer–Tropsch synthesis 127
        2.6.6 SNG 134
    2.7 Chemical recuperation 138

Part II Steam Reforming Technology

3   Technology of Steam Reforming 143
    3.1 Early developments 143
    3.2 Steam reforming reactors 146
        3.2.1 Role of catalyst 146
        3.2.2 The tubular reformer 149
        3.2.3 Scale-up of steam reforming technology 153
        3.2.4 Plant measurements 154
        3.2.5 Reformer temperature measurements 157
    3.3 Modelling of steam reforming reactors 159
        3.3.1 Two-dimensional reactor model 162
        3.3.2 Heat transfer in the two-dimensional model 168
        3.3.3 Heat transfer parameters in syngas units 171
        3.3.4 Pressure drop 176
        3.3.5 Convective reformers 178
        3.3.6 Tubular reformer furnace chamber 181
        3.3.7 Tubular reforming limits of operation 187
        3.3.8 Micro-scale steam reforming reactors 189
    3.4 Modelling of the catalyst particle 191
        3.4.1 Catalyst particle model 192
        3.4.2 Effective diffusion coefficients 195

		3.4.3 Simulation of a hydrogen plant reformer 197
	3.5	Reaction kinetics 199
		3.5.1 Industrial rates and the scale-down problem 199
		3.5.2 Intrinsic kinetics. Steam reforming of methane 204
		3.5.3 Steam reforming of higher hydrocarbons 210
		3.5.4 CO_2 reforming 212
4	Catalyst Properties and Activity 213	
	4.1	Catalyst structure and stability 213
		4.1.1 Reactions with the support 213
		4.1.2 Activation and nickel surface area 216
	4.2	Nickel surface area 219
		4.2.1 Measurement of nickel surface area 219
		4.2.2 Nickel surface area and catalyst preparation 224
		4.2.3 Sintering 224
	4.3	Catalyst activity 227
		4.3.1 Group VIII metals 227
		4.3.2 Non-metal catalysts 228
		4.3.3 Thermal reactions – catalytic steam cracking 230
5	Carbon and Sulphur 233	
	5.1	Secondary phenomena 233
	5.2	Carbon formation 233
		5.2.1 Routes to carbon 233
		5.2.2 Carbon from reversible reactions 241
		5.2.3 Principle of equilibrated gas 247
		5.2.4 Principle of actual gas and steady-state equilibrium 252
	5.3	Steam reforming of higher hydrocarbons 257
		5.3.1 Whisker carbon in tubular reformer 257
		5.3.2 Catalyst promotion 260
		5.3.3 "Gum formation" in prereformers 264
		5.3.4 Carbon from pyrolysis 270
		5.3.5 Regeneration of coked catalyst 273
	5.4	Sulphur poisoning of reforming reactions 275
		5.4.1 Chemisorption of hydrogen sulphide 275
		5.4.2 Chemisorption equilibrium 277
		5.4.3 Dynamics of sulphur poisoning 281
		5.4.4 Regeneration for sulphur 282
		5.4.5 Impact of sulphur on reforming reactions 285
	5.5	Sulphur passivated reforming 288
	5.6	Other poisons 293
6	Catalysis of Steam Reforming 295	
	6.1	Historical perspective 295
	6.2	The role of step sites 298

6.3   Geometric or electronic effects..305
  6.4   Metal activity. Micro-kinetics 307
  6.5   The parallel approach 311
Appendix 1 Enthalpy of formation...313
Appendix 2 Chemical equilibrium constants..................................317
Notation and Abbreviations...323
References...331
Author index...357
Subject index..369

Part I
Basic Principles

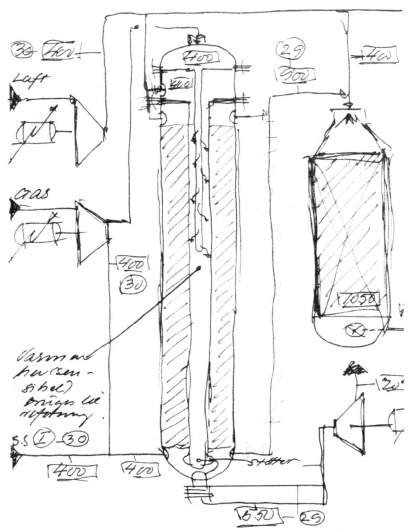

Sketch by Haldor Topsøe, 20 Dec. 1989

1 Routes to Syngas

1.1 General trends

1.1.1 Towards focus and sustainability

Synthesis gas (syngas) is a mixture of hydrogen, carbon monoxide and carbon dioxide. It may also contain nitrogen as applied for the ammonia synthesis. Syngas is a key intermediate in the chemical industry. It is used in a number of highly selective syntheses of a wide range of chemicals and fuels, and as a source of pure hydrogen and carbon monoxide. Syngas is playing an increasing role in energy conversion [418].

Synthesis gas can be produced from almost any carbon source ranging from natural gas and oil products to coal and biomass by oxidation with steam and oxygen. Hence it represents a key for creating flexibility for the chemical industry and for the manufacture of synthetic fuels (synfuels).

Figure 1.1 Conversion via syngas.

The conversion via syngas results in products plus heat (Figure 1.1). In most plants, the heat is utilised for running the plant. As an alternative, the heat may be exported, but that is not always necessary.

The present use of syngas is primarily for the manufacture of ammonia (in 2006, 124 million tonnes per year) and of methanol (in 2005, 33 million tonnes per year), followed by the use of pure hydrogen for hydrotreating in refineries as shown in Table 1.1.

The main commodity products based on natural gas are shown in Table 1.1 [420]. It is evident that the chemical conversion of natural gas (approximately $7 \cdot 10^9$ GJ/y) is marginal to the total natural gas production ($3.07 \cdot 10^{12}$ Nm^3/y [78] or $1.17 \cdot 10^{11}$ GJ/y assuming a lower heating value (LHV) equal to 38 MJ/Nm^3). Recent trends in the use of syngas are dominated by the conversion of inexpensive remote natural gas into liquid fuels ("gas to liquids" or "GTL") and by a possible role in a future "hydrogen economy" mainly associated with the use of fuel cells. These trends imply, on the one hand, the scale-up to large-scale GTL plants (more than 500,000 Nm^3 syngas/h) and, on the other hand, the scale-down to small, compact syngas units for fuel cells (5–100 Nm^3 syngas or H$_2$/h). These forecasts create new challenges for the technology and for the catalysis.

Table 1.1 Main chemical products based on natural gas.

Product	Yearly prod. (mil. t/y)	Energy consumpt. (GJ/t)	Thermal LHV Practical (%)	Efficiency Ideal (%)	CO$_2$ (t/t)	Main technology
Ammonia	124	29	65	89	16^a	Syngas/synthesis
Ethylene	75	15^b	62	93	0.65	Steam cracking C$_2$H$_6$
Propylene	53	-	-			Steam cracking C$_2$H$_6$
Methanol	32	28	72	84	0.28	Syngas/synthesis
Hydrogen	20	12.6	84^c	92	0.9	Steam reforming
Synfuels	18^d	67	60	78	1.18	Syngas/synthesis

a) incl. CO$_2$ converted into urea
b) data kindly provided by F. Dautzenberg, ABB Lummus, 2005
c) CH$_4$ used for reaction heat; no steam export
d) excl. 3 million tonnes per year under construction

The data in Table 1.1 show that the practical efficiencies for natural gas conversion into products are approximately 80% of the ideal values expressed as:

$$\eta_{ideal} = \frac{\text{LHV product/mol}}{\text{LHV methane/mol}} \qquad (1.1)$$

For endothermic reactions (ethylene, hydrogen), the LHV of the fuel providing the reaction heat should be added to the nominator.

The world energy production is dominated by fossil fuels as the main energy source. It amounted to 88% in 2007 with oil responsible for 37%

[78]. The energy consumption is growing fast in Asia, and China has become the world's second largest consumer of oil, after the USA. The proved reserves of oil are concentrated in the Middle East (61%) and those of natural gas are also in the Middle East (41%), followed by Russia (23%) [78]. Coal is more evenly distributed between the continents.

Apart from the large reserves (Middle East, Russia), natural gas is present as associated gas in oil fields. However, as many fields are far from the marketplace and often off-shore, the gas there is called remote gas or stranded gas [102]. Part of the associated gas is reinjected to enhance the oil recovery, but unfortunately still a significant fraction is flared for convenience. The flared gas amounts to close to 5% of the total natural gas production (corresponding to about 1% of total world CO_2 production from fossil fuels) [263] [420].

So far, the proven reserves for oil have followed the increase in production as expressed by the reserves/production ratio (R/P ratio) staying at about 40 for oil over the last 20 years; however, at a steadily increasing cost of exploration and production. A big fraction of the reserves is present as oil sand (tar sand) and other non-conventional sources under active development [78]. This means that at the present world production, the oil reserves known today will be used up within about 40 years. This figure should be considered with care. It does not include reserves still to be found and it does not include the changes in consumption (for instance the growth in Asia). Furthermore, the R/P ratio for oil varies from region to region, being above 80 in the Middle East and below 20 in North America.

The R/P ratio (2007) for natural gas is about 60 and 122 for coal [78]. The total R/P for fossil fuels (based on oil equivalent) is less than 100 years. These figures emphasise the need for flexibility in the energy network and the need for alternative fuels. Oil is the most versatile of the fossil fuels with high energy density and it is easily transported.

The power industry is very flexible to feedstocks and it is feasible to transport coal over long distances to big centralised power plants close to deep water harbours. Natural gas is transported to the marketplace in pipelines over still longer distances or as liquified natural gas (LNG).

The automotive sector represents a special challenge as the energy conversion is strongly decentralised. So far oil-derived products have been the solution, but in view of the limited reserves of oil, a number of alternative fuels are being considered, such as liquefied petroleum gas (LPG), natural gas, methanol, dimethylether (DME), ethanol, biodiesel, synfuels and hydrogen. Biofuels represent a "sustainable" response to liquid fuels. It may be based on ethanol and biodiesel derived from conventional agricultural products or from synfuels via gasification of biomass. The alternative fuels may be blended with conventional fuels or used directly in internal combustion engines (ICE) or fuel cells. In Western Europe alternative fuels may amount to 20% of energy sources by 2020.

Globalisation has caused companies to concentrate on core business and critical mass. It has resulted in a restructure of the chemical industry into two types of focused companies [190]: the molecule suppliers (commodities and fine chemicals) and the problem solvers (functional chemicals like additives and pharmaceuticals). Each type has its own characteristics as reflected by the role of the catalyst [418].

The most important parameter for large-volume chemicals is production costs (variable and fixed costs). The variable costs are related to the feed costs, the use of energy, process selectivity and environmental costs.

Four trends have characterised plants for commodity chemicals:

- Location of cheap raw materials;
- Economy of scale;
- More integrated plants; and
- CO_2 footprint (tonnes CO_2 per tonne product).

Plants are moved to locations where raw materials are cheap. As illustrated in Figure 1.2, the ammonia production is hardly feasible at natural gas prices typical for Europe and USA (3–4 USD/GJ with high seasonal variations) [420]. As a result, new plants for commodity chemicals are built at locations (Middle East, Trinidad, Nigeria, West Australia...) with low natural gas prices (0.5–1 USD/GJ). It means that the use of natural gas as feedstock may not be feasible where there is a big market for natural gas as fuel.

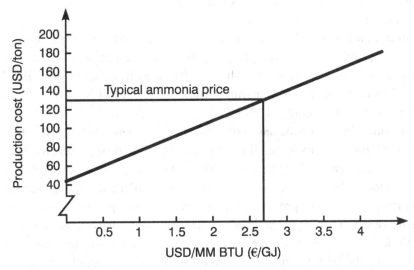

Figure 1.2 Ammonia production costs [420]. Reproduced with the permission of Springer.

Plants have become larger to take advantage of the economy of scale. The economy of scale can be expressed by:

$$\text{Cost}_1 = \text{Cost}_2 \left(\frac{\text{capacity}_1}{\text{capacity}_2} \right)^n \quad (1.2)$$

n typically varies between 0.6–0.9.

The economy of scale means choice of different technologies as they may be characterised by different values of n.

Today, ammonia plants are built with capacities up to more than 3000 metric tons per day (MTPD) and methanol plants are being considered at capacities of 10000 MTPD. This corresponds to the size of synthetic fuel plants based on FT synthesis (35,000 bpd). At the same time, as plants become bigger, there is a trend to minituarise chemical process plants and take advantage of mass production, the economy of numbers competing with the economy of scale. This is one of the key issues in the "hydrogen economy" and the application of fuel cells. Micro-structured process equipment components such as heat exchangers, and new reactor concepts are becoming available. Plants have also become more integrated to minimise energy consumption.

It can be shown that the plant costs for a variety of processes correlate with the energy transfer (heat transfer, compression) within the process scheme [289]. As an example, the energy consumption of ammonia production has decreased over the last 50 years from about 40 GJ/t to 29 GJ/t corresponding to a thermal efficiency (LHV) of 65% or 73% of the theoretical minimum [169] [420].

Commodity plants depend on steady improvement and sophistication of the technology. Even small improvements of the process scheme may show short payback times. On the other hand, the uncertainties associated with new technology may easily outbalance the economic advantage of a new process. Improvement of one process step might easily result in less favourable performance of another process step. The high degree of integration means that the weakest part of the chain may determine the performance of the entire plant. As an example, there is a need for more coke-resistant catalysts and often deactivation phenomena determine the process layout and the optimum process conditions to be applied [404]. It is evident that catalyst life, i.e. on-stream factor, is crucial for large-scale commodity plants in contrast to batch-wise manufacture of fine chemicals.

A few days' production stop because of a catalyst failure may be crucial for the plant economy. It means that secondary phenomena such as catalyst deactivation are important issues. For large-scale operation, economic arguments will limit the minimum space time yield to approximately 0.1 tonne product/m^3 at a typical catalyst life of 5 years [289]. This corresponds to a catalyst consumption of less than 0.2 kg cat/t product. For ammonia synthesis a typical figure is 0.03 kg cat/t NH_3.

These risks mean that it has become more expensive to develop new process technology. New technology must be demonstrated to a larger extent – not only the basic principles, but also the solutions to a series of secondary problems [400] [418].

Many well-established processes are approaching their theoretically achievable efficiency, selectivity, etc. (refer to Table 1.1), but new challenges have been introduced by objectives for sustainable growth formulated by society. This has not only led to the introduction of new products, but also necessitated the development of new processes. Environmental challenges represent major room for breakthroughs in the catalytic process industry.

For any process scheme, it is essential at an early stage to establish the overall mass balance and to estimate the $\Delta \mathcal{P}$ as simply being the difference between the price of products and the price of feedstocks [418]. Hence, there has been a trend to develop processes using cheaper raw materials. The gain in $\Delta \mathcal{P}$ could, however, be lost by lower selectivity or higher investments. Selectivity is crucial to achieving a high $\Delta \mathcal{P}$. Low selectivity and conversion per pass result in low concentrations in process streams and hence more expensive separation systems.

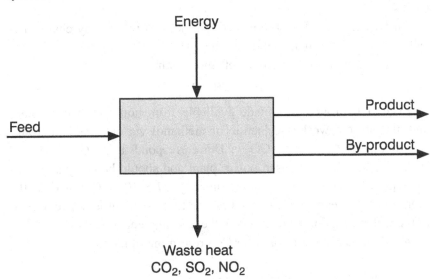

Figure 1.3 Simplified mass balance.

It may be argued that energy efficiency is of less importance when natural gas is cheap, but high energy efficiency means small feed pretreat units and reduced requirements for utilities and hence less investments.

Moreover, high efficiency means less CO_2 production. As illustrated in Figure 1.3, the $\Delta \mathcal{P}$ calculation should consider also the energy consumption and the by-products which may easily have a negative value. This may be expressed by the so-called E-factor [458] expressing the amount of by-product produced per kg of product.

The emissions may have great negative value. This may be reflected by the costs of carbon capture and storage (CCS) (refer to Section 1.4.2). The CO_2 emission expressed as a C-factor [106] [420] (tonnes CO_2 per tonne product, refer to Table 1.1) may become an important process parameter in the future. CO_2 emissions are often directly related to the energy consumption of the process.

As an example, a reduction of the energy required to produce ammonia from natural gas of 1 GJ/t means a reduction of CO_2 emissions of around 8.5 million tonnes CO_2/y, worldwide [416]. In many ammonia plants about 80% of the CO_2 is reacted with ammonia to urea from which it is, of course, liberated to the atmosphere when the urea is used as fertiliser.

On the other hand, CO_2-consuming processes will hardly change the picture from non-carbon containing fuel [416].

As an example, consider the methanol synthesis:

$$CO_2 + 3H_2 \leftrightarrow CH_3OH + H_2O \tag{1.3}$$

Even if hydrogen was made available from non-carbon containing fuel, the present world production of methanol via this reaction would only amount to 40 million t CO_2/y. This corresponds to the CO_2 emission from a 4000 MW coal-based power plant and should be compared with the total CO_2 emissions of approximately 27.5 10^9 t CO_2/y (7.5 10^9 t carbon/y). It means that CO_2 as a reactant will have little impact on the CO_2 problem. Again, the products will eventually end as CO_2.

A similar argument is valid for CO_2 reforming of natural gas.

1.1.2 Direct or indirect conversion

An important challenge in "C_1 chemistry" is to circumvent the syngas step by a direct conversion of methane into useful products. Still, yields are far from being economical [238] [287] [307]. The methane molecule is very stable, with a C-H bond energy of 439 kJ/mol; hence methane is resistant to many reactants. Electrophillic attack requires superacidic conditions, and radical abstraction of a hydrogen atom by a reactant Q requires that the Q-H bond exceeds 439 kJ/mol [130]:

$$CH_4 + Q- \leftrightarrow CH_3 + QH \qquad (1.4)$$

This is feasible when Q is an oxidising agent. However, the product often has much weaker C-H bonds than methane, which implies that it is difficult to eliminate further reactions leading to complete oxidation.

$$CH_4 \xrightarrow{O_2} CH_xO \xrightarrow{O_2} CO_2 + 2H_2O \qquad (1.5)$$

The direct conversion of CH_4 into methanol may have a high selectivity, but at a low conversion per pass. For example, Zhang et al. [537] reported a selectivity of 60% at a conversion of 12–13%. This corresponds to a yield of about 7.5%. This low yield per pass results in a large recycle ratio and a difficult separation associated with a low partial pressure of the product. This is illustrated by simple calculations in Figure 1.4 [410].

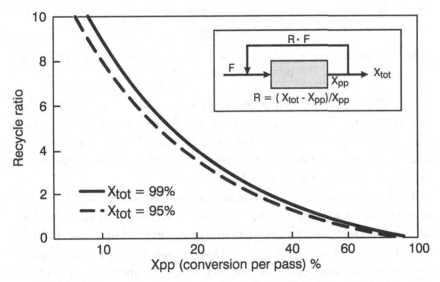

Figure 1.4 Recycle ratio and conversion. A selectivity of 95% at a conversion per pass of 5% means a large recycle ratio of approximately 10, and hence a difficult separation due to low partial pressure of the product. Reproduced with the permission of Springer [410].

A simple kinetic analysis of consecutive first-order reactions [410] may illustrate the problem. The data in Figure 1.5 show that the higher the ratio k_2/k_1, the lower the yield of the intermediate B.

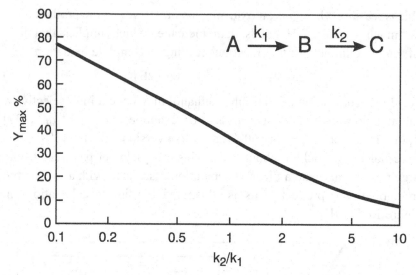

Figure 1.5 Consecutive reactions. Reactivities and maximum yields. Reproduced with the permission of Springer [410].

The direct oxidation of methane to methanol or formaldehyde has been a "dream reaction" for a long time [537]. Attempts include gas-phase reaction, catalytic reactions, and use of other oxidants than air. Selectivities may be high, but at a lower conversion per pass resulting in yields being inferior for industrial use.

High selectivity and conversion may not be sufficient. A process using superacid activation (Catalytica) for converting CH_4 via methyl bisulphate into methanol has the potential of achieving a high selectivity of 95% at a conversion of 90% [345]. However, the process would require a large sulphuric acid plant (1 mol SO_3/mol methanol) and a unit for concentrating a large recycle of diluted acid [410].

Other attempts have aimed at creating a carbon-carbon bond from methane, although most natural gas sources contain a fraction of ethane and other lower alkanes.

Most work in direct conversion has focused on the oxidative conversion of methane into ethylene [196] [307]. It has proven to be more promising than high-temperature pyrolysis of methane into primarily acetylene. However, the process suffers from ethane being a significant part of the products (low $\Delta \mathcal{P}$) and that above 20% of the converted methane is oxidised to carbon oxides. Under industrial conditions C_{2+} yields are less than 20% at a conversion of 24–35% per pass. As a result, the process scheme ends up being rather complex, meaning that the oxidative coupling is not economically feasible with the present low selectivities to C_2 hydrocarbons. Although the reaction scheme is elegant, the principles behind Figure 1.4 may explain why yields in oxidated coupling have never passed an apparent ceiling [292].

Catalytic partial oxidation at high temperature and ultra-short residence time over noble metals gauze has shown formation of olefins and oxygenates [206]. The feasibility of this route is still to be analysed. The indirect route via methanol appears to be a more promising route for olefins (see Section 2.6).

Direct conversion of methane to higher hydrocarbons without the assistance of oxygen is not favoured by thermodynamics. This constraint can be circumvented in a two-step process via carbides, but so far yields have been insignificant [279].

Other studies have explored the direct conversion of methane into benzene [420]. Selectivities of 70% were obtained close to equilibrium conversion at 600°C (12%).

From a thermodynamic point of view [145] the manufacture of synthetic transportation fuels should aim at a minimum change of the hydrogen content of the feedstock to that of the product (typically around H/C=2). It means that, in principle, it is more efficient to convert natural gas to paraffinic diesel than to aromatic rich gasoline. For coal the indirect conversion via syngas appears less efficient than the direct hydrogenation routes. However, these theoretical considerations should be supplemented with an analysis of the process steps and selectivities involved [145].

The main advantage of the indirect routes via syngas is the very high carbon efficiency. As an example, a modern methanol synthesis loop based on natural gas may operate with more than 50% conversion per pass having a selectivity of 99.9% and a carbon efficiency above 95% (refer to Section 2.6.2). The synthesis gas routes are highly efficient as illustrated in Table 1.1, but they are capital intensive because they involve exchange of energy in the reformers and heat recovery units, as illustrated in Figure 1.6 [413].

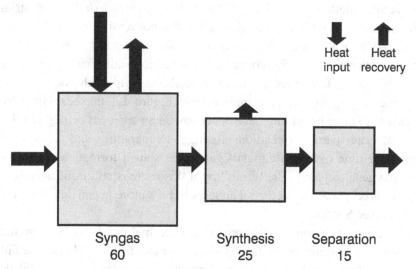

Figure 1.6 Indirect conversion of natural gas (numbers indicate the relative investments) [413]. Reproduced with the permission of Elsevier.

Syngas manufacture may be responsible for approximately 60% of the investments of large-scale gas conversion plants based on natural gas. Therefore, there is great interest in optimising process schemes based on steam reforming and autothermal reforming as well as in exploring new routes for the syngas manufacture.

1.2   Manufacture by steam reforming of hydrocarbons

1.2.1   Reactions and thermodynamics

Steam reforming is the reaction between steam and hydrocarbons into a mixture of hydrogen, carbon monoxide, carbon dioxide and unconverted reactants. Steam may be replaced by carbon dioxide as reactant. The reforming reactions are accompanied by the water-gas-shift reaction.

The term "steam reforming" should not be confused with catalytic reforming used for the conversion of paraffinic hydrocarbons to high octane hydrocarbons such as iso-alkanes and aromatics. A better term may be "oxygenolysis" [381] [389] as the reaction involves the breakage of C-H and C-C bonds by means of oxygen containing species.

A complete steam reforming reaction scheme can thus be written as:

Table 1.2 Steam reforming reactions.

Reaction		$-\Delta H^0_{298}$ $\frac{kJ}{mol}$
R1	$CH_4 + H_2O \leftrightarrow CO + 3H_2$	-206
R2	$CH_4 + 2H_2O \leftrightarrow CO_2 + 4H_2$	-165
R3	$CH_4 + CO_2 \leftrightarrow 2CO + 2H_2$	-247
R4	$CO + H_2O \leftrightarrow CO_2 + H_2$	41
R5	$C_nH_m + nH_2O \leftrightarrow nCO + (n + 0.5m)H_2$	<0

Although all reactions may describe specific operating conditions, only two out of the first four reactions are independent from a thermodynamic point of view, since the other two can be established as linear combinations of the two selected ones. Catalytic studies indicate that it is steam reforming of methane to carbon monoxide and the water-gas-shift reactions that are the independent reactions in addition to the steam reforming of higher hydrocarbons as the last reaction. This set of reactions (R1, R4, and R5 in Table 1.2) will consequently be used in the following.

The term "steam reforming" is also used for the reaction between steam and alcohols (methanol and ethanol) as well as liquid-phase reactions with carbohydrates or biomass (see Section 1.4).

The last reaction R5 in Table 1.2 is the reverse Fischer–Tropsch synthesis, but the conversion of higher hydrocarbons can be considered irreversible at normal reforming temperatures. The higher hydrocarbons react on the metal surface to C_1 components or stay as carbonaceous deposits. At temperatures above 600–650°C, the catalytic reactions may be accompanied by thermal cracking.

The symbol "↔" indicates that a reaction is reversible, i.e. at a given temperature the reaction will not have full conversion. Usually only the last reaction is considered irreversible, so parts of methane and steam will remain in the mixture at outlet conditions.

Preparation of heat and mass balances for synthesis gas processes thus requires methods to calculate mass and heat balances and chemical equilibrium.

For a reaction system like the one shown above in Table 1.2 it is convenient to define the thermodynamic reference state as the enthalpy and free energy of formations as an ideal gas at 25°C (298.15 K). This definition allows direct calculation of heat duty in the enthalpy calculation without having to distinguish between the parts for heating and reaction.

Different functions may be used to represent the ideal gas heat capacity or enthalpy as a function of temperature; values in Appendix 1 are based on a fourth-degree polynomial for the enthalpy of formation, where the zero-order coefficient has been adjusted to obtain the enthalpy of formation at 25°C.

$$H_i = \sum_{k=1}^{5} E_k \cdot T^{k-1} \tag{1.6}$$

Appendix 1 shows the coefficients in the enthalpy polynomials for a number of characteristic synthesis gas reaction key components and a table with actual values of the ideal enthalpies of formation as a function of temperature.

The preparation of mass balances requires calculation of chemical equilibrium by solution of the following Equation (1.7) for the coupled reactions between all components:

$$K_{eq,j} = \prod_{i=1} a_i^{v_{i,j}} \tag{1.7}$$

where $v_{i,j}$ is the stoichiometric coefficient for component i in reaction j.

The right-hand side includes the activity, which rigorously is defined as:

$$a_i = \frac{f_i^{gas}}{f_i^o} = \frac{y_i \cdot \varphi_i \cdot P}{f_i^o} \tag{1.8}$$

in which f is the fugacity, f° the reference fugacity and φ the fugacity coefficient, which describes the deviation from an ideal gas. Preparation of syngas is carried out at high temperature and a modest pressure so the assumption of an ideal gas is acceptable when the operation is not close to the dew point of the mixture. This implies that the fugacity coefficient φ usually is set to unity. The equilibrium constant, K_{eq}, is a function of temperature only, so the conversion in the steam reforming of methane reaction is favoured by a low pressure.

As the reference state for all components in the reactions is defined as an ideal gas at 25°C (298.15K) and 1.01325 bar, the value that must be used for f° is 1.01325 bar. If another unit of measurement for pressure is used, the reference pressure must be changed correspondingly so that the activity becomes independent of pressure. If a component is a solid, such as carbon in carbon formation, the activity, a, is unity, since the reference state for the Gibbs free energy of carbon is also defined as the solid state.

The temperature equation for the equilibrium constant is derived from thermodynamics using the Gibbs energy of formation, G°, the enthalpy of formation H°, and the temperature dependence as derived from:

$$RT\ln(K_{eq,j}) = -\Delta G_j^o$$

$$\frac{d\ln(K_{eq,j})}{dT} = \frac{\Delta H_j^o}{RT^2} \tag{1.9}$$

After insertion of Equation (1.6) the final equation for the equilibrium constant is:

$$\ln(K_{eq,j}) = C_{1,j} \cdot \ln(T) + \frac{C_{2,j}}{T} + C_{3,j} + C_{4,j} \cdot T + C_{5,j} \cdot T^2 + C_{6,j} \cdot T^3 \tag{1.10}$$

The basic reaction properties for some key reactions are shown in Table 1.3.

Table 1.3 Basic reaction properties for steam reforming reactions. Formation data at 25°C. Data from [137] [375].

	$CH_4 + H_2O \leftrightarrow$ $CO + 3H_2$	$CO + H_2O \leftrightarrow$ $CO_2 + H_2$	$CH_4 + CO_2 \leftrightarrow$ $2CO + 2H_2$	$C_2H_6 + 2H_2O \leftrightarrow$ $2CO + 5H_2$
$\Delta H^0 = \sum_i v_i H_i^o$ (kJ/mol)	206.15	-41.16	247.31	347.27
$\Delta G^0 = \sum_i v_i G_i^o$ (kJ/mol)	142.12	-28.52	170.64	215.53
$\Delta S^0 = \dfrac{\Delta H^0 - \Delta G^0}{T}$ (kJ/mol/K)	0.2148	-0.0424	0.2572	0.4419

The definition of the standard state as the ideal gas of formation at 25°C (298.15K) and 1.01325 bar implies that the absolute values of entropy will not fulfil the third law. The heat of reaction is defined as the heat of formation, but with the opposite sign so it is written as $-\Delta H^0_{298}$.

Tables with equilibrium constants at selected temperatures for the main synthesis gas reactions are shown in Appendix 2.

The reforming reaction involving two stable molecules as methane and water is strongly endothermic and it leads to formation of more molecules. This means that the affinity for the reaction ($-\Delta G°$) is established by the entropy term. The basic thermodynamic reaction properties as a function of temperature are shown for the methane reforming reaction in Figure 1.7.

It is seen that reaction enthalpy and entropy are weak functions of temperature, but also that they are positive in the entire interval implying that other heat sources are necessary for conversion. The free energy decreases strongly with temperature due to the corresponding increase in

TΔS giving more favourably equilibrium. Hence, the steam reforming reaction is entropy-driven.

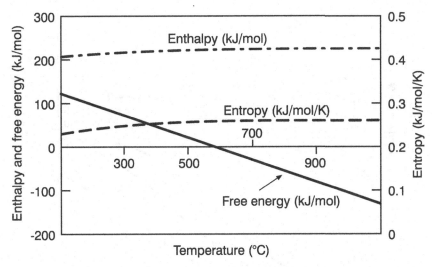

Figure 1.7 Steam reforming. Thermodynamic functions.

Conversion at equilibrium is calculated by solving the coupled set of equations for K_{eq} for the two independent reactions R1 and R4 in Table 1.2.

A simplified method to find the conversions in the two reactions is available as will be shown below, but a general method which can solve any chemical equilibrium problem is preferred. For this purpose two methods may be used. The first is minimisation of the Gibbs free energy [316], whereas the other one is the solution for conversions [468]. The first one may be attractive from a theoretical point of view and it is readily combined with phase equilibrium, but the last one is preferred in catalysis, since no combination of reactions may proceed in all cases. The set of equations in Table 1.2 may be solved using the Newton–Raphson method with the conversions as independent variables. Some of the components (higher hydrocabons or oxygen) may almost disappear in the final mixture so it is necessary to handle elimination of reactions with almost complete conversion.

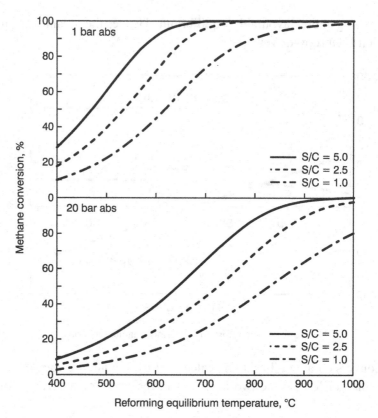

Figure 1.8 Steam reforming and methane conversion [415]. Reproduced with the permission of Elsevier.

Methane conversions as a function of temperature in the combined methane reforming and shift reactions are shown in Figure 1.8. It is seen that conversion is strongly favoured by high temperature, low pressure, and high steam-to-carbon ratio.

> Example 1.1
> A simplified method [276] may be used to solve the chemical equilibrium for the reactions R1, R4 and R5 in Table 1.2 and find the conversions ξ_1 and ξ_4 in the reactions R1 and R4.
> Initially all higher hydrocarbons are converted quantitatively to CO and H_2 using Equation R5, whereafter the mole amounts in the initial

mixture are converted according to the ordinary stoichiometric conversion equations below:

In	Out	y
$y^o_{CH_4}$	$y^o_{CH_4} - \xi_1$	$y_{CH_4} = (y^o_{CH_4} - \xi_1)/(1+2\xi_1)$
$y^o_{H_2O}$	$y^o_{H_2O} - \xi_1 - \xi_4$	$y_{H_2O} = (y^o_{H_2O} - \xi_1 - \xi_4)/(1+2\xi_1)$
y^o_{CO}	$y^o_{CO} + \xi_1 - \xi_4$	$y_{CO} = (y^o_{CO} + \xi_1 - \xi_4)/(1+2\xi_1)$
$y^o_{CO_2}$	$y^o_{CO_2} + \xi_4$	$y_{CO_2} = (y^o_{CO_2} + \xi_4)/(1+2\xi_1)$
$y^o_{H_2}$	$y^o_{H_2} + 3\xi_1 + \xi_4$	$y_{H_2} = (y^o_{H_2} + 3\xi_1 + \xi_4)/(1+2\xi_1)$
1	$1+2\xi_1$	

The bottom line is the total amount before and after conversion. The normalisation is in principle with respect to the total amount of preconversion of higher hydrocarbons; here it is simply assumed that the sum of the initial mole fractions is unity.

Assuming ideal gas the mole fractions can now be inserted in the two equations for K_{eq}:

$$K_{eq,ref} = \frac{y^3_{H_2} \cdot y_{CO} \cdot P^2}{y_{CH_4} \cdot y_{H_2O}}$$

$$K_{eq,shf} = \frac{y_{H_2} \cdot y_{CO_2}}{y_{CO} \cdot y_{H_2O}}$$

Considering the shift reaction alone results in a second-order equation in the conversion ξ_4, and this equation can be solved analytically after reformulation of the equilibrium equation as:

$$\{K_{eq,shf} - 1\}\xi_4^2 - \{K_{eq,shf}(y^o_{CO} + y^o_{H_2O}) + (y^o_{H_2} + y^o_{CO_2})\}\xi_4 +$$
$$\{K_{eq,shf} y^o_{CO} y^o_{H_2O} - y^o_{H_2} y^o_{CO_2}\} = 0$$

or

$$A\xi_4^2 + B\xi_4 + C = 0$$

The roots are:

$$\xi_4 = \frac{-B \pm \sqrt{B^2 - 4AC}}{2A} = \frac{-B \pm D}{2A}$$

Both roots are real, but only the smallest one is used, since the largest one (by experience) may give negative mole fractions. $-B$ is always positive, so the solution is in principle only relevant for $-D$, but since B and D numerically may have the same size, it is appropriate to use the $+$ sign for the other root and then calculate the final root by using the fact that the product of the two roots is C/A.

The conversion in the shift reaction alone is then:

$$\xi_4 = \frac{2C}{-B+D}$$

This implies that for any value of ξ_1 the conversion in the shift reaction can be found analytically. The reforming reaction with known ξ_4 is a fourth-order equation in ξ_1. It can be solved analytically, but it is more practical to implement a combined method to solve the two equations [276]. This method uses bisection for the determination of ξ_1 with insertion of the correct value of ξ_4 for each step by solution of the quadratic equation. The equations are highly non-linear, so it is necessary to know the conversion limits. The lower limit may be set to zero here. It need only be considered if the reverse reaction must be considered. The upper limit is the minimum concentration of either CH_4 or H_2O+CO_2.

The method is suitable for implementation in a spreadsheet and can be written as:

1. Calculate the equilibrium constants for the two reactions from the data in Appendix 2.
2. Specify the lower and upper boundaries as $\xi_{1,min}$ and $\xi_{1,max}$.
3. Guess a new value of ξ_1 as the average of the minimum and maximum values.
4. Calculate new initial mole fractions assuming the conversion ξ_4 in the shift reaction is 0.
5. Find the conversion in the shift reaction from the analytical solution.
6. Calculate new mole fractions using both conversions.
7. Calculate the reaction quotient for the steam reforming reaction by inserting the new calculated mole fractions in the equilibrium expression.
8. If the reaction quotient is larger than the equilibrium constant, the conversion is too large, since the reaction quotient increases

with conversion. Set the upper conversion limit to the calculated value of ξ_1 and jump to step 3.
9. If the reaction quotient is smaller than the equilibrium constant, the conversion is too small. Set the lower conversion limit to the calculated value of ξ_1 and jump to step 3.

The iteration is slow but safe and continues until the desired accuracy has been reached.

The steam reforming reactions are fast reactions due to the high temperature and the presence of the catalyst implying that the actual conversion will be close to the equilibrium conversion.

Figure 1.9 shows the conversions in the methane steam reforming and shift reactions, if they are treated as independent reactions. The conversion in the endothermic steam reforming is increasing with temperature, whereas the opposite is the case for the shift reaction.

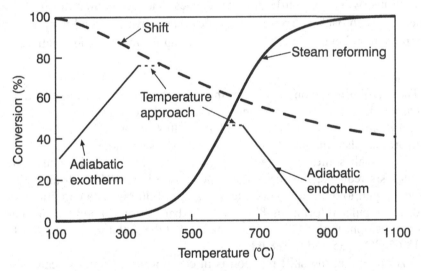

Figure 1.9 Temperature, conversion and temperature approach to equilibrium for steam reforming and shift.

Using actual conversions as a measure of reaction extent is not convenient, since they must be relative to equilibrium conversions. Instead the so-called "temperature approach" to equilibrium is used, which is the horizontal distance between the actual outlet temperature and the corresponding equilibrium temperature. This is illustrated in Figure 1.9 by use of the adiabatic reaction paths. The temperature approach to equilibrium is portrayed by horizontal dotted lines.

It is seen that an appropriate definition of the temperature approach is:

$$\Delta T_{app,ref} = T_{exit} - T_{eq}$$
$$\Delta T_{app,shf} = T_{eq} - T_{exit}$$
(1.11)

T_{eq} is the temperature on the equilibrium curve having the same conversion or the same reaction quotient as the actual outlet gas.

By this definition the temperature approach will always be positive provided the reaction path is below the equilibrium curves.

It is true that if two of the reactions in the reaction set R1 to R4 in Table 1.2 are at equilibrium, the other two will also be at equilibrium. If, however, one of the reactions has an approach larger than zero, the other reactions will have different approaches and some may even have negative approaches. This also signifies that conversion and temperature approach in fact only has a practical meaning for the limiting reaction.

Example 1.2

The equilibrium conversion in the steam reforming of methane to carbon monoxide and hydrogen is determined from the two equilibrum expressions shown in Example 1.1. In this second example a large hydrogen plant making 100,000 Nm^3/h of H_2 is considered.

Although some hydrogen must be present in the feed to keep the catalysts reduced, the feed is assumed to be pure CH_4 with an actual flow equal to 500 mol/s and a temperature of 500°C. The condition out of the tubular reformer is 875°C and 31 bar (absolute) and the dry gas composition (mole %) has been measured at the outlet as: H_2 70.36, CO 13.65 CO_2 7.35 and CH_4 8.64.

Assume that the shift reaction is in equilibrium in order to calculate the unknown amount of water and then calculate the temperature approach to the steam reforming of methane reaction.

The equilibrium constant for the shift reaction at 875°C is found in Appendix 2 to be 0.82838. The equation for the shift reaction can be rearranged as a function of the dry mole fractions and the unknown water mole fraction as:

$$K_{eq,shf} = \frac{y_{dry,H_2} \cdot y_{dry,CO_2}(1-y_{H_2O})}{y_{dry,CO} \cdot y_{H_2O}}$$

Insertion of the specified dry mole fractions and the equilibrium constant gives:

$$y_{H_2O} = 31.38 \, mole\%$$

The wet mole fractions are now calculated, and these and the pressure are then inserted into the steam reforming equilibrium reaction to calculate the reaction quotient. The result is:

$$Q_{ref} = 544 \Rightarrow T_{eq,ref} = 850°C \Rightarrow \Delta T_{ref} = 25°C$$

In addition to the steam reforming reactions *carbon* may also be formed according to the reactions in Table 1.4.

Table 1.4 Carbon-forming reactions.

Reaction		$-\Delta H_{298}^0 \; \frac{kJ}{mol}$
R6	$CH_4 \leftrightarrow C + 2H_2$	-75
R7	$2CO \leftrightarrow C + CO_2$	172
R8	$CO + H_2 \leftrightarrow C + H_2O$	131
R9	$C_nH_m \rightarrow nC + 0.5mH_2$	<0

Two forms of carbon may be found. The first is ordinary graphite, but carbon may also be found in a whisker structure on the catalyst, which has another free energy of formation as will be discussed in Chapter 5.

The reaction properties are seen in Table 1.5 below. A table with equilibrium constants is found in Appendix 2.

Table 1.5 Basic reaction properties for carbon-forming reactions.
Data from [137] [375] for carbon as graphite. Data from [425] for carbon as whisker
(refer to Section 5.3.1).

	$CH_4 \leftrightarrow$ $C + 2H_2$	$CH_4 \leftrightarrow$ $C_{whisker} + 2H_2$	$2CO \leftrightarrow$ $C + CO_2$	$CO + H_2 \leftrightarrow$ $C + H_2O$
$\Delta H^\circ = \sum_{i=1}^{C} v_i H_i^\circ$ (kJ/mol)	74.85	89.62	-172.46	-131.30
$\Delta G^\circ = \sum_i v_i G_i^\circ$ (kJ/mol)	50.79	58.19	-119.85	-91.32
$\Delta S^\circ = \dfrac{\Delta H^\circ - \Delta G^\circ}{T}$ (kJ/mol/K)	0.0807	0.1054	-0.1765	-0.1341

(See notes to Table 1.3)

1.2.2 Product gas composition

The methane steam reforming reaction R1 in Table 1.2 results in a H_2/CO ratio close to 3. Steam can be replaced by CO_2, resulting in a H_2/CO ratio close to 1. The addition of oxygen in ATR and POX gives a lower H_2/CO ratio.

The H_2/CO ratio can be varied over a wider range, as illustrated in Figure 1.10, as the reforming reactions are coupled to the shift reaction. In the manufacture of hydrogen, the reforming process is followed by water gas shift carried out in the presence of an iron and/or copper catalyst at low temperatures (210–330°C) to ensure complete conversion of carbon monoxide (refer to Section 1.5). The conversion of methane is restricted by the thermodynamics of the reforming reactions. The endothermic steam (and CO_2) reforming reactions must be carried out at high temperature and low pressure to achieve maximum conversion, as illustrated in Figure 1.8.

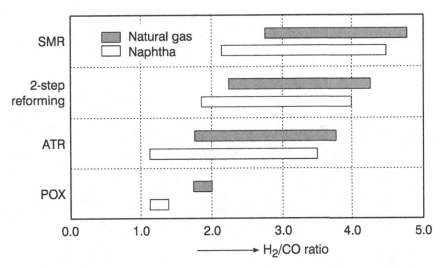

Figure 1.10 H_2/CO ratios from various syngas processes [415]. Reproduced with the permission of Elsevier. SMR: steam methane reforming; ATR: autothermal reforming (see Section 1.3.2); POX: partial oxidation (see Section 1.3.1).

As an example, modern hydrogen plants are normally designed for low steam-to-carbon ratios (refer to Section 2.2), although high steam-to-carbon ratios (4–5 molecules of H_2O/C atom) would result in higher conversion of the hydrocarbons. However, a low steam-to-carbon ratio (typically 2.5 or less) reduces the mass flow through the plant, the steam production, and thus the equipment sizes. The lowest investment is therefore generally obtained for plants designed for low steam-to-carbon ratios.

In principle, a low steam-to-carbon ratio increases the amount of unconverted methane from the reformer (Figure 1.8), but this can be compensated for by increasing the reformer outlet temperature, typically to 920°C. In synthesis plants, the unconverted methane flows downstream with the synthesis gas. Unconverted methane thus implies a larger syngas unit and results in restrictions on recycle ratios in the synthesis because of accumulation of the inert methane in the synthesis gas, which reduces the partial pressures of the syngas components.

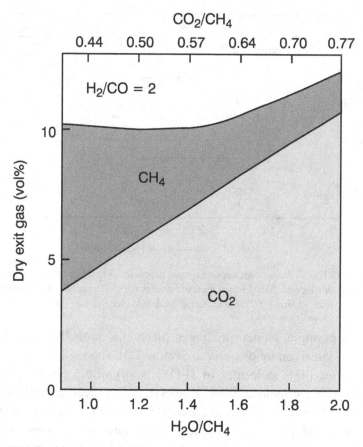

Figure 1.11 Combined steam and CO_2 reforming for producing a syngas with $H_2/CO=2$. 25 bar abs. $T_{exit}=950°C$ [415]. Reproduced with the permission of Elsevier.

Stoichiometric reforming according to Equations R1 and R3 in Table 1.2 at H_2O/CH_4 or CO_2/CH_4 ratios of 1 is rarely feasible [399], because it would result in incomplete conversion at the pressures that are economical for industrial syngas plants (20–50 bar). This is also true for mixed CO_2/H_2O reforming as illustrated in Figure 1.11 [415]. The low-pressure manufacture of reducing gas for direct ore reduction is one exception (refer to Section 2.4.3).

Other thermodynamic constraints are related to the risk of carbon formation when too little oxidant is present (refer to Chapter 5).

The requirements to the composition of the syngas vary with the synthesis in question, as shown in Table 1.6.

Table 1.6 Syngas composition for various processes [415].

Process	Stoichiometric composition	Co-reactants
Ammonia	$M = \dfrac{H_2}{N_2} = 3$	
Methanol	$M = \dfrac{H_2 - CO_2}{CO + CO_2} = 2$	
DME from hydrocarbons	$M = \dfrac{H_2 - CO_2}{CO + CO_2} = 2$	
DME from coal gas	$M = \dfrac{H_2}{CO} = 1$	
High-temp. Fischer–Tropsch	$M = \dfrac{H_2 - CO_2}{CO + CO_2} = 2$	
Low-temp. Fischer–Tropsch	$M = \dfrac{H_2}{CO} \cong 2$	
SNG	$M = \dfrac{H_2 - CO_2}{CO + CO_2} = 3$	
Acetic acid	CO	Methanol
Higher alcohols	$M = \dfrac{H_2}{CO} = 1$	Olefins
Industrial hydrogen	99.99 H_2	
Hydrogen for PEMFC	<50 ppm CO	
Reducing gas (iron ore)	$\dfrac{CO_2 + H_2O}{H_2 + CO + CO_2 + H_2O} \leq 0.05$	

1.2.3 Thermodynamics of higher hydrocarbons

Higher hydrocarbons are steam reformed according to Reaction R5 in Table 1.2. In natural gas mixtures and light hydrocarbon mixtures the components are identified as such, but in heavier hydrocarbon feedstocks such as naphthas, kerosene and diesel, the chemical compounds and composition are usually unknown. Instead, such mixtures are characterised by the following measurements:

- A distillation curve with initial and final boiling point. It is converted to a true boiling-point (TBP) curve;
- Specific gravity;
- The carbon to hydrogen (C/H) weight ratio;
- PNA distribution: Paraffin (alkane), Naphthene (cycloalkane), and Aromatic;
- Lower heating value (LHV);
- Refractive index;
- Liquid viscosity.

Except for the LHV, no reaction properties are thus measured. Correlations are available to estimate reaction properties using conventional refinery pseudo-component characterisation methods, but it is advantageous to calculate a mixture of real chemical compounds, which represent the measurements, as well as possible to be able to calculate accurate mass and energy balances over the plant.

This is carried out selecting a number of components, whose type is chosen approximately proportional with the PNA distribution and the C/H ratio. The components are selected in such a way that their normal boiling points are distributed equally on the measured TBP curve. For each additional measurement an equation is set up using proper mixing rules and the total equation system is solved using appropriate uncertainty factors. After solution of the equations, the resulting composition of the mixture will give accurate mass and energy balances in the syngas process.

1.2.4 The tubular reformer

The negative heat of reaction of the reforming reaction and the high exit temperatures at typical process conditions mean that heat must be supplied to the process, typically in a fired reactor, the tubular reformer [389] as shown in Figure 1.12. The product gas will normally be close to the equilibrium of Reactions R1 and R4 in Table 1.2 with an "approach to equilibrium" of about 5–10°C and 0°C for the two reactions, respectively.

Figure 1.12 Photo of a tubular reformer.

To illustrate the strong endothermicity, adiabatic steam reforming of methane (carried out with $H_2O/CH_4=2.5$ at a pressure of 20 bar abs and a feed temperature of 500°C) will result in a temperature drop of approximately 12°C for each 1% of methane converted.

The overall heat requirement can be estimated from enthalpy tables (see Appendix 1) when the product gas composition is known from equilibrium calculations.

Example 1.3

The heat input for a tubular reformer in a hydrogen plant (100,000 Nm^3/h) should be estimated. The outlet equilibrium properties of the feed were calculated in Example 1.2, so this example is a continuation of this with the purpose of calculating the heat balance over the tubular reformer.

The total inlet and outlet atomic balances for C and O are set up. The C balance can be used to calculate the total outlet flow by use of the wet composition found in Example 1.2. Similarly, the O balance can be used to calculate the water inlet flow as follows:

$$C: \quad CH_{4,in} = CH_{4,out} + CO_{out} + CO_{2,out}$$
$$O: \quad H_2O_{in} = CO_{out} + 2CO_{2,out} + H_2O_{out}$$

gives

$$F_{out} = 2459 \, mol/s$$
$$H_2O_{in} = 1250 \, mol/s$$
$$CH_{4,out} = 146 \, mol/s$$
$$H_{2,out} + CO_{out} = 1417 \, mol/s$$

The steam-to-carbon ratio in the feed is thus: $H_2O/CH_4=2.5$ and the total conversion of CH_4 is 71%.

If all CO in the product can be converted to H_2 in the subsequent shift reaction system, the total amount of product is 114,338 Nm^3/h of which 100,000 Nm^3/h or 88% in this case is recovered as final product.

A total heat balance over the tubular reformer can now be established by using the calculated component flows and the corresponding enthalpies of formation of the feed components at 500°C and the product components at 875°C from Appendix 1 as:

Inlet: $-224.88 \cdot H_2O_{in} + (-51.75) \cdot CH_{4,in} = -306975 \, kJ/s$

Outlet: $-209.61 \cdot H_2O_{out} + (-25.47) \cdot CH_{4,out} + (-83.88) \cdot CO_{out} +$
$(-351.90) \cdot CO_{2,out} + 25.20 \cdot H_{2,out} = -198500 \, kJ/s$

$Q_{added} = -198500 - (-306975) = 108475 \, kJ/s \cong 108 \, MW$

If an average heat of reaction equal to 225 kJ/mol (valid at 700°C) is used, it is found that in this case 75% of the total heat added is used in the reaction and the rest is used for heating of feed and product. The split between the two parts changes, of course, with H_2O/CH_4 ratio and other operating conditions.

The catalyst is normally nickel on a stable support (refer to Chapter 4). The catalyst properties are dictated by the severe operating conditions, including temperatures of 450–950°C and steam partial pressures of up to 30 bar. The intrinsic activity of the catalyst depends on the nickel surface area.

The catalyst is placed in a number of high-alloy reforming tubes placed in a furnace as shown in Figure 1.13. Tubular reformers are designed with a variety of tube and burner arrangements (as shown in Section 3.2.2). Such reformers are built today for capacities up to more than 300,000 Nm³ of H_2 (or syngas) /h.

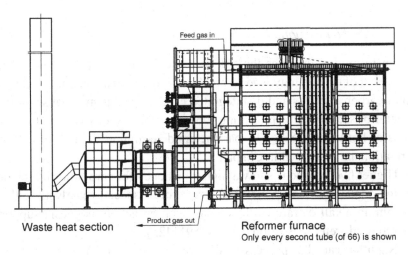

Figure 1.13 Reformer furnace and waste heat section [425]. Reproduced with the permission of Wiley.

Heat transfer takes place primarily (>95%) by radiation from the furnace gas and in a sidewall-fired furnace also from the furnace walls. The remaining transfer is by convection.

About 50% of the fuel combustion heat is transferred through the tubes for the reforming reactions and for heating up the gas to the exit temperature. The remaining combustion heat is recovered in the waste heat section (Figure 1.13).

It is possible to increase the amount of heat transferred to the process gas in the reformer from about 50% to about 80% of the supplied heat when using a convective heat exchange reformer in which the flue gas as well as the hot product gas are cooled by heat exchange with the process gas flowing through the catalyst bed. This results in a more compact piece of equipment [171]. However, in all types of heat-exchange reformers, the heat exchange is by convection, and this generally leads to lower average heat fluxes to protect the construction materials than in reformers with radiant heat transfer. Therefore, in principle the fired tubular reformers may appear the most economic solution for large-scale operation, but the convective reformer may be applied in combination with the tubular reformer for more efficient heat recovery by "chemical recuperation" [426] [427]. This is further discussed in Section 2.2 and Chapter 3.

1.2.5 Carbon formation. Higher hydrocarbons

Steam reforming involves the risk of carbon formation (Table 1.4). The formation of carbon may lead to breakdown of the catalyst and the build-up of carbon deposits and disintegrated catalyst pellets may cause partial or total blockage of the reforming tubes resulting in development of hot spots or hot tubes [389]. The parameters determining the risk of carbon formation are discussed in Chapter 5.

Higher hydrocarbons show a higher tendency for carbon formation on nickel than does methane and, therefore, special catalysts either containing alkali or rare earths or based on an active magnesia support are required (refer to Section 5.3.2) [389] [425].

Naphtha can be processed directly in the tubular reformer, as practiced in many industrial units, but the control of the preheat temperature and heat flux profile may be critical [384]. This is a severe constraint as the heat required in the tubular reformer and hence the reformer costs may be reduced by increasing preheat temperature. However, the preheater may then work as a "steam cracker" producing olefins from higher hydrocarbons in the feed [532] (refer to Section 4.3.3). The olefins easily form carbon in the reformer. Apart from the pressure, the conditions in the tubular steam reformer and in the preheater are not far from that of a steam cracker in an ethylene plant.

These constraints are removed when using an adiabatic prereformer [394] [415] as illustrated in Figure 1.14. The prereforming catalyst is typically a highly active nickel catalyst. This catalyst also works as an effective sulphur guard for the tubular reformer and downstream catalysts, by removing any traces of sulphur still left after the desulphurisation section.

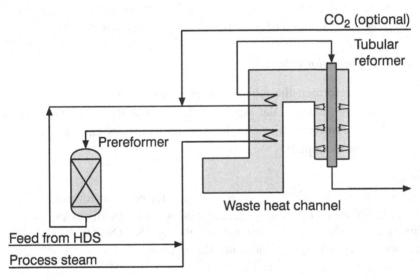

Figure 1.14 Flow diagram of process with tubular reformer with prereformer [394]. Reproduced with the permission of Elsevier.

With proper desulphurisation, it is possible to convert naphtha and heavy distillate feedstocks such as kerosene and diesel in a prereformer into syngas with no trace of higher hydrocarbons [389] [405].

All higher hydrocarbons are converted in the prereformer in the temperature range of 350–550°C, and the reforming and shift reactions are brought into equilibrium. After a prereformer, it is possible to preheat to temperatures around 650°C, thus reducing the size of the tubular reformer. A revamp of an ammonia plant [457] by installing a prereformer and larger preheater involved the increase in the reformer inlet temperature from 520°C to 650°C. This resulted in reduced fuel consumption.

With low catalyst activity, the thermal cracking route (pyrolysis) may also take over in the reformer tube [389]. This is the situation in case of severe sulphur poisoning or in attempts to use non-metal catalysts with low activity. The risk of carbon formation depends on the type of hydrocarbon with the contents of aromatics being critical. Ethylene formed by pyrolysis results in rapid carbon formation on nickel (refer to Section 5.2). Ethylene may also be formed by oxidative coupling if air or oxygen is added to the feed – or by dehydration of ethanol.

1.2.6 Non-tubular reforming

In a tubular reformer, the tube diameter is selected from the mechanical considerations leaving the space velocity (catalyst volume) as a dependent parameter. This so-called tubular constraint can be illustrated by the simple example:

> **Note**
> For given tube length, L, flow, F, and transferred heat (reformer duty), Q, the number of tubes, n, is determined by the tube diameter,: d_t, the average heat flux,: q_{av} and the space velocity,: SV ($Nm^3/h/m^3$ cat). For constant inlet and outlet conditions, this means:
>
> $$Q = n\pi d_t L q_{av}$$
>
> $$Q \approx F = SV \cdot n \frac{\pi}{4} d_t^2 \cdot L$$
>
> from which :
>
> $$q_{av} \approx SV \cdot d_t$$
>
> (1.12)

The tubular constraint (the last equation) can be made less restrictive by convective heat exchange reformers. This may involve the use of catalysed heat transfer surfaces [188] [397] [488] in the form of plate type reformers and multi-channel reformers. This is discussed further in Section 3.3.8.

Another approach is to decouple the heat transfer and the reaction. This includes reheat schemes [403] [455] [548] in which the process gas is heated in a heater followed by reforming reaction in an adiabatic reactor as illustrated in Figure 1.15. However, many steps are required to reheat the gas because of the strong endothermicity of the reaction.

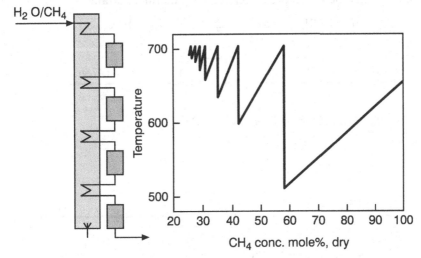

Figure 1.15 Reheat scheme for steam reforming [403]. Reproduced with the permission of Japan Petr. Inst.

A variation of the reheat process scheme is the use of a circulating catalyst bed using one bed for reaction and the other for heating up the catalyst [549]. This is also applied in other fluidised petrochemical processes [439]. However, for steam reforming the recirculation rate would be very high. Moreover, catalyst dust in downstream heat exchangers would result in methane formation by the reverse reforming reaction (methanation). Other attempts have aimed at utilising the high heat transfer in fluidised beds and supplying the heat by an external heater [10]. Other suggestions [262] have dealt with supplying the heat by addition of a CO_2 acceptor (CaO, etc.) to the fluidised bed. The heat from the formation of carbonate is almost sufficient for the reforming reaction [262].

An alternative to the reforming process may be the use of a cyclic process [438] as illustrated in Figure 1.16. Hydrogen is generated by reacting steam with a metal (Cu, Fe, etc.). The resulting metal oxide is reduced by reaction with methane-forming steam and CO_2 at a pressure well suited for sequestration. The scheme involves a number of constraints relating to heats of reaction. The addition of air is necessary to ensure that the overall reaction becomes thermoneutral.

$$CH_4 + 1.32H_2O + 0.34O_2 = 3.32H_2 + CO_2 \qquad \Delta H^o_{298} = 0 \qquad (1.13)$$

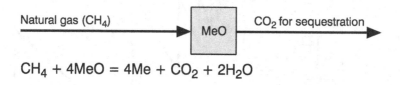

$$CH_4 + 4MeO = 4Me + CO_2 + 2H_2O$$

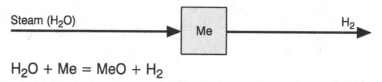

$$H_2O + Me = MeO + H_2$$

Figure 1.16 Cyclic process for CO_2-free hydrogen [414]. Reproduced with the permission of Balzer.

1.3 Other manufacture routes

1.3.1 Partial oxidation

An alternative approach to steam reforming is to add oxygen to the feed and hence gain the necessary heat by internal combustion. It means that the steam formed by the combustion is condensed in the process stream instead of leaving as water vapour in the flue gas from the fired reformer. Hence, the higher heating value of the fuel is recovered in the partial oxidation schemes in contrast to the lower heating value in the steam reforming process.

Partial oxidation can be carried out in three different ways as illustrated in Figure 1.17:

- Non-catalytic partial oxidation (POX);
- Autothermal reforming (ATR);
- Catalytic partial oxidation (CPO).

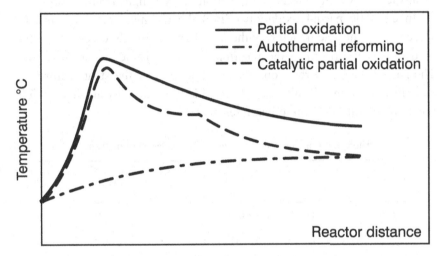

Figure 1.17 Syngas by partial oxidation. Temperature profiles for various routes to syngas by partial oxidation [398]. Reproduced with the permission of Elsevier.

The non-catalytic partial oxidation [486] (POX, Texaco, Shell) needs high temperature to ensure complete conversion of methane and to reduce soot formation. Some soot is normally formed and is removed in a separate soot scrubber system downstream of the partial oxidation reactor. The thermal processes typically result in a product gas with $H_2/CO=1.7-1.8$. Gasification of heavy oil fractions, petcoke, coal and biomass may play an increasing role as these fractions are becoming more available and natural gas (NG) less available.

The autothermal reforming (ATR) process is a hybrid of partial oxidation and steam reforming using a burner and a fixed catalyst bed for equilibration of the gas. This allows a decrease in the maximum temperature and hence the oxygen consumption can be lowered. On the other hand, the high temperature in POX units may result in lower contents of methane and carbon dioxide in the product gas than in ATR.

Soot formation in ATR can be eliminated by addition of a certain amount of steam to the feedstock and by special burner design. Steam can hardly be added to the non-catalytic processes without the risk of increased soot formation because of the resulting lower temperature. This means less flexibility for the composition of the syngas from POX units.

In catalytic partial oxidation (CPO) the chemical conversions take place in a catalytic reactor without a burner. In all cases of partial oxidation, some or all of the reactions listed in Table 1.7 are involved. The partial oxidation reactions are accompanied by the steam reforming and shift reactions (Table 1.2). The oxidation reactions are irreversible under all conditions of practical interest.

Table 1.7 Reactions occurring in partial oxidation of methane.

Reaction		$-\Delta H^\circ_{298} \; \dfrac{kJ}{mol}$
R10	$CH_4 + 0.5O_2 \rightarrow CO + 2H_2$	36
R11	$CH_4 + O_2 \rightarrow CO + H_2O + H_2$	278
R12	$CH_4 + 1.5O_2 \rightarrow CO + 2H_2O$	519
R13	$CH_4 + 2O_2 \rightarrow CO_2 + 2H_2O$	802

The overall reaction in a partial oxidation reactor is strongly exothermic and no heat must be supplied to the reactor. The partial oxidation reactions may be accompanied by cracking of the hydrocarbons or by oxidative dehydrogenation into non-saturated compounds including olefins, poly-aromatics and soot. The control of the heat balance and the formation of by-products are main considerations in the design of partial oxidation reactors.

Two-step reforming features are a combination of tubular reforming (primary reformer) and oxygen-fired secondary (autothermal) reforming. In this concept the tubular reformer is operating at less severe operation, i.e. lower outlet temperatures (refer to Section 2.6.2).

1.3.2 Autothermal reforming

The ATR technology was pioneered by SBA and BASF in the 1930s [413], by Topsøe and SBA in the 1950s [331], and later Topsøe alone [107] [111] developed the technology at first for ammonia plants and later for large-scale gas-to-liquid plants. Today, ATR is a cost-effective technology for the synthesis gas section for a variety of applications [187].

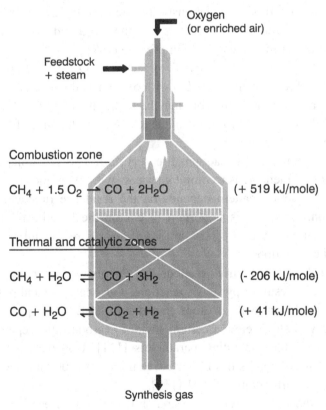

Figure 1.18 ATR reactor.

The ATR reactor consists of a burner, a combustion chamber, and a fixed catalyst bed placed in a compact refractory lined vessel [107] as illustrated in Figure 1.18. Irrespective of whether the burner is thermal or catalytic or whether a fixed or a fluidised catalyst bed is used, the product gas will be determined by the thermodynamic equilibrium at the exit temperature, which in turn is determined by the adiabatic heat balance.

The feedstocks are hydrocarbons, steam, and either oxygen or air (or a mixture thereof). Optionally, carbon dioxide may be added to the hydrocarbon stream (refer to Section 2.4.2). The mixture of hydrocarbon and steam is preheated and mixed with oxygen in the burner. An adiabatic prereformer may be advantageous to eliminate thermal cracking of higher hydrocarbons in the preheater.

A turbulent diffusion flame ensuring intensive mixing is essential to avoid soot formation. The burner is designed to avoid excessive metal temperatures in order to ensure long lifetime [109].

Typically, the molar ratio of oxygen (as O_2) to carbon in the hydrocarbon feed stream is 0.5–0.6 with oxygen as the oxidant [111] and the temperature of the flame core may be higher than 2000°C. Hence, the design of the combustion zone is made to minimise transfer of heat from the flame to the burner [107].

Thermal combustion reactions are very fast. The sub-stoichiometric combustion of methane is a complex process with many radical reactions [466]. The reaction pattern depends on the residence time/temperature distribution. Hence, it is important to couple the kinetic models with CFD simulations by post processing [466] or by direct coupling in more advanced calculations.

The sub-stoichiometric combustion involves the risk of soot formation as a result of pyrolysis reactions with acetylene and polycyclic aromatic hydrocarbons as soot precursors [107] [465]. The soot formation will start below a certain steam-to-carbon ratio depending on pressure and other operating parameters [111]. However, the data in Table 1.8 shows results from a soot-free pilot test (100 Nm^3 NG/h) at a low steam-to-carbon ratio of 0.21 [111].

The methane steam reforming and shift reactions (Reactions R1 and R4 in Table 1.2) also occur thermally in the combustion zone, but far from equilibrium for the reforming reaction, whereas the shift reaction remains close to equilibrium.

Table 1.8 ATR pilot test at low H_2O/C. $H_2O/C=0.21$, $O_2/C=0.59$, P=24.5 bar, $T_{exit}=1057°C$. Product gas analysis vol% [111].

H_2	56.8
N_2	0.2
CO	29.0
CO_2	2.9
CH_4	1.0
H_2O	10.1
	100.0%

$H_2/CO=1.96$, selectivity to H_2+CO (dry gas): 95%.

Soot precursors and residual methane are converted by steam reforming and shift reactions in the catalyst bed. These reactions are in equilibrium in the gas leaving the catalyst bed and the ATR reactor. The catalyst size and shape is optimised to have sufficient activity and low pressure drop to achieve a compact reactor design. The catalyst must be able to withstand the high temperature without excessive sintering or weakening and it should not contain components volatile at the extreme conditions. A catalyst with nickel on magnesium alumina spinel has proven to fulfil these requirements [107] [111].

Specific studies [3] have been made on the use of ATR for synthesis gas production for very large methanol and FT plants, considering the limitation by other parts of the plant, e.g. boilers, compressors, air separation units, etc.

1.3.3 Catalytic partial oxidation

In catalytic partial oxidation (CPO), the reactants are premixed and the reactions proceed on the catalytic reactor without a burner, in contrast to ATR [184]. The principle of a fixed-bed catalytic partial oxidation (CPO) is illustrated in Figure 1.19.

CPO with no flame was practiced in the 1950s [355] at low pressure and later by Lurgi at pressure [231]. The reaction was started by an ignition catalyst. The addition of oxygen to a fluidised bed steam reformer has been studied by a number of groups [53] [429] including pilot scale operation [180] and lately with hydrogen membranes [100] [341].

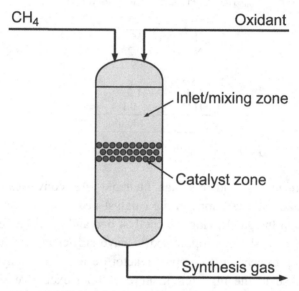

Figure 1.19 Principle of catalytic partial oxidation (CPO).

Extensive studies of CPO reactions were carrried out by Lanny Schmidt et al. [229] [443] using a millisecond fixed-bed reactor. It was possible to produce syngas over a rhodium monolith at residence times of milliseconds [229]. Platinum was less active than rhodium. It was shown [242] that the reactions take place in an oxidation zone as combined surface/gas-phase reactions followed by a steam reforming zone with equilibration of the steam reforming and shift reactions. It was also possible to convert liquid hydrocarbons [164], ethanol [434] and biomass [444] in the millisecond reactor.

The fixed-bed CPO technology has been studied at pilot scale [41] and at pressure. When operating at 20 bar [2] [41] and $O_2/NG=0.56$, it was possible to achieve stable conversions close to thermodynamic equilibrium.

The direct CPO reaction R10 in Table 1.7 appears to be the ideal solution for methanol and Fischer–Tropsch syntheses, as it provides a H_2/CO molar ratio of 2 and has a low heat of reaction:

$$CH_4 + 0.5 O_2 \rightarrow CO + 2H_2 \qquad (1.14)$$

It is often considered a "dream reaction" with H_2/CO ratios lower than two at low conversions, but industrial utilisation would imply expensive recycle of non-converted methane as also discussed for direct conversion of methane (refer to Section 1.1.2). Other studies which claim high yields at low temperatures may be misleading [105].

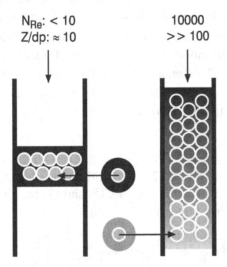

Figure 1.20 CPO catalyst and operating conditions [410]. In a flat bed reactor the catalyst temperature may easily be significantly higher than the gas temperature, whereas the catalyst and gas temperatures may follow each other in a reactor operating at high Reynold numbers. Reproduced with the permission of Springer.

Note
High yields of carbon monoxide and hydrogen were claimed in CPO studies at reactor temperatures as low as 300°C [105] with product gas compositions corresponding to equilibrium of the reforming reaction at 600–700°C. However, it could be estimated [153] [410] that the catalyst surface might have had an over-temperature of 300–400°C because of strong film diffusion due to low Reynolds number N_{Re}. Therefore, the

product gas most likely reflects the gas equilibrium at the catalyst surface. At low N_{Re} in a flat reactor operating with low conversion, the catalyst temperature may easily be higher than the gas temperature and close to that resulting from the adiabatic temperature increase at full conversion as illustrated in Figure 1.20. In an integral reactor operating at high N_{Re}, the temperatures of the gas and the catalyst will increase simultaneously through the bed.

Measurements of conversions at CPO conditions reported in the literature are seldom showing conversions relative to the equilibrium conversions. Yields are often expressed as selectivities to hydrogen and carbon monoxide although the product gas in most situations is close to equilibrium for the reforming and shift reactions. Therefore selectivity data should be supplemented by calculation of approach to equilibrium, ΔT_{ref} (Equation 1.11). Reported selectivities may be misleading. Even sophisticated catalyst compositions lead to equilibrated product gas as illustrated in Example 1.4 below.

Example 1.4
As an example, one study [27] claimed high selectivity to $CO+H_2$ using a sophisticated catalyst at 775°C. Equilibrium calculations are carried out for a stoichiometric mixture of CH_4 and O_2 reacting at 775°C and ambient pressure according to:

$$CH_4 + 0.5 O_2 \leftrightarrow CO + 2H_2 \quad K_{eq,ox} = \frac{P_{H_2}^2 \cdot P_{CO}}{P_{O_2}^{0.5} \cdot P_{CH_4}} = 2.7210^{11}$$

From an equilibrium point of view all oxygen is reacted, but due to the presence of a catalyst, equilibrium is also established in the steam reforming and shift reactions:

$$CH_4 + H_2O \leftrightarrow CO + 3H_2 \quad K_{eq,ref} = \frac{p_{H_2}^3 \cdot p_{CO}}{p_{CH_4} \cdot p_{H_2O}} = 96.06$$

$$CO + H_2O \leftrightarrow CO_2 + H_2 \quad K_{eq,shf} = \frac{p_{H_2} \cdot p_{CO_2}}{p_{CO} \cdot p_{H_2O}} = 1.1603$$

Values of the equilibrium constants are calculated using the data given in Appendix 2. It is seen that full conversion can be assumed in the oxygen reaction, so after solving the two remaining equations for the unknown conversions using the principle in Example 1.1, the following equilibrium composition is found: CH_4 3.53, H_2 62.08, H_2O 2.24, CO 30.87 and CO_2 1.29. This corresponds to a total conversion of CH_4 equal to 90% and selectivities to CO and H_2 equal to:

$$CH_{4,CO} = \frac{CO}{CO+CO_2} 100 = 96\%$$

$$CH_{4,H_2} = \frac{H_2}{H_2+H_2O} 100 = 97\%$$

This is in fine agreement with measurements such as those published in [27] for CPO over transition metal catalysts. It should be noted that adding an inert makes the equilibrium even more favourable due to the decrease in partial pressures.

As shown in Table 1.8, the ATR process at low steam carbon ratio is very close to fulfilling $H_2/CO=2$ at selectivities above 90% for CO→H_2.

Figure 1.21 The dark reactor shows a CPO pilot (210 Nm^3/h CO+H_2 and 25 bar) in front of a tubular reformer of twice the capacity. The size of the CPO catalyst bed is indicated.

Therefore, the product gas from CPO at high conversions will be close to the thermodynamic equilibrium of the steam reforming and water-gas-shift reactions [242]. For adiabatic operation, the exit temperature is determined from a heat balance based on inlet flows and temperatures assuming that all oxygen is consumed.

If so, the same exit temperature is required in ATR and CPO to achieve similar syngas compositions for given feedgas streams. As the CPO requires a lower preheat temperature (approximately 250°C) than ATR (approximately 650°C), it means that CPO inherently has a higher oxygen consumption per tonne of product than does ATR.

Hence, CPO appears less feasible for large-scale operation. However, the compactness of the CPO reactor as illustrated in Figure 1.21 makes it attractive for smaller decentralised units (hydrogen, fuel cells, see Section 2.3).

1.3.4 Air-blown technologies and membranes

The partial oxidation routes have the advantages of eliminating the expensive heat transfer surfaces of the steam reforming processes, but they suffer from the necessity of an expensive oxygen plant which with the present technology will be based on cryogenic separation of air (Air Products, Air Liquide, Linde) [25]. The air separation units can be built at capacities of approximately 3500 tonnes O_2/d and above, but the investment may amount to 40% of the syngas unit.

The costs of oxygen plants have created a potential for air-blown technologies. However, with few exceptions (secondary reforming for ammonia: refer to Section 2.5; CO_2-free hydrogen for power: refer to Section 2.2; and fuel cells: refer to Section 2.3), the amount of nitrogen in the syngas is prohibitive for recycle syntheses, because of a huge accumulation of nitrogen. The use of air instead of oxygen results in large gas volumes and consequently big feed/effluent heat exchangers and compressors [2] [172] [408]. This may not be feasible for large-scale plants provided that the big purge streams (with low heating value) cannot be used for export of energy as steam or electricity.

One attempt to solve these problems has been the development of syngas technology based on oxygen ion selective membranes [95]. The membrane materials are non-porous and are composed by mixed metal oxides (perovskite, etc.) that conduct oxygen ions and electrons through the oxygen-deficient lattice structure. At high temperature, 700–1100°C, it is possible to achieve the required diffusivities [95] [99]. The challenge is the tightness of ceramic scale and metal/ceramic joints. The technology resembles that for solid oxide fuel cells (SOFC), however, without the transport of electrons involved. This means that the driving force depends on $\ln[p_{O_2,1}/p_{O_2,2}]$ in contrast to a gas diffusion driven membrane with the driving force proportional to $\sqrt{p_{O_2,1}/p_{O_2,2}}$.

With the high driving force across the membrane, there is no need for compression of air to the process pressure. However, as illustrated in Figure 1.22 there is still a need for a big feed/effluent heat exchanger in the "ceramic membrane reforming" scheme. One design (Air Products) is based on planar membranes with air and process gas flows in microchannels [95] [317]. This ITM (Ion Transport Membrane) technology has been demonstrated at pilot scale (28 Nm^3 syngas/h) involving temperatures and pressures up to 900°C and 28 bar, respectively.

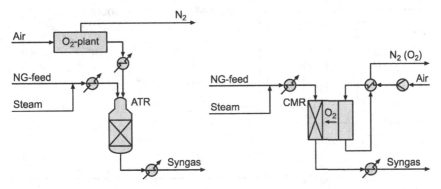

Figure 1.22 Comparison of ceramic membrane reforming (CMR) and ATR [409]. Reproduced with the permission of ACS.

Other studies [99] [341] apply ion-selective membranes in fluid bed reactor concepts. The ITM technology may have a higher potential for air separation [71] as an alternative to the cryogenic technology.

1.3.5 Choice of technology

The choice of technology is dictated by the need for high conversion, the requirements of the syngas composition, and by the scale of operation.

For methanol, tubular reforming would be cheapest at low capacities, whereas the autothermal reforming is favoured at large capacities. This is caused by the economy of scale being different for the tubular reformer and the oxygen plant as illustrated in Figure 1.23 [408].

For the intermediate range representing the capacity of world-scale methanol plants, a combination of tubular reforming and a secondary oxygen blown reformer is normally the optimum solution. In this concept, the tubular reformer is operating at less severe conditions, i.e. lower outlet temperature (see Section 2.6.2).

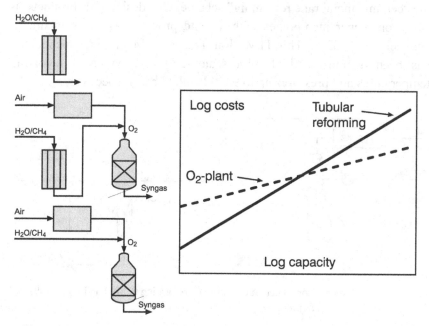

Figure 1.23 Impact of economy of scale [408]. Reproduced with the permission of Elsevier.

For very large scale, autothermal reforming and POX remain the preferred choice. ATR has lower oxygen consumption than POX. POX has slightly lower contents of methane and CO_2 in the product gas. CPO has a higher oxygen consumption than ATR and it may be difficult to scale up the premixing of reactants. The oxygen membrane technology still operates at the maximum temperature of 900°C, which means higher content of non-converted methane.

For small-scale operation, convective reformers are preferred due to compactness. The same is true for CPO units, in particular when an air-blown process is acceptable or when oxygen is available. The oxygen membrane technology may have a potential in such situations.

1.4 Other feedstocks

1.4.1 Alcohols, oxygenates

The thermodynamic constraints described for steam reforming disappear when methanol is used as feed [411]. The reaction takes place over a copper catalyst being active above 200°C and at the same time, the catalyst is not active for the methanation reaction (reverse reforming reaction R1 in Table 1.2). This means that a methane-free gas can be produced at low temperatures and at high pressures and that full conversion to CO_2 (and CO) hydrogen is achieved as illustrated in Figure 1.24. The heat of reaction is less than for steam reforming of hydrocarbons (Table 1.9). In contrast, the use of nickel catalysts results in methane-rich gases [125] [151] [392] and an overall exothermic process.

Table 1.9 Steam reforming of alcohols and DME.

Reaction		$-\Delta H^o_{298} \; \frac{kJ}{mol}$
R14	$CH_3OH + H_2O \leftrightarrow CO_2 + 3H_2$	-50
R15	$CH_3OH \leftrightarrow CO + 2H_2$	-91
R16	$C_2H_5OH + 3H_2O \leftrightarrow 2CO_2 + 6H_2$	-173
R17	$(CH_3)_2O + H_2O \leftrightarrow 2CO + 4H_2$	-205
R4	$CO + H_2O \leftrightarrow CO_2 + H_2$	41

Methanol reforming (decomposition) over Cu/Zn/Al catalysts [83] [258] [343] is a well-established technology [61] [260], mainly used for small hydrogen plants (less than 1000 Nm^3/h). Since the amount of heat required per mole of hydrogen is far less than for steam reforming of natural gas, the equipment becomes much cheaper than the tubular reformer. On the other hand, the heat of evaporation on a mass basis of methanol is about four times higher than that of naphtha.

A methanol reformer is typically a reactor heated by electricity or indirectly by circulating heating oil [61]. A more advanced and compact scheme [183] applies steam condensing on the outside of catalyst tubes.

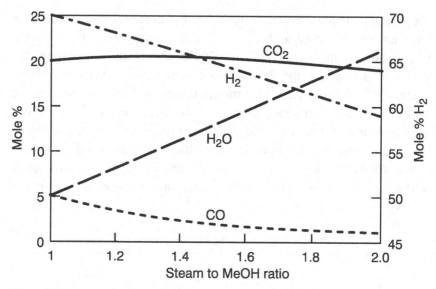

Figure 1.24 Steam reforming of methanol. Equilibrium composition (P=5 bar, T=280°C) [411]. Reproduced with the permission of Royal Soc. Chem.

The optimum choice of operating conditions [63] is around a steam-to-methanol ratio of 1.5 and a temperature in the range 250 to 300°C. The pressure does not influence the reaction rate, but very high pressures limit the equilibrium conversion, which otherwise is above 99%.

Methanol may be a suitable alternative in areas with expensive hydrocarbons considering the simplicity of the methanol reforming unit

[414]. Methanol from biomass may represent a sustainable route to hydrogen [485] (see Section 2.2).

DME is easily converted over Cu/Zn/Al catalysts [32] [191] [483] with little change in the layout of the plant [63]. Without water addition, decomposition of DME over a Cu/Al/Zn catalyst results in a syngas with $H_2/CO=2$. This route has been applied for small-scale supply of syngas, eliminating storage of carbon monoxide. By using methyl formate as feed, a syngas with $H_2/CO=1$ is achievable.

The three different methods of methanol (and DME) conversion are compared in the example below:

Table 1.10 Methanol conversion. Typical conditions.

Process	Cat.	H_2O/C	T_{exit} (°C)	P_{exit} (bar)	% Conv.	Product gas (vol.%)				
						H_2	CO	CO_2	CH_4	H_2O
MeOH Reforming	Ni	1.5	350	20	100.0	0.7	0.0	8.5	24.7	66.1
MeOH Cracking	Cu	0.	350	20	99.3	66.6	33.3	0.	0.	0.
MeOH Cracking	Cu	1.5	285	20	99.8	64.5	2.0	20.1	0.	13.4
DME Cracking	Cu	1.5	285	20	81.1	69.5	19.9	9.9	0.	0.7

The interest in fuel cells for automotive applications has resulted in a large number of investigations of reforming of methanol [286] [501] for on-board reforming or for distributed units for hydrogen production. Compact units have been studied [183] using micro-channel or plate reformers [94] [340] [442] or a combination with selective hydrogen membranes [26] [40].

As copper catalysts are sensitive to explosure to air during shutdown and start-up of on-board reformers, there has been an interest in palladium catalysts for the methanol reforming process [103].

Ethanol has attracted interest as a feedstock for syngas [419], thereby coupling biotechnology to classic catalysis [369]. However, steam reforming of ethanol is not as simple as the conversion of methanol, because ethanol is easily dehydrated to ethylene, being a coal precursor (refer to Chapter 5). Moreover, Reaction R16 in Table 1.9 involves the breakage of a carbon-carbon bond. This means that copper catalysts are not suitable, as they are poor catalysts for hydrogenolysis [462]. On the other hand, Group VIII metals like nickel being active for hydrogenolysis are also active for carbon formation (refer to Chapter 5). Therefore, there are many attempts to identify catalyst with stable performance [220] [359] [530]. It appears that noble metal catalysts (Ru, Rh) [280] [298] [359], promoted Co catalysts [299] [302], or bi-metallic catalysts such as Ni,Cu [539] look promising. Autothermal reforming of ethanol and other alcohols is an option [138] [147] [441], as it is for methanol.

The increased production of biodiesel leaves glycerol as a by-product which has been considered as a source for hydrogen by reforming [9].

Dumesic *et al.* [128] [248] found that aqueous phase reforming (APR) is a promising route for converting oxygenated hydrocarbons such as simple alcohols, ethylene glycol, glycerol and carbohydrates (sorbital, glucose, etc.) into hydrogen and carbon dioxide.

$$C_nH_{2n}O_n + nH_2O \leftrightarrow 2nH_2 + nCO_2 \qquad (1.15)$$

The reaction is carried out at temperatures in the range 150–265°C over a platinum catalyst or bi-metallic catalysts [248] [249] (Pt, Ni Pt,Co) including a non-noble metal catalyst, Ni,Sn [247] [456]. Although there is high thermodynamic potential at the reaction conditions for the formation of methane and other hydrocarbons, the experimental data show reasonable selectivities for hydrogen. The addition of Sn to Ni almost eliminated the formation of methane [456].

Although carbohydrates ($C_nH_{2n}O_n$) contain a lot of hydrogen, this is bound to oxygen, meaning that the "effective" hydrogen-to-carbon ratio is zero:

$$C_nH_{2n}O_n \rightarrow nC + nH_2O \qquad (1.16)$$

This is illustrated in Table 1.11, comparing with other molecules.

The liquid phase reforming by Dumesic *et al.* solves the problem by extracting oxygen as carbon dioxide, but then making hydrogen (Equation 1.15).

Table 1.11 Gross formula from the very low H/C ratios is obtained.

Molecule	Molecule – H_2O	H/C
Methane, CH_4	CH_4	4
Heptane, C_7H_{16}	C_7H_{16}	2.3
Ethylene, C_2H_4	C_2H_4	2.0
Acetylene, C_2H_2	C_2H_2	1.0
Methanol, CH_4O	CH_2	2.0
DME/C_2H_6O	C_2H_4	2.0
Ethyl-glycol $C_2H_6O_2$	C_2H_2	1.0
Glycerol, $C_3H_8O_3$	C_3H_2	0.7
Glycose, $C_6H_{14}O_6$	C_6H_2	0.3
$C_nH_{2n}O_n$	C	0

1.4.2 Coal, gasification

Coal conversion was considered in the 1970s as a reaction to the oil crisis. The target was the manufacture of substitute natural gas (SNG) via coal gasification and methanation (refer to Section 2.6.6). SNG was never introduced at large scale before oil prices decreased. The conversion of coal to liquid fuels was not feasible except under special circumstances (South Africa) because of the high investments in coal gasification. Coal gasification was also applied in combined cycle power plants (IGCC) in which the "syngas" was burned in a gas turbine followed by a steam turbine.

With the increasing use of coal, in particular in China [78], coal gasification has come into focus again [268]. Clean coal conversion is an issue in the US with its large coal reserves – however with an increasing demand on CO_2 capture and storage (CCS). Similar plans are considered in a number of countries. Gasification technologies have the advantage that CO_2 will be available at much higher pressures than envisaged by normal combustion. Gasification is the basis for using coal, petcoke, oil sand (tar sand) and biomass for the production of syngas. These fuels differ in properties and composition. Coals are normally classified in four groups, characterised by their content of fixed carbon and related heating value as listed in Table 1.12 [230].

The volatile matter increases from less than 8% in antracite to more than 27 wt% in lignite. In addition, the content of water may vary from less than 5 wt% in antracite to about 60% in German brown coal. Nitrogen (0.5–2%) will be converted into ammonia. The sulphur content may typically vary from 0.5–5 wt%. Sulphur will be converted to COS and H_2S. Sulphur will poison downstream synthesis catalysts and must be removed. Chlorine is normally below 1 wt%. Chlorine may cause corrosion problems in downstream equipment. Chlorine will react with ammonia from the nitrogen and deposition of ammonia chloride may foul waste heat boilers and limit their operating temperature [230].

Table 1.12 Classification of coal (Higman *et al.* [230]).

	Fixed carbon[a] (wt%)	Heating value[b] (MJ/kg)
Antracite	>92	36-37
Bituminous carbon	78-92	32-36
Subbituminous carbon	73-78	28-32
Lignite (brown coal)	65-73	26-38

a) Dry and ash-free. b) HHV.

The ash may contain a number of components which are volatile at high temperatures such as As, Hg. The ash content may typically be around 10% in anthracite coal, but up to 40% in some coals considered for gasification [230]. The properties of the ash (melting point, etc.) are important parameters.

The C/H ratio in coal varies around 14 wt/wt in lignite to 25 wt/wt in antracite as shown in Table 1.13.

Coke made from coal consists mainly of fixed carbon plus the ash. It is of high value for use in blast furnaces because of its strength.

Petroleum coke, petcoke and liquid refinery resid feedstocks are used for gasification. Petcoke is made from heavy residues in refineries by the coking process (delayed coking). Resids and petcoke are characterised by a low C/H ratio of 7–10 wt/wt and high content of sulphur 1–7 wt%.

These feedstocks typically contain vanadium (300–3500 mg/kg) which is volatile as V_2O_5 and nickel.

Table 1.13 H/C ratios of various fuels.

	C/H (wt/wt)	H/C (atom/atom)
Methane	3	4
Naphtha C_7H_{16}	5.3	2.3
Biomass	Approximately 9	1.3
Resid/asphalt	7–10	1.7–1.2
Lignite	14–17	0.9–0.7
Bitum. coal	14–19	0.9–0.10
Antracite	25	0.5

Orimulsion is another liquid fuel being an emulgation of water and bitumen from the Orinoco fields in Venezuela. Like resids, it contains sulphur, vanadium and nickel. Oil sand resids are heavy hydrocarbons recovered from deposits in sandstone, for instance in Alberta, Canada.

The different characteristics of these feedstocks determine the technology being optimum for the gasification to syngas. The feed is characterised by the atomic amounts of C, H, O, N, S and halogens, X. Oxygen will usually be in deficit so it must be added to the gasifier.

The main gasification reactions are shown in Table 1.14.

Table 1.14 Gasification of carbon. Main gasification reactions.

Reaction		$-\Delta H^o_{298} \dfrac{kJ}{mol}$
R18	$C + 0.5O_2 \rightarrow CO$	111
R19	$C + O_2 \rightarrow CO_2$	394
R20	$C + H_2O \rightarrow CO + H_2$	-131
R21	$C + CO_2 \rightarrow 2CO$	-172
R22	$C + 2H_2 \rightarrow CH_4$	75
R23	$H_2 + 0.5O_2 \rightarrow H_2O$	242
R24	$S + H_2 \rightarrow H_2S$	20

It is seen that combustion to CO produces 28% of the lower heating value of carbon. The remaining heating of the reactor can be used to drive the endothermic water gas reaction C [230]. In addition to these full conversion reactions the following six side reactions in Table 1.15 can also take place:

Table 1.15 Gasification of carbon. Equilibrium side reactions.

Reaction		$-\Delta H^\circ_{298}$ $\frac{kJ}{mol}$
R1	$CH_4 + H_2O \leftrightarrow CO + 3H_2$	-206
R4	$CO + H_2O \leftrightarrow CO_2 + H_2$	41
R25	$N_2 + 3H_2 \leftrightarrow 2NH_3$	92
R26	$N_2 + 2CH_4 \leftrightarrow 2HCN + 3H_2$	-285
R27	$CO + H_2O \leftrightarrow HCOOH$	26
R28	$H_2S + CO_2 \leftrightarrow COS + H_2O$	-29

The conversions in these reactions depend on the actual design of the coal gasifier. They are usually specified using experimentally determined temperature approaches to equilibrium.

The H_2/CO ratio in the syngas relates to the H/C ratio in the feed [230] (see Table 1.13) and to the addition of steam.

However, a simple adiabatic heat balance should be corrected for heat capacity of ash and at high temperature also heat of melting, heat losses, etc.

The optimum operation for syngas is: 1) under pressure to minimise compression of the larger syngas volume to the synthesis pressure; 2) at a low surplus of oxygen to minimise CO_2 production and to reduce oxygen costs; and 3) at high temperature to ensure complete combustion of tar components to minimise the formation of methane.

The conditions for maximising $CO+H_2$ production are illustrated in Figure 1.25.

These ideal conditions should be balanced by the constraints given by solving a number of secondary problems caused by the ash and other impurities as well as low temperature.

The gasification technologies are normally classified into 3 groups: moving bed gasifiers; fluid bed gasifiers; and entrained flow gasifiers. The main characteristics are summarised in Table 1.16. Typical raw gas compositions are shown in Table 1.17.

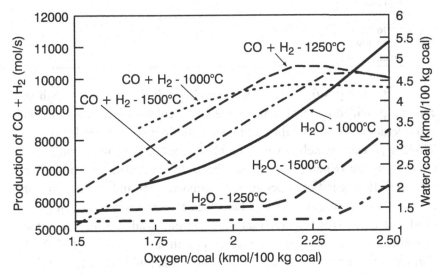

Figure 1.25 Gasification of carbon yields of CO+H$_2$. Discrete value of T. CO + H$_2$ refer to the left y-axis and H$_2$O to the right y-axis.

In the moving bed gasifier [230], the coal is added in the top via a hopper. It has a large consumption of oxygen and steam. The coal moves downwards by gravity in counter-current with the gas stream. This results in high efficiency and low oxygen consumption, but the low gas exit temperature means high methane (9–12 vol% dry gas) and tar in the product gas. Methane is a disadvantage for most syntheses, but an advantage for substitute natural gas. Tar results in difficulties in gas cleaning and recovery of water from the process condensate.

Table 1.16 Coal gasifiers. Main characteristics.

	Ash	T_{exit} (°C)	P (bar)
Moving bed:			
Lurgi–Sasol	Dry	650	25–30
BGL	Slagging	>650	
Fluid bed:			
HT Winkler	Dry	900–1050	10
Entrained flow:	Slagging	1250–1600°C	20–80
GE (Texaco)			
Shell			
Prenflo			
E Gas (Dow)			
Siemens (GSP)			

The dry-ash gasifier is the preferred choice if the coal has a high ash content, provided the coal lumps have sufficient strength for the moving bed operation. If not, the fluidised bed technology may be preferred. The Winkler process, which is very flexible to coal quality, was used for the German lignite (Rheinbraun). It has been demonstrated for 10 bar. The entrained flow gasifiers operate with grained coal and oxygen in con-current flows. The gasification takes place at high temperature and short residence time. This requires more oxygen but results in high feedstock flexibility and a tar-free product gas with high yields of $CO+H_2$. Furthermore, pressures up to 80 bar can be applied. Hence, entrained flow gasifiers are typically preferred for manufacture of syngas.

Table 1.17 Coal gasification. Typical composition of raw gas (vol%). Coal – Illinois No. 6 [461].

	Lurgi Dry ash	BGL	GE (Texaco)	Shell
H_2	16.1	26.4	29.8	30.0
CO	5.8	46.0	41.0	60.3
CH_4	3.6	4.2	0.3	-
CO_2	11.8	2.9	10.2	1.6
H_2S	0.5	0.6	1.0	
COS	-	0.1	0.1	0.1
N_2	0.1	3.8	0.7	0.1
H_2O	61.8	16.3	17.1	3.6
NH_3	0.3	0.3	0.2	2.0
HCN				0.1

The various gasifiers differ in method of feeding and in waste heat recovery [230]. The GE (Texaco) gasifier feeds coal in a waste slurry (paste) which means that oxygen is used for evaporation of water. However, the slurry feed system allows a much higher pressure (up to 70–80 bar) than possible for gasifiers with dry feeding systems.

The cooling of the exit gas takes place in a simple water quench. The Shell and Prenflo gasfiers are a result of a joint optimisation of the Koppers Totzek atmospheric gasifier. They apply dry feeding of coal and heat recovery in a downstream boiler. This is in principle more efficient, but the boiler is expensive and subject to fouling. The Prenflo gasifier is now marketed with the option of having a water quench and GE is studying systems for dry feeding. In this way, the gasification technologies are approaching each other.

The GE (Texaco) and the Shell gasifiers have been used for decades for gasification of natural gas and liquid hydrocarbons (heavy oil, resid).

The E-gasifier [230] applies a coal slurry feed as the GE gasifier, but with a two-stage feeding where the second stage has a function as "chemical quench". The tar in the product gas is recycled to the first combustion stage. The Siemens gasifier [230] is a further development of the Schwarze Pumpe gasifier (GSP). It operates in process steps close to an optimum by using dry feeding and a two-step heat recovery using a boiler followed by a quench, thus avoiding the low-temperature problems of a boiler. There is still room for improving the gasification technology.

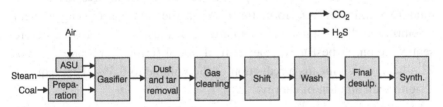

Figure 1.26 Coal conversion. Principal flow sheet.

The gasifier is the heart of a coal-based plant, but as shown in Figure 1.26, it includes a number of units for preparation of the coal feed as well as for scrubbing the raw product gas for removal of dust followed by a number of catalytic reactors and adsorption masses for removal of impurities and units for removal of CO_2 and recovery of sulphur. As a rule of thumb, the investments for a coal-based synthesis plant is around three times higher than those for a natural gas-based plant. This does not include the costs of CO_2 sequestration.

Carbon capture and storage (CCS) may well dictate the acceptance of coal conversion in the future [30]. Capture of CO_2 by absorption using the wet scrubbing processes is well-known technology in the process industry (refer to Section 1.5.3). CO_2 is already stored when used for re-injection in oil fields for enhanced oil recovery. In nature, CO_2 has been stored in geological cavities for millions of years with no sign of leakages and there appears to be plenty of capacity for geological sequestration. However, the experience is limited and there are a number of concerns in relation to leakages, chemical reactions, pollution of ground water, etc. There is a need for more research and demonstration projects to create a solid basis for public acceptance. The CO_2 capture is the most expensive part of CCS [315]. The costs of the absorption process is almost proportional to the partial pressure of CO_2. For power plants, three options for CO_2 capture are considered: post combustion (flue gas); pre-combustion (as in IGCC power plant); and oxy-fuel (using oxygen for combustion).

Post-combustion capture of CO_2 in the flue gas is expensive and reduces the efficiency of a power plant, whereas gasification under pressure yields a higher CO_2 partial pressure with lower cost of loss in efficiency. Oxy-fuel is an attempt to increase the partial pressure of CO_2 (and to eliminate NO_x formation). A gasification-based syngas plant resembles an IGCC plant. The captured CO_2 must be compressed before sequestration, at best to a super-critical liquid (critical pressure of CO_2, 73.8 bar), which means a reduction of the volume to 0.3%. This requires a significant amount of energy.

Example 1.5
A 4000 MW coal-based power plant has an efficiency equal to 40% based on the LHV of the fuel. The coal can be assumed to be pure C and it has an LHV equal to 27 MJ/kg.
The total amount of feed is:

$$C_{feed} = \frac{4000}{0.4 \cdot 27} = 370.4 \, kg/s = 30.9 \, kmol \, carbon/s$$

This is also the total emission of CO_2 in the flue gas from the plant. ($P_{CO2}=0.2$ bar).

If this amount is to be compressed for CCS, a simplified expression for the minimum work is (see Section 2.1.3):

$$W_{compression} \cong C_{feed} RT \ln\left(\frac{73.8}{0.2}\right) = 30.9 \cdot 8.3144 \cdot 298.15 \cdot 5.91 = 452 MW$$

In a gasification plant the CO_2 is available at 1 bar after the CO_2 wash. This means a reduction in compression energy to 329 MW or 0.24 GJ/tonne CO_2 based on the amount of C in the feed.

The example illustrates that CCS means that the efficiency of an IGCC power plant will decrease from 43% to 35% [418]. The capture costs are already a part of syngas-based plants, but the energy for CO_2 compression is similar to that of IGCC plants.

1.4.3 Biomass

Sequestration of CO_2 should not be an issue of gasification of biomass to syngas. Biomass covers a wide field of materials ranging from vegetable biomass such as wood, straw, grain, black liquor from the paper industry, animal biomass and various waste. Gasification of biomass may be an alternative to biochemical routes [419]. However, there are still problems to be solved.

Biomass differs from coal and heavy refinery products [230]. It has a smaller heating value (10–20 MJ/kg), low sulphur and, in particular straw has a high content of chlorides. The ash will typically have a low melting point and be aggressive. The fixed carbon is below 15%. It means that it is difficult to achieve high temperatures in gasification without using excess oxygen in the air stream. The relatively low temperatures (8–900°C) means that the product gas will have a high content of tar components which may be difficult to remove. At present, a number of gasifiers are under development. It appears that fluid bed gasifiers [127] [230] are being preferred (Carbona, Foster Wheeler, Güssing).

Biomass can be co-gasified in entrained coal gasifiers, provided there are no milling (grinding) problems.

A special entrained flow gasifier was developed by Chemrec for gasification of black liquor from the paper industry. If successful, there is a huge potential for this niche market [478].

Biomass has another disadvantage as feedstock which is related to its very low energy density (approximately 3.7 GJ/m^3), which is almost ten times lower than that of oil. It means that there are severe logistics problems. Except for grain, it is not economic to transport biomass over long distances. This may be solved by using a two-step procedure [225] with a decentralised flash pyrolysis [133] to a bio-oil which can be gasified in gasifiers similar to those used for heavy refinery products [230]. It was also shown that bio-oil and solid biomass could be gasified in the millisecond CPO unit [435]. Another catalytic route being explored is gasification (steam reforming) of biomass to hydrogen and carbon dioxide in super critical water (T>374°C, P>221 bar) [445]. Potassium carbonate improves the rate [76]. Other attempts have used nickel catalysts [132].

1.5 Gas treatment

1.5.1 Purification

Most of the catalysts for syngas manufacture and downstream syntheses are sensitive to poisoning – in particular sulphur. Therefore, there is a need to purify the feedstock for the reforming processes and to purify the raw gas from gasification of heavy feedstocks which cannot be purified before gasification. Modern process plants take full advantage of high catalyst activity by operating at more severe and economic conditions. This has sharpened the requirements to purification. Characteristic reactions are seen in Table 1.18.

Table 1.18 Reactions for gas purification.

Reaction		$-\Delta H^\circ_{298} \dfrac{kJ}{mol}$
R29	$H_2S + Cu \leftrightarrow Cu-S + H_2$	
R30	$H_2S + ZnO \leftrightarrow ZnS + H_2O$	63
R31	$SO_2 + 3H_2 \leftrightarrow H_2S + 2H_2O$	207
R32	$COS + H_2O \leftrightarrow H_2S + CO_2$	30
R33	$RSH + H_2 \leftrightarrow RH + H_2S$	
R34	$HCN + H_2O \leftrightarrow NH_3 + CO$	50

Sulphur in natural gas is present mainly as hydrogen sulphide, lower mercaptanes and thioethers. Hydrogen sulphide is easily absorbed on zinc oxide (Table 1.18). The uptake takes place in three zones as illustrated in Figure 1.27 [391]. A layer of complete conversion – with solid state diffusion as rate-determining step – to bulk zinc sulphide is followed by a front for bulk saturation with gas diffusion in the pore system as rate-determining step and a chemisorption front adsorbing the traces of sulphur passing through the absorption fronts.

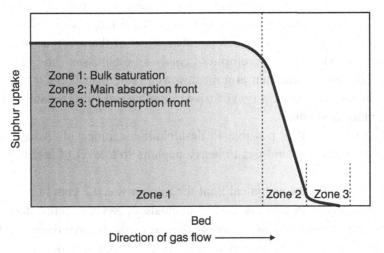

Figure 1.27 Sulphur profile in a zinc-oxide bed.

It means that the H_2S/H_2O ratio of the gas leaving the reactor is determined by the chemisorption equilibrium and not the bulk phase equilibrium.

Commercial zinc-oxide absorbents are a compromise of saturation level (kg S/m^3 bed), porosivity (high diffusivity), and high chemisorption capacity (purity, surface area). In practice, this means that sulphur is removed to the ppb range provided the gas is dry. The ZnO reactor typcially operates at 350–400°C, but is effective also at low temperature, although with longer absorption front and lower saturation value.

Although lower mercaptanes are partly withheld by zinc oxide, a hydrodesulphurisation step is normally installed prior to the zinc-oxide vessel. The hydrogenation catalyst is a Co(Ni)Mo/Al_2O_3 catalyst operating at 350–400°C and it converts organic sulphur into hydrogen sulphide (Table 1.18). The HDS step needs a (recycle) stream of hydrogen for the reaction. Without hydrogen present, thioethers and thiophenes will pass unconverted through the reactor. The hydrogen-to-feed ratio varies with the feedstock in question. A recycle hydrogen stream may often contain CO_2, which may result in H_2O by reverse shift over the HDS catalyst. This may have an impact on the equilibrium over the zinc oxide (Table 1.18). Carbon oxides may also cause a lower activity of the HDS catalyst.

A zinc-oxide mass containing copper may take care of traces of sulphur passing through the zinc oxide by establishing the chemisorption equilibrium over copper, which is independent of the presence of water (Table 1.18). The prereformer catalyst establishes the H_2S/Ni chemisorption equilibrium at a much lower value than H_2S/Cu (refer to Section 5.4). However, it is expensive to use the prereformer catalyst as a desulphurisation mass.

In this way, it is possible to desulphurise a range of hydrocarbon feedstocks from natural gas to heavy naphtha to a level of less than 10 ppb.

In the 1960s, the analytical limit for sulphur was 0.2 ppm [126] [391] and this stayed in process design manuals of several contractors for years. Surface science and better analytical methods have demonstrated that real limits are in the low ppb range (refer to Section 5.4).

Chlorine, which will be present as HCl or organic chlorine compounds, is a poison, in particular for copper catalysts [391] and it may cause stress corrosion in the equipment. Chlorine can be removed by promoted alumina. Chlorine may be present in certain refinery off-gases and in landfill gas. Chlorine may also originate from failure in the water purification system. If so, it will pass the guard bed and should be captured by a guard in the low-temperature shift reactor (see Section 1.5.2).

Ammonia is not a poison for metal catalysts (Ni, Cu, Fe), but it may deactivate acidic zeolite catalysts and HDS catalyst. It can be removed over a low-temperature bed with zeolite.

The raw dust-free syngas from purification of coal and heavy hydrocarbons contains sulphur as COS and H_2S (Table 1.18) according to the equilibrium (see Appendix 2). COS can be removed by a promoted zinc oxide [246].

1.5.2 Water gas shift

The water-gas-shift (WGS) reaction R4 in Table 1.2 is used to adjust the H_2/CO ratio of the syngas (refer to Figure 1.10). For ammonia and hydrogen plants, as much CO as possible should be converted into H_2. This involves the WGS reaction. The reaction is exothermic, meaning that low temperatures are required for high conversion. The WGS accompanies the steam reforming and the partial oxidation reactions, and the equilibrium is established at the high exit temperatures, but the WGS processes operate at lower temperatures. Many materials are active for WGS [214] (even rusty surfaces in sample lines may be active and cause misleading analyses).

Depending on the operating conditions, three different types of WGS processes are applied [232] [303] [391]. High-temperature shift (300–500°C) over a robust catalyst is used for primary conversion. Medium temperature shift (200–330°C) is used for special purposes. Low-temperature shift (185–250°C) is used to achieve maximum conversion. Sour gas shift (350°C) is used to operate under high sulphur conditions and low H_2/CO (raw coal gas, etc).

The *high-temperature shift* process is typically carried out in adiabatic reactors at an inlet temperature above 300°C and with a temperature increase up to 500°C. The catalyst is a robust Fe-Cu-Cr catalyst [68]. Chromium, which prevents sintering, is present as an iron chromium internal spinel [175] [361]. The activity is improved by promotion with a low percent copper, which will be present as small metallic crystallites on the iron chromium spinel [232] [266]. This will also inhibit the formation of hydrocarbons at low H_2/CO ratios [96]. It was shown [235] [391] that the activity for this reaction could be related to the phase transition into iron-carbide, which is a Fischer–Tropsch catalyst:

$$5Fe_3O_4 + 32CO \leftrightarrow 3Fe_5C_2 + 26CO_2 \qquad (1.17)$$

The transformation point can be determined by means of the "principle of equilibrated gas" [235] (refer to Section 5.2.3).

The catalyst is robust towards sulphur [67] [68]. However, this can hardly be utilised as the product gas leaving the steam reformer is practically sulphur-free and because the raw syngas from a coal gasifier will have the potential for carbide formation due to the low H_2/CO ratio (see below).

Low-Temperature Shift (LTS) catalysts are copper-based catalysts which operate at temperatures as low as 185–225°C [232] [303] [391]. They may in principle operate at even lower temperatures, but limited by the dew point of the process gas. The commercial catalysts are typically based on $Cu/ZnO/Al_2O_3$. The active phase is copper in close connection with zinc oxide. Although the activity relates to the copper surface area, the role of the zinc oxide is still being discussed [216] [232], as is the reaction mechanism [207] [337].

Another approach is the use of a $Cu/ZnO/Cr_2O_3$ catalyst [232] [69] [329].

The LTS catalyst resembles the methanol synthesis catalyst and promotion is required to eliminate the by-product formation of methanol [97]. The activation of copper catalysts is strongly exothermic and should be carried out with care. The phase transitions during activation have been followed through *in situ* measurements using EXAFS and XRD [124].

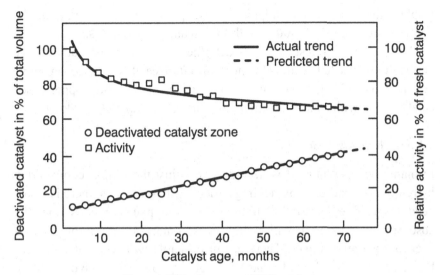

Figure 1.28 Performance of LTS catalyst.

Sulphur is a poison for copper catalysts, but due to the presence of zinc oxide, the catalyst has a high sulphur capacity.

The LTS catalysts lose activity during operation due to sintering. After a large decrease, the activity levels off as shown in Figure 1.28. The sintering is accelerated by chlorine apparently promoting transport of copper while migrating through the catalyst bed [391].

The performance of the LTS catalyst can be analysed by a model including the absorption and chemisorption parts for the uptake of sulphur and chlorine as well as the progression of sintering as illustrated in Figure 1.28.

It may be advantageous to carry out the WGS reaction on the raw syngas from gasification of coal and heavy hydrocarbons by a so-called *sour shift catalyst*. This allows removal of CO_2 and H_2S in the same wash system (see Section 1.5.3). It requires a catalyst that is sulphur-tolerant and capable of working at low H_2O/C ratios. The conventional iron-based HTS catalyst can operate in the presence of sulphur, but it requires addition of significant amounts of steam to eliminate the problem of the carbide formation reaction. This problem is solved by using a molybdenum sulphide-based catalyst [168] [232]. The catalyst is promoted and is based on alumina support. It requires the presence of sulphur to maintain the sulphide phase [245]. It can operate in the temperature range 200–500°C with low steam/dry gas ratios.

The recent interest in compact fuel cell units for automotive applications has led to development of WGS catalysts based on noble metals [134] [277], which in contrast to the iron- and copper-based catalysts can withstand exposure to air during start-up and shutdown.

1.5.3 Acid gas removal

In many syngas plants, it is necessary to adjust the syngas composition by removal of carbon dioxide. In gasification-based plants, there is also a need to remove hydrogen sulphide. In large plants, this is done by absorption processes in which the gas is contacted with a solvent in a packed absorption tower. There are in principle two types of processes based on physical and chemical absorption, respectively. Main commercial processes are listed in Table 1.19 [281] [284]. The physical absorption processes (Recticol, Selexol) are driven by ordinary gas solubility. The solvent circulation rate, (size of tower, heat exchange) depends on the total quantity of gas and hence they are best suited for syngases with high contents of CO_2 and H_2S. In contrast the design of the chemical absorption processes (MEA, Benfield) is related to the concentration of CO_2 (and H_2S) and they are the preferred choice for gases with low partial pressures. The chemical processes require energy to release CO_2. This is normally done by heating. The selection of process depends on parameters such as selectivity (co-absorption of useful gas components), required purity, etc.

Table 1.19 Absorption processes for acid gas removal.

	Solvent
Chemical	
MEA	Methanol amine
DEA	Diethanol amine
Benfield	Potassium carbonate
Physical	
Rectisol	Methanol
Selexol	Dimethylether of polyethylene glycol

The physical wash processes can be designed so that H_2S and CO_2 are absorbed separately, meaning that both gases can be recovered at high concentration. The CO_2 stream may be compressed for CCS (refer to Section 1.4.2). The H_2S stream is converted either into elementary sulphur by the Claus process or into concentrated sulphuric acid by the WSA process [291].

2 Syngas Applications

2.1 Thermodynamic framework for syngas processes

A syngas process can be divided into the following basic elements:

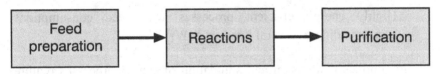

Figure 2.1 Process steps in syngas plants.

Where:

- "Feed preparation" includes unit operations, such as heating, purification of feedstock and addition of steam;
- "Reactions" are the steam reforming reactions;
- "Purifications" include unit operations for adjustment of product composition in agreement with downstream requirements and removal of undesired compounds such as water, if the product must be dry.

Simulation of syngas processes are performed using flow sheeting [82], where the individual unit operations are connected by streams and the heat and mass balances for the entire process are solved sequentially. Well-proven commercial tools and solution methods for many processes are available as described in [82], but to design catalytic processes both simplified and rigorous reactor models are key items and for this purpose a dedicated tool such as the one described in [116] can be used.

Rigorous reactor models are the subject of Chapter 3, but process simulation relies on thermodynamic properties, and this will be described below.

Syngas processes are usually characterised by:

- Large production volumes;
- Gas-phase catalytic processes;
- High pressure (up to 100 bar);
- Temperatures ranging from 0°C in condensation and up to 2500°C in combustion;
- Steam generation, steam superheating and condensation as water; and
- Highly energy efficient processes with feed consumptions corresponding to several hundred MW.

This implies that design and simulation of syngas processes require thermodynamic properties with an accuracy corresponding to what is used in an electric power plant.

2.1.1 Syngas properties

To calculate thermodynamic properties for hydrocarbon mixtures and gases, well-proven methods based on modifications of the Soave–Redlich–Kwong (SRK) or Peng–Robinson (PR) equations of state [353] [470] and with parameter extensions for hydrogen [212] are available. Such methods are used within refining. These EOS models are, however, not sufficiently accurate for steam and water, where a method as accurate as the one described in the International Steam Tables [515] must be used. Although syngas processes are carried out at high temperatures, where an ideal gas model even at the higher pressures gives sufficient accuracy, properties at low temperature and process parts with vapour and liquid phases show significant deviation from an ideal gas.

It would be highly desirable for reasons of thermodynamic consistency and to avoid energy discontinuities to use just one thermodynamic model representing the properties of all parts of a process; but no consistent thermodynamic model can with a sufficient accuracy calculate the thermodynamic properties of all relevant syngas mixtures. This is not only due to the absence of adequate mixing rules for mixtures of steam/water and other compounds, but also that water has molecular properties requiring an elaborate model as the one in the Steam Tables. New formulations, such as the CPA equation of state [528] that has been used to model the H_2O-H_2 system, are promising. An alternative is to use the principle of corresponding states [316], but calculating all properties of syngas mixtures from water will require a significant number of additional parameters.

A practical solution is therefore to calculate thermodynamic properties by combining the steam/water model from the Steam Tables with the properties of the other components as calculated from the SRK equation of state.

For a gas phase, this can in practice be carried out by first calculating the properties using the SRK equation of state for the complete mixture and then interpolate with the pure steam properties, so that for a mixture without water, the SRK equation state is used, whereas for pure steam, the steam formulas are used.

The advantage of using such a method is that the impact on the reaction properties is small and that heat duties connected with condensation and steam generation are in agreement with the Steam Tables.

Example 2.1
Example 1.3 resulted in an overall tubular reformer duty equal to 108 MW when using ideal gas thermodynamic properties. If the non-ideal gas is considered by use of the method described above, the duty increases to 109 MW.

The heat balance over the steam reformer itself is thus almost unchanged when real gas properties are used. The reason is the moderate pressure and the high temperature, but at a higher pressure or for mixtures close to the dew point the difference increases.

The liquid present in syngas processes is almost pure water with dissolved gases. Other liquid components may be present, but only in trace amounts. The properties of a liquid mixture are calculated by mixing the properties of water with those of the dissolved gases or of other liquid components. The heat of mixing can usually be neglected. A special case arises if a liquid hydrocarbon (such as naphtha) is used as feedstock, where the thermodynamic properties may be calculated using the SRK equation of state. Liquid naphtha and water form two liquid phases, but this will only be present in abnormal situations in a syngas plant, since steam will not be added till after evaporation of the naphtha.

The *phase equilibrium* method used in simulation of syngas processes must thus be able to calculate the properties of the following mixtures:

1. Water/steam
2. Gases, such as H_2, CO, CH_4, N_2, Ar and lower hydrocarbons present in natural gas
3. Trace amounts of organic compounds and liquids originating in other parts of a plant and recycled to the syngas section
4. By-products such as NH_3 that may be formed in the catalytic reactors
5. NH_3, CO_2, H_2S and HCN that may dissociate in water-forming ionic compounds
6. Hydrocarbon liquids such as naphthas
7. Coal gasification
8. CO_2 removal
9. Purification of biofuels

Obviously no thermodynamic phase equilibrium method can handle all these mixtures.

The compounds in 1–5 may be present in the entire syngas plant, whereas those in 6–9 are in specific process sections only, where they may be simulated using their own appropriate thermodynamic method. Liquid hydrocarbons (6) are simulated using the above-mentioned SRK equation of state and gasification (7) takes place at such a high temperature that an ideal gas assumption is appropriate. Purification and in particular CO_2 removal require specific packages, usually only available to the licensor.

As discussed above no equation of state can accurately calculate phase equilibrium for the components in 1–5 at all relevant conditions; instead activity coefficient models must be used. In such models, the characteristic basic property used by the phase equilibrium algorithms is the K-value, defined as $K_i=y_i/x_i$ where y_i and x_i are the mole fractions in the vapour and liquid phases, respectively.

The K-values in dilute water mixtures are divided into an equation for condensable components and an equation for non-condensable components. Whether the component is condensable or not is determined by the actual application temperature compared with the critical temperature for the component in question.

The rigorous equation for the K-value for a *condensable* component (1 and 3) is established from the equality of the fugacities in the two phases as [316] [353]:

$$K_i = \frac{\gamma_i P_i^{sat} \varphi_i^{sat}}{\varphi_i^V P} \exp\left\{\frac{v_i^{sat}}{RT}\left(P - P_i^{sat}\right)\right\} \quad (2.1)$$

The main condensable component is water, and it is seen that in this case the K-value equation converges towards Raoult's law.

All other liquid components are present in dilute solutions in water, which require an appropriate activity coefficient model for the dilute component. Here the UNIFAC/UNIQUAC model [197] is selected, where the activity coefficient is calculated as the sum of a combinatorial part and a residual part. The residual part uses fitted binary parameters valid for the components present in the liquid mixture. If no binary parameters are available – which is often the case in syngas preparation – the UNIFAC method may be used to calculate the residual part.

For a *non-condensable* component (2), a saturation pressure cannot be calculated and the rigorous equation for the K-value for a non-condensable component is thus described by Henry's constant as:

$$K_i = \frac{\gamma_i^* H_i^{sat}}{\varphi_i^V P} \exp\left\{\frac{v_i^{-\infty}}{RT}\left(P - P^{sat,solvent}\right)\right\} \quad (2.2)$$

The activity coefficient γ^* converges towards unity at infinite dilution in order to be in agreement with the definition of Henry's constant. In water mixtures this is a good assumption. The Poynting correction factor consequently uses the partial molar volume at infinite dilution in the liquid mixture. The Henry's law constants are correlated using scaled particle theory [353] [447]. Parameters fitted to experimental data are available for the common gases: O_2, H_2, N_2, CO, CO_2, Ar, CH_4, He and H_2S. It should be noted that the experimental data for the solubility of

CO are "disturbed" at higher temperatures (>100°C) by chemical reactions in the liquid phase forming HCOOH, so the CO correlation is actually extrapolated low-temperature solubility data [447]. The solubilities of the two components NH_3 and HCN are obtained by fitting to experimental data.

For the weak *electrolytes* (4+5), a number of reactions may take place in the liquid such as:

$$\begin{align} NH_3 + H_2O &= NH_4^+ + OH^- \\ CO_2 + H_2O &= HCO_3^- + H^+ \\ H_2O &= H^+ + OH^- \\ HCO_3^- &= CO_3^{--} + H^+ \\ NH_3 + HCO_3^- &= NH_2COO^- + H_2O \\ HCN &= H^+ + CN^- \\ H_2S &= H^+ + HS^- \end{align} \qquad (2.3)$$

The impact of these liquid phase reactions on the phase equilibrium properties is thus an increased solubility of NH_3, CO_2, H_2S and HCN compared with the one calculated using the ideal Henry's constants. The reason for the change in solubility is that only the compounds present as molecules have a vapour pressure, whereas the ionic species have not. The change thus depends on the pH of the mixture. The mathematical solution of the physical model is conveniently formulated as an equilibrium problem using coupled chemical reactions. For all practical applications the system is diluted and the liquid electrolyte solution is weak, so activity coefficients can be neglected.

Data for the equilibrium constants can be found in [52] [176] [177] [209] [269] [324]. Calculations show that the pH is very sensitive to the equilibrium constants justifying the assumption to neglect activity coefficients, since they are very difficult to determine accurately.

Example 2.2
The natural gas in the H_2 example (1.3) may contain 1% N_2, part of which is reacted in the tubular reformer to NH_3. It is assumed that the temperature approach to equilibrium is 200°C corresponding to an equilibrium temperature equal to 1075°C. The chemical equilibrium

corresponds to 64 ppm NH_3 in the dry gas after the tubular reformer. When condensing the stream, all NH_3 will remain in the liquid due to the excess of CO_2, and NH_3 will thus have to be removed in another way. If only Henry's law had been used, nearly all NH_3 would go to the vapour phase after the separator.

2.1.2 Synthesis process properties

The methods described in the previous chapter can in addition to the syngas processes also be used to simulate processes for manufacturing SNG, hydrogen, and reducing gas and for other processes such as fuel cells. The requirement is that no further liquid separation is used. Production of pure CO uses cryogenic separation requiring accurate thermodynamic properties at such conditions.

The following chapters also describe processes for manufacturing ammonia, methanol and derivatives such as DME and the Fischer–Tropsch synthesis. These processes are characterised by operation at a high pressure to favour chemical equilibrium, so the fugacity coefficients in Equation (1.8) are not unity. The product is usually separated out by condensation. As for a syngas preparation process, it would be preferable to use thermodynamically consistent and continuous models, which can be used not only to predict chemical equilibrium out of the reactors but also phase equilibrium. For reasons of continuity and consistency, however, and in particular since production of ammonia was one of the first high-pressure processes, a similar thermodynamic approach as described for syngas processes with separate models for the gas and liquid phases is used [328]. An equation of state is hence used to predict gas-phase properties including non-ideal chemical equilibrium, whereas an activity coefficient model is used to predict phase equilibrium using Equations (2.1) and (2.2).

For synthesis processes it should be added that basic chemical equilibrium data such as the free energy and enthalpy of formations originally have been derived from experimental data using a specific method to correct for non-ideal gas fugacity coefficients. This must be taken into consideration when selecting an appropriate method. This is the case for methanol, where a generalised method only as a function of pure component critical temperature and pressure was used to derive the ideal gas-phase equilibrium properties. It is also evident that accurate chemical equilibrium data are not easy to find in the literature and in many cases it is necessary to perform measurements. This is in particular the case for any by-product formation. To perform such measurements highly qualified personnel with experimental skills are required.

A thermodynamic model to simulate ammonia synthesis is described in [117]. It includes the use of the so-called Martin–Hou equation of state for the gas phase, whereas a simple Van Laar model is used to model the phase equilibrium including the solubility of the gases.

For methanol synthesis and its derivatives, such as DME, a UNIQUAC model may be used to predict phase equilibrium, but other models from [468] may be used as well. The key point is that parameters must be fitted to experimental data.

For Fischer–Tropsch a model similar to the one used for simulating ordinary refinery mixtures is used, but any organic by-product must be taken into consideration.

2.1.3 Process analysis

Process energy balances are established using the first law of thermodynamics. It has the following form for a steady-state flow process where kinetic and potential energy contributions are neglected:

$$\Delta H = Q + W \qquad (2.4)$$

For an adiabatic unit operation, such as an adiabatic catalytic reactor where transferred heat, Q, and shaft work, W, both are zero, the enthalpy change is thus zero, since the basis for the enthalpy is the heat of formation. A heat loss may be taken into consideration.

When simulating a process, only relative enthalpies have significance, but when a process is evaluated, it is also necessary to relate the properties of a stream to the amount of energy which can be released or consumed when the stream is converted to a standard state. Such a concept is used for pricing fuels, where the usual reference state for syngas mixtures is the lower heating value, LHV. It is defined as: "The heat released when the component in its standard state at 25°C and atmospheric pressure is reacted with oxygen to give CO_2, H_2O, SO_2, N_2, and halogens X_2, all in their gaseous state at 25°C and at a total pressure equal to 1.01325 bar abs."

The general reaction scheme to calculate LHV can thus be written as:

$$C_{n_C}H_{n_H}S_{n_S}N_{n_N}X_{n_X}O_{n_O} + (n_C + 1/4 n_H + n_S - 1/2 n_O)O_2 \rightarrow$$
$$n_C CO_2 + 1/2 n_H H_2O + n_S SO_2 + 1/2 n_N N_2 + 1/2 n_X X_2 \quad (2.5)$$
$$LHV = n_C \cdot H^o_{CO_2} + 1/2 n_H \cdot H^o_{H_2O} + 1/2 n_N \cdot H^o_{N_2} + n_S \cdot H^o_{SO_2} + 1/2 n_X \cdot H^o_{X_2}$$

where n_i is the individual atomic amounts and H^o the corresponding enthalpies of formation.

A total energy balance for a syngas plant may be established using the LHV, but it would be better to use the higher value, HHV, to avoid negative LHV values for liquid water or combustion efficiencies above 100, but this is not standard for syngas processes, where the LHV is used.

For use in process evaluation, it is important to note that both heating values only consider the absolute enthalpy conversion without considering where and how energy is degraded in the process. Such an analysis requires the second law of thermodynamics, and the combination of the first and second laws introduces the concepts of the ideal work and lost work [148] [468]. The ideal work is the thermodynamic work a mixture can carry out, when it is transferred from one state to another. It is closely related to the Gibbs free energy of formation, the difference being that all heat transfer is carried out reversibly at the temperature of the surroundings, T_s, so that the first law can be written as:

$$W = \Delta H - Q \quad \text{or}$$
$$W_{ideal} = \Delta H - T_s \Delta S \quad (2.6)$$

Hence, ideal work considers not only ideal shaft work, but also the work that can be extracted by reversibly transferring all heat quantities to and from the surroundings at the temperature T_s. Note that the change in properties caused by the reactions are automatically included, since the reference state for enthalpy is the enthalpy of formation.

The lost work can be calculated as:

$$W_{lost} = W - W_{ideal} = T_s \Delta S - Q \quad (2.7)$$

In addition to the concept of lost work, it would be advantageous to have an absolute value of the ideal work a stream can carry out. For this purpose the concept of exergy (originally defined as chemical availability by Gibbs, see e.g. [148]) is used. Such an 'objective' thermodynamic basis will in particular provide a more rigorous basis for calculating overall process thermodynamic efficiencies. In analogy with ideal work, the exergy is defined as the ideal work that can be obtained from a stream when it is converted from an actual state -1- to a standard state of the surroundings. Exergy is then the ideal work for the conversion to the standard condition, except that the sign is reversed:

$$Ex_1 = -(\Delta H_{1 \to s} - T_s \Delta S_{1 \to s}) \tag{2.8}$$

The standard state for exergy is here selected as the one used for the HHV, which is the concentrations the components have in dry air at atmospheric pressure and 25°C [115] [482]. This implies that the exergy of dry air and liquid water at atmospheric pressure and 25°C both are zero. Other reference states could have been selected. This concerns in particular the one chosen for CO_2, where it may seem more appropriate to select the concentration in a characteristic flue gas, but such a choice requires definition of a 'standard' flue gas composition and this is not convenient, since it depends on the amount of excess air used or if enriched air has been used in the combustion process. The state corresponding to HHV is convenient, although a flue gas even at 25°C consequently will have a positive exergy.

The change in exergy from state 1 to state 2 is thus the connected ideal work performed with the change:

$$\Delta Ex_{1 \to 2} = Ex_2 - Ex_1 = \Delta H - T_s \Delta S \tag{2.9}$$

The calculation of exergy is carried out in two steps:

1. The first contribution is the Gibbs free energy of formation at 25°C of the mixture from the compounds in the surroundings. This is similar to the LHV defining Equation (2.5), considering the work required to transfer all components to 1.01325 bar, which is the standard stated for the Gibbs free energy. This term consequently includes the entropy of mixing for transferring compounds from

atmospheric pressure to the actual pressures the components have in air.
2. The second contribution is the transfer of the mixture from 1.01325 bar to the actual temperature and partial pressures in the mixture. This is carried out using calculated values of the residual changes in entropy and enthalpy, ΔH^{res} and ΔS^{res}.

Example 2.3
In the hydrogen plant, Example 1.3, the exergies of the process inlet and outlet streams of the tubular reformer can be used to calculate lost work from:

$$\Delta Ex_{1 \to 2} = Ex_2 - Ex_1 = 525.3 \text{MW} - 451.6 \text{MW} = 73.7 \text{MW}$$

The ideal work increases not only due to the temperature increase, but also since the free energy increases in the reaction. This exergy increase must, however, be delivered by heat exchange with a warmer fluid on the outside of the reaction space and this heat is usually supplied by combustion. The increase in exergy should be compared with the enthalpy duty which is 109 MW.

Here, it is assumed that the fuel is pure CH_4 and it is burned in air at a ratio between actual and theoretical air $\lambda=1.15$ and that the temperatures of both fuel and air are 25°C. The resulting adiabatic flue gas temperature is 1825°C. The actual temperature of the resulting flue gas depends on how the heat is transferred to the reformer. Figure 2.2 below shows how the combustion efficiency using either exergy or LHV depends on the actual flue gas temperature after the required amount of heat to the process (duty) has been subtracted. The efficiency is defined as the ratio between the duty or exergy on the process side and the difference on the combustion side.

It is seen that both efficiencies become zero at the adiabatic combustion temperature, since an infinite amount of fuel must be used here, but also that the largest efficiency gain is obtained by decreasing the flue gas temperature in the high end. It is also seen that the increase in exergy efficiency is levelling off when the flue gas exit temperature decreases, whereas the LHV shows an almost straight-line dependence. The reason is that although the heat content of the flue gas at low temperature appears large, the ideal work of it is small. The increase below 100°C is caused by condensation of water.

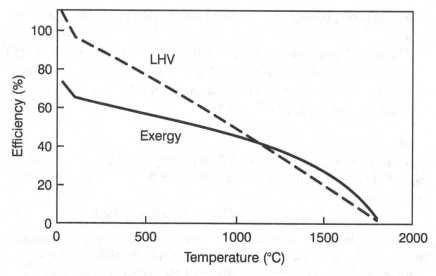

Figure 2.2 Steam reforming. Combustion efficiency as a function of flue gas temperature.

> The efficiency may be seen as a Carnot efficiency
>
> $$\eta_{Carnot} = 1 - \frac{T_{cold}}{T_{hot}}$$
>
> where the adiabatic combustion temperature is the temperature of the hot reservoir, but the temperature of the cold reservoir is an (undefined) average temperature of the reforming system. If the flue gas temperature is 875°C, similar to the outlet of the tubular reformer, the exergy change on the combustion side is 149.45 MW and the efficiency 49%. If this is interpreted as a Carnot efficiency, it corresponds to a cold reservoir temperature approximately equal to 790°C.

It is evident that optimisation of syngas processes must concentrate on combustion and decrease in the flue gas temperature. Lost work analyses help to quantify the losses, but commercial syngas processes are already highly optimised processes where the amount of fuel is minimised to the extent possible. In this case it is usually construction materials and catalysts that exhibit the constraints in process optimisation together with local conditions such as availability and price of fuel and cooling water. In process development exergy analysis is important and it may be supplemented with other tools such as pinch analysis of the heat exchanger networks aiming at minimising the temperature differences in the heat recovery part [301] [473]. Lost work analysis has also been extended to a detailed reactor analysis of the tubular reformer [335].

2.2 Hydrogen

2.2.1 Routes to hydrogen

Hydrogen is an important raw material for the chemical and the refinery industries, and it may play a future role in the energy sector. The total hydrogen consumption amounted to about 50 million tonnes per year (approximately $550\ 10^9\ Nm^3/y$) in 2004. The present use of manufactured hydrogen is primarily for the production of ammonia and methanol (approximately 51% in mixtures with nitrogen or carbon oxides) followed by hydrotreating in refineries (44% including co-production). Pure hydrogen is also used for a number of hydrogenation reactions (4% of total consumption). Other present uses (1%) of hydrogen are related to the food industry, the semi-conductor industry, and the metallurgical industry (for instance direct reduction of iron ore).

Specifications for reformulated gasoline have meant less aromatics and olefins and constraints on light hydrocarbons and sulphur. New legislation for diesel requires deep desulphurisation to 10–50 ppm S. This is done by reacting the sulphur compounds with hydrogen into hydrogen sulphide (hydrotreating), which is removed from the hydrocarbon stream. The requirement for sulphur removal may be accompanied by a wish to remove aromatics.

In general, these trends result in an increasing atomic ratio H/C of the fuels approaching two [161], while available oil resources become heavier with higher contents of sulphur and metals. This has created a large requirement for more hydrotreating (HDS, HDN, HDM) and hydrocracking.

Traditionally, a major part of the hydrogen consumption in refineries was covered by hydrogen produced as a by-product from other refinery processes, mainly catalytic reforming ("plat-forming"). However, there is a fast growing need for increased hydrogen production capacity in refineries. This need is being met mainly by installation of steam reforming-based hydrogen plants. The conversion of oil sand (tar sand) in Canada has also resulted in large hydrogen plants based on steam reforming of natural gas [360].

Hydrogen as an "energy vector" was discussed in the wake of the energy crisis in the 1970s [480]. Today, "hydrogen economy" is on the political agenda [251]. Hydrogen is claimed to replace hydrocarbons and to provide a clean fuel with no carbon emissions for use in stationary and mobile applications as well. Fuel cells may play a key role for both situations. Steam reforming or CPO of natural gas and liquid fuels (logistic fuels) may be essential to the introduction of a hydrogen economy including the use of hydrogen-driven fuel cells [418].

Hydrogen is an energy carrier, not a fuel. Hydrogen can be manufactured via a variety of routes [237] [251] [414]. The choice depends on size of production and cost of available feedstocks. The most important method is steam reforming of hydrocarbons followed by gasification of coal, oil sands (tar sands), etc. Gasification of heavy oil fractions may become more important as these fractions are becoming more available because of falling demand. Some refineries have installed gasification units for power production and co-generation of hydrogen [349]. For small-scale production, investments are dominating and simple equipment may be preferred over high energy efficiency. Still electrolysis of water and other sources accounts for less than 5% of the production.

Hydrogen production from coal was considered in the US Future Gen Program as co-production in ICCC power plants as illustrated in Figure 2.3 (Section 2.6.4). Carbon dioxide is removed by sequestration (refer to Section 1.4.2). This is feasible only in large centralised plants.

Plants supplying hydrogen for the build-up of a future hydrogen infrastructure are faced with a dilemma when based on fossil fuels [418]. Centralised large-scale hydrogen production is penalised by significant costs for compression and transportation.

Hence, it might appear more feasible to manufacture hydrogen decentralised at gas stations, but CO_2 sequestration (refer to Section

1.4.2) appears feasible only with large-scale production. Without CO_2 sequestration, it may be better to use natural gas directly in the car.

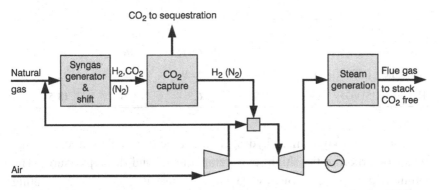

Figure 2.3 Hydrogen by air-blown reforming for CO_2-free power production [414]. Reproduced with the permission of Balzer.

The ultimate solution for decentralised production of hydrogen should be based on sustainable methods such as electrolysis or steam reforming of biofuels. Still, the economy is not attractive [326].

CO_2 sequestration from hydrogen plants based on steam reforming of hydrocarbons suffers from high capture costs as CO_2 ends up strongly diluted in the flue gas. It is essential that CO_2 is captured at as high a pressure as possible, as CO_2 absorption from flue gas will be very energy requiring (refer to Section 1.4.2).

This is illustrated in Figure 2.3, showing a scheme for large-scale power production based on air-blown reforming of natural gas coupled to CO_2 sequestration [414].

2.2.2 Hydrogen by steam reforming of hydrocarbons

Steam reforming of natural gas and other hydrocarbons remains the cheapest route to large-scale commercial production of hydrogen [251] [414] [426]. The costs are increased if CCS is included as illustrated in Table 2.1.

Table 2.1 Production costs for hydrogen with/without CO_2 capture/storage. Data from IEA 2005 [251].

	Large	Large with CCS		Small
H_2 pressure (bar)	78	78	Liq.	340
LHV efficiency (%)	73.2	61.1	46	46.7
Investment (USD/GJ/y)	9.4	11.9	21	42
H_2 costs (USD/GJ)	5.5	6.7	9.6	13.5

From thermodynamics, hydrogen manufacture is favoured by high steam-to-carbon ratio, high exit temperature and low pressure. The optimum pressure is typically 25 bar. For use in typical hydrotreating units (30–170 bar) the hydrogen product must be further compressed. A high steam-to-carbon ratio means that a surplus of steam has to be heated in the reformer. This results in a larger reformer and high energy consumption. Therefore, modern hydrogen plants [414] are normally designed for low steam-to-carbon ratios (1.8–2.5 mol/C-atom). A low steam-to-carbon ratio reduces the mass flow through the plant and thus the size and costs of equipment. The size of the reformer can further be reduced by increasing the inlet temperature. This is possible by using an adiabatic prereformer (refer to Section 1.2.5).

A low steam-to-carbon ratio results in a more energy efficient plant and thus lower operating costs. A low steam-to-carbon ratio increases the amount of unconverted methane from the reformer, but this is compensated for by increasing the reformer outlet temperature, typically to 900–950°C.

The non-converted methane is recovered in a unit for pressure swing adsorption (PSA) for cleaning hydrogen to a purity of 99.999% H_2. The PSA off-gas is then used as fuel for the reformer as shown in Figure 2.4 [414] [426].

Syngas Applications 89

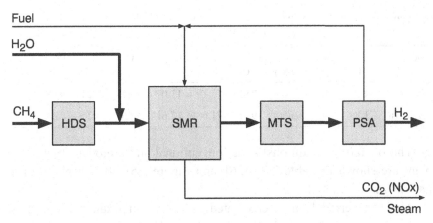

Figure 2.4 Hydrogen plant. Simplified scheme [426]. Reproduced with the permission of ACS.

Figure 2.5 H_2 plant based on steam reforming of natural gas. The tubular reformer is shown in the background with the PSA unit to the right (capacity: 39,000 Nm^3 H_2 eqv./h).

Table 2.2 Typical conditions for hydrogen plants using steam reforming of methane (100,000 Nm³ H₂ /h).

Design	H_2O/CH_4	P (bar)	T_{inlet} (°C)	T_{exit} (°C)	H_2	CO	CO_2	CH_4	Process duty MW
Previous	4.5	30	500	850	75.19	11.07	10.49	3.24	97.8
Modern	1.8	30	650	920	71.08	17.61	4.56	6.75	85.4

The operating conditions for a conventional and a modern hydrogen plant are shown in Table 2.2 [426], and Figure 2.5 shows a photo of a natural gas-based hydrogen plant.

The transferred heat is calculated as shown in Example 1.3. It is evident that the reformer is significantly smaller than for the previous design.

The net reaction can be written as:

$$CH_4 + 2H_2O(liq) \rightarrow CO_2 + 4H_2 \qquad -\Delta H^{\circ}_{298} = 253 \text{ kJ/mol } CH_4$$

The basic thermodynamic data of the feed and product are shown in Table 2.3 below.

It is seen in Table 2.3 that the exergy balance is fulfilled, whereas the LHV balance is not due to the liquid water. If the balance should be fulfilled, it is necessary to set the LHV of the liquid water to -88 kJ/mol CH_4 corresponding to the heat of evaporation.

In Table 2.3 is shown the product value in (%) calculated as the value of the product divided by the total value of the feed and the required amount of heat added for reaction. The actual energy efficiency may be calculated from the LHV of the product divided by the actual energy consumption. It is a very large efficiency equal to 86%. A theoretical efficiency may be calculated by dividing the energy efficiency by the product value resulting in 94%. This corresponds to assuming that the theoretical minimum consumption is the contributions from CH_4 and reaction as shown in [414] [426]. Although it may be discussed which is the proper efficiency, it is seen that today's processes for manufacturing of hydrogen are operating very close to the theoretical minimum energy consumption.

Table 2.3 Thermodynamic data for energy analysis of hydrogen production based on 1 mole of CH_4 or 4 mole H_2. The practical energy consumption corresponds to 12.6 GJ/1000 Nm^3 H_2 [414] [426].

	LHV (kJ/molCH$_4$)	Exergy (kJ/molCH$_4$)
CH_4	802	830
H_2O(liq)	0	0
Reaction heat (25°C)	253	125
4 mole H_2	967	941
CO_2	0	20
$H_2 - CO_2$ separation	0	6
Product value (%)	92	99
Practical energy consumption	1130	
Energy efficiency (%)	86	

If hydrogen has to be supplied at higher pressure or as liquid hydrogen, the efficiency decreases as illustrated in Table 2.1. The compression of hydrogen from 1 bar into liquid hydrogen uses energy corresponding to 1/3 of the LHV of the hydrogen to be compressed [222].

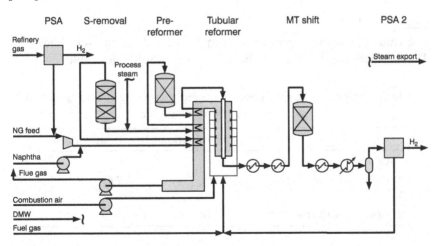

Figure 2.6 Process layout of a typical multi-feedstock hydrogen plant [414]. Reproduced with the permission of Balzer.

A typical process layout of a feedstock flexible hydrogen plant operating at 25 bar on refinery gas, natural gas and naphtha is given in Figure 2.6. Refinery gas containing large amounts of hydrogen is sent to a pressure swing adsorption (PSA) unit where pure hydrogen is extracted. The off-gas from the PSA, containing non-converted methane, is compressed and used as feed in the hydrogen plant. In this way, low-grade refinery gas is used as feed to a hydrogen plant and thereby substituting more expensive natural gas or naphtha.

Methanol can be added directly to the shift converter of a natural gas-based hydrogen plant in which the methanol is converted into hydrogen simultaneously with the shift reaction. This offers a cheap opportunity to boost the capacity of the hydrogen plant [265].

2.2.3 The steam export problem

Hydrogen production often results in a by-product of steam which in many cases should be minimised [414].

In ammonia plants and other petrochemical plants, the steam produced in the waste heat section of the reformer can be used to drive the synthesis compressor, but in hydrogen plants there is little need for steam. Hence, many hydrogen plants have a significant export of steam.

Example 2.4
The value of steam depends on the requirements from the surroundings. The thermodynamic value can be seen from the following table:

Table 2.4 Thermodynamic value of steam. It is seen that the exergy of the generated steam is 40% of the enthalpy input.

	Enthalpy (kJ/mol)	Exergy (kJ/mol)
H_2O(liq)	0	0
H_2O(liq) boiling at 100 bar (312°C)	23.54	7.43
H_2O(steam) at 100 bar (312°C)	47.18	19.03

It is possible to reduce the steam production from a hydrogen plant based on tubular steam reforming [208]. Introduction of a prereformer with reheat increases the thermal efficiency for reforming from 50% to about 60%. Another part of the flue gas heat content can be used for preheating of combustion air.

Convective reformers result in less waste heat. The flue gas as well as the product gas is cooled by heat exchange with the process gas flowing through the catalyst beds, so that they leave the reformer at about 600°C. The amount of waste heat is reduced from 50% in the conventional design to about 20% of the fired duty in the heat exchange reformer. This means that the steam generated from the remaining waste heat just matches the steam needed for the process, so that export of steam can be eliminated.

Figure 2.7 Topsøe package hydrogen plants (2 x 5,000 Nm^3/hr) at Air Liquide, Belgium.

However, the heat exchange is primarily by convection which generally leads to lower average heat fluxes (and hence bigger heat transfer surfaces) than in tubular reformers with radiant heat transfer (refer to Section 3.3.5). This is compensated by elimination of the reformer furnace and a smaller waste heat section, which results in a more compact process unit. Convective reformers are industrially proven and are preferred for smaller units due to their compactness (Figure 2.7).

Convective reforming can also improve the productivity of the fired reformer by utilising the hot product gas for supplementary heat input to the process, i.e. by "chemical recuperation" of the heat in the process gas instead of raising steam [426] [427]. This can be done in various ways. In the Topsøe Bayonet Reformer (TBR) the product gas may simply be passed through the reformer tube in a bayonet type arrangement, i.e. counter-current to the flow through the catalyst bed. In this way, the capacity of the reformer furnace can be increased by 25% and the steam export reduced by 30% [426].

Another approach [426] [522] is the installation of a parallel convective reformer heated by the hot process gas from the tubular reformer as illustrated in Figure 2.8.

With this approach, it is possible to increase the capacity of a hydrogen plant by 30% with the same tubular reformer or reduce the steam export by 40%.

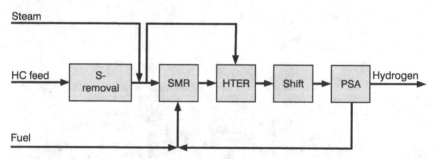

Figure 2.8 SMR: steam methane reforming. HTER: Haldor Topsøe Convective Reformer. Hydrogen plant with chemical recuperation [426]. Reproduced with the permission of ACS.

2.2.4 Membrane reforming

Thermodynamics require high exit temperatures to achieve high conversion of methane. This is in contrast to the potential of the catalyst showing activity even below 400°C (refer to Section 3.5). Steam reforming at low temperatures would allow the use of cheaper construction materials and the use of low-temperature heat to drive the reaction. This has led to efforts to circumvent constraints by the use of a selective hydrogen membrane installed in the catalyst bed with extraction of hydrogen though a palladium membrane [1] [273] [414] [471], thus pushing the reforming equilibrium to higher conversion. Reactor simulations and experiments [1] have shown that the reformer exit temperature can indeed be reduced to below 700°C while maintaining the same conversion.

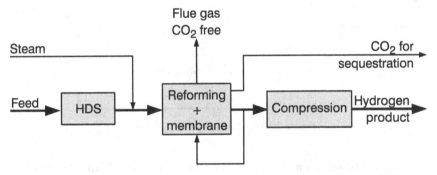

Figure 2.9 Membrane reforming. Hydrogen plant with CO_2 sequestration [414]. Reproduced with the permission of Balzer.

As shown in Figure 2.9, this results in hydrogen at lower pressure and CO_2 at high pressure. The latter is good for CO_2 sequestration and a low supply pressure may be acceptable for fuel cells.

Still, there is a need to develop stable thin Pd membranes (thickness of a few microns) and a reliable manifold system.

2.2.5 Hydrogen via catalytic partial oxidation (CPO)

Steam reforming remains the most economic and efficient technology, even in very small scale (50 Nm^3/h). However, other parameters may play a role as well for small units, such as simplicity, compactness and (for automotive units) short start-up time. Air-blown catalytic partial oxidation fulfils these requirements, in particular for fuel cell applications where it is normally acceptable that the hydrogen stream contains nitrogen. A CPO plant has a simpler steam and heat recovery system than a steam reforming plant, but an air compressor is needed, which makes the technology less suited for high-pressure operation [408] [410].

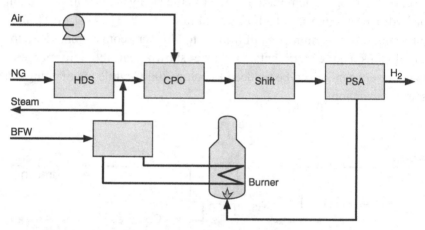

Figure 2.10 Hydrogen by CPO (20 bar).

Figure 2.10 shows a CPO process scheme for pure hydrogen using a PSA unit for purification, just as for a reforming-based scheme (Figure 2.4). The PSA off-gas has to be burned in a heater producing export steam. In conclusion, a CPO plant for making hydrogen at 20 bar would be close to 50% more expensive. The main prospects of the CPO technology are linked to fuel cells (Section 2.3) and small-scale plants.

2.3 Fuel cells

2.3.1 Fuel processing system

Most fuel cells involve electrochemical oxidation of hydrogen on the anode. There are only few examples of direct conversion of other fuels on the anode. In high-temperature fuel cells, it is possible to convert the fuel to hydrogen inside the cell utilising the heat from the electrochemical reaction [411], but otherwise it is necessary to convert the primary fuel outside the stack into a hydrogen-rich gas which is fed to the anode. A wide range of feedstocks is considered for conversion in fuel cells including hydrocarbons from natural gas to diesel, alcohols, syngas from gasification, landfill gas, etc. The coupling of fuel processing with the fuel cell operation is essential to achieve high plant efficiencies [411].

The layout of the fuel processing system depends on the type of fuel cell as summarised in Table 2.5.

Table 2.5 Characteristics of fuel cells.

	T_{exit} (°C)	External Reformer	H_2 purification	Internal reforming
PEMFC polymer membrane	80	+	50 ppm CO	-
PAFC phosphoric acid	200	+	0.05% CO	-
MCFC molten carbonate	600-650	(+)	None	+
SOFC solid oxide	600-1000	(+)	None	+

Fuel cells are developed for modular design mainly for small-scale applications (1–250 kW). A small-scale hydrogen plant of 1000 Nm^3/h would correspond to a fuel cell size of approximately 1 MW. A scale-down of large-scale hydrogen plant designs as described in Section 2.2 will not be economic. It is a challenge to reduce piping and instruments and to develop integrated units suited for mass production. Micro-structured process components such as heat exchangers and new reactor concepts are becoming available [282].

For small-scale units (about 10–50 kW), for instance providing hydrogen for automotive fuel cells, the choice of technology is dictated by parameters such as simplicity and quick response to transients.

The low-temperature fuel cells (PEMFC and PAFC) require pure hydrogen, as carbon monoxide is a poison to the platinum anode. A PSA unit as used in large-scale hydrogen plants (Figure 2.4) cannot be used at the small scale in question, and the final purification is made either by methanation (as used in ammonia plants, Section 2.5) or by preferential partial oxidation (PROX) of carbon monoxide to carbon dioxide over a noble metal catalyst. A typical process scheme for a PEMFC unit is shown in Figure 2.11 with natural gas as feed [62]. The convective (heat exchange) reformer is followed by shift and PROX reactors. The anode off-gas is used as fuel for the reformer.

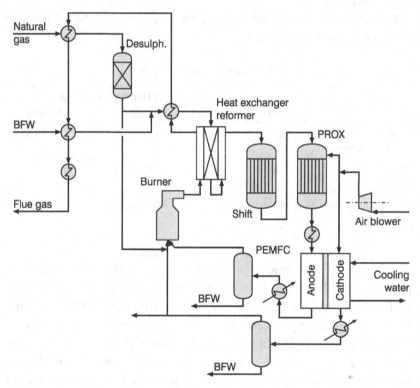

Figure 2.11 PEMFC (J.B. Hansen [62]). Fuel processing system. Reproduced with the permission of the author.

The catalytic process steps may appear trivial, but it is a challenge that most units must have simple start-up and shutdown procedures allowing exposure of the hot catalysts to air. This has resulted in the development of noble metal catalyst for most of the steps.

The reformer is the key element of the fuel processing system. The special requirements have led to new concepts for steam reforming and partial oxidation [282] [488]. One example is micro-channel reactors (refer to Section 3.3.8) which use catalysed hardware on heat exchange surfaces [188] or split heat transfer and reforming reactions [455]. These types of fuel processes have been developed for a variety of fuels including natural gas [188], diesel [253] and methanol [61] [182].

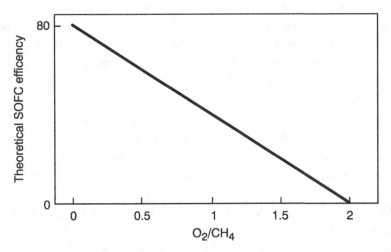

Figure 2.12 Efficiency and degree of combustion. Internal reforming is represented by $O_2/CH_4=0$, CPO at 0.5, and total combustion by 2.

A simple CPO process scheme (refer to Section 2.2.5) may be preferred due to its compactness. However, in principle the partial oxidation route results in smaller efficiency than the steam reforming route as illustrated in Figure 2.12. The reaction with oxygen results in an additional heat production which is highest for complete combustion and not existing for steam reforming. This additional heat production can only be converted into electricity with the Carnot efficiency.

2.3.2 Internal reforming

With high-temperature fuel cells (SOFC, MCFC), it is possible to convert the free energy of methane (or hydrogen and other fuels) directly into electric energy with the theoretical voltage expressed by the Faraday equation ($\Delta G° = -n\mathcal{F}E$) where n is the number of electrons participating in the electrochemical reaction, E the voltage and $\mathcal{F}$ the Faraday constant. It means that the ideal electric efficiency of a fuel cell, η_{ideal}, is expressed simply by $\Delta G°/\Delta H°$ for the overall combustion reaction [411].

The maximum useful work which can be obtained by combustion of natural gas is equal to the exergy of methane being 830 kJ/mol (1 bar, 25°C), as shown in Table 2.3. The LHV is almost the same, 802 kJ/mol, since the entropy change is very small. This means that almost as much energy can be obtained as heat as would have been obtained as work if the process had been carried out reversibly. The subsequent conversion of the thermal energy to mechanical energy is limited by the Carnot efficiency.

Figure 2.13 shows η_{ideal} versus temperature for various fuels electrochemically converted in the cells [411]. For methane, η_{ideal} is close to 100% and independent of temperature (reflecting that the entropy of the combustion reaction is almost zero). η_{ideal} for hydrogen fuel is less than 100% and decreases strongly with temperature ($\Delta S°<0$).

For methanol (with $\Delta S°>0$), η_{ideal} is above 1 and increases with temperature. However, the advantage of using methanol is partly lost because methanol is manufactured from natural gas with an efficiency of approximately 66% (LHV) (refer to Section 2.6).

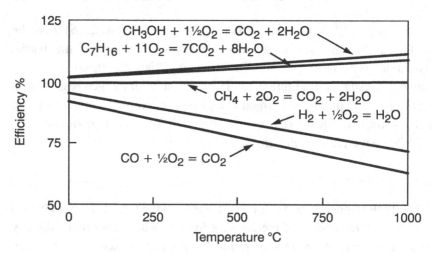

Figure 2.13 Ideal fuel cell efficiencies for various fuels [411]. Reproduced with the permission of Royal Soc. Chem.

These ideal efficiencies show the upper limits of the practical efficiencies using the corresponding fuels. The ideal voltage in the cell is calculated from ΔG°, but this reversible voltage is reduced due to the actual activities of the reactants and products in the cell as expressed by the Nernst equation, and by the polarisation on the cathode and anode and the internal resistance of the electrolyte when current is drawn from the cell.

The advantage of converting the hydrocarbons in the fuel cell rather than converting the fuel to hydrogen first is evident. If steam reforming of natural gas takes place in an externally fired reformer, there is a loss in efficiency because of the high temperature created in the flame, which is not utilised fully for work because the waste heat can only be recovered via the Carnot cycle as shown in Example 2.3.

This is illustrated in Figure 2.14 showing the exergy changes going from methane to electricity directly or via hydrogen [418] [420]. With direct internal reforming, high-temperature fuel cells (SOFC, MCFC) may achieve electric system efficiencies of 46–47% [195] [204], whereas hydrogen-driven PEMFC are reported to have efficiencies in the range 28–38% [62].

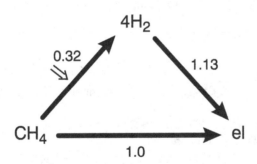

Heat input by combustion

Figure 2.14 Fuel cell conversion of methane into electric power [418] [420]. Exergy changes at ideal conditions, 250°C. By the indirect route, $4H_2$ represents an exergy of 1.13, but at the expense of the exergy of CH_4 used as feed + fuel (0.32). Reproduced with the permission of Taylor & Francis.

It is essential that the reforming reaction is coupled to the electrochemical reactions on the anode [123]. If the reforming reaction in the MCFC took place outside the anode chamber, but at the same temperature (650°C), complete conversion of methane would not be possible. The methane conversion at equilibrium would be less than 90% at MCFC conditions (but close to 100% at SOFC conditions). Therefore, for an MCFC without the coupling, either an external reformer with high exit temperature or a recycle of non-converted methane is required.

The internal reforming can be carried out either as anode chamber reforming or as stack heat integrated reforming in the cooling channels of the stack. The catalyst may be placed as pellets in the anode chamber or in cooling channels, but a more elegant solution is catalysed hardware where the surfaces of the anode chamber are covered with a thin layer of a reforming catalyst [401].

The internal reforming was pioneered by Baker *et al.* [339] and strongly demonstrated by the cell voltage data shown in Figure 2.15.

The stack performance is almost identical for the operation with gas from external and internal reforming, respectively. However, the external reforming requires fuel for the fired reformer and energy for the cooling of the stack by recycle of cathode air.

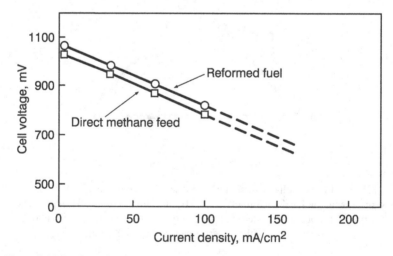

Figure 2.15 Stack performance with and without internal reforming. MCFC stack. Paetsch *et al.* [339].

The internal reforming means that the heat produced by the electrochemical reactions in the stack can be absorbed by the steam reforming reaction [12]. This results in a strongly reduced demand for external cooling by recycling of air to the cathode [402].

Example 2.5
The coupling of the catalytic steam reforming and the electrochemical reactions of hydrogen and oxygen at the electrodes takes place as follows:

$CH_4 + 2H_2O \leftrightarrow CO_2 + 4H_2$ $\quad -Q_R \quad\quad -\Delta H^o_{923} = -188 \text{ kJ/mol } CH_4$ (a)

$4H_2 + 2O_2 \leftrightarrow 4H_2O$ $\quad\quad Q_C + W_E \quad -\Delta H^o_{923} = 987 \text{ kJ/4mol } H_2$ (b)

$CH_4 + 2O_2 \leftrightarrow CO_2 + 2H_2O \quad -Q_R + Q_C + W_E \quad -\Delta H^o_{923} = 799 \text{ kJ/mol } CH_4$ (c)

Values for ΔH^o are shown at MCFC conditions at 650°C. In the equations W_E is inserted as the electric energy, Q_R as the heat of reaction for the steam reforming reaction, and Q_C as the electrochemically produced heat.

For full conversion of the reforming reaction the relation between, Q_R, Q_C and W_E are given by [401]:

$$+\Delta H^o(a) = Q_R$$
$$-\Delta H^o(b) \cdot \alpha = W_E + Q_C$$

where α is the fuel utilisation of hydrogen corresponding to the extent of reaction (b).

In order to maintain the temperature of the cell, the heat required for the reforming reaction must be less than the electrochemically produced heat:

$$Q_R \leq Q_C \Rightarrow W_E \leq -\Delta H^o(a) - \Delta H^o(b) \cdot \alpha$$

The maximum fuel cell efficiency, η_{max}, leaving enough heat for the reforming reaction is then obtained from:

$$\eta_{max} = \frac{W_E}{-\Delta H^o(c)} = \frac{-\Delta H^o(a) - \Delta H^o(b) \cdot \alpha}{-\Delta H^o(c)}$$

For $\alpha=1$, $\eta_{max}=1$, since reaction c is obtained by adding reactions a and b. For $\alpha=0.75$, which is a typical design value, $\eta_{max}=0.69$ corresponding to a cell voltage of 0.95 V using Faraday's law. This is well above practical MCFC cell voltages and hence, the heat balance is not a limiting factor for an MCFC based on internal reforming. With a typical fuel cell voltage of 0.75 V, a minimum fuel utilisation $\alpha=0.46$ is required to maintain the temperature of the cell [401]. Similar results are obtained for an SOFC.

2.3.3 Process schemes for SOFC

A process scheme of a natural gas-based SOFC plant is shown in Figure 2.16 [62]. The fuel processing scheme is simpler than that for the PEMFC shown in Figure 2.11. An adiabatic prereformer has been added to convert higher hydrocarbons in the natural gas, which might otherwise easily be cracked in the preheater (E2) or in the stack leading to coke deposits (refer to Chapter 5). The anode off-gas is partly recycled, partly combusted for providing heat for preheating.

A comparison [62] of the PEMFC and SOFC schemes in Figure 2.11 and Figure 2.16 shows electric efficiencies of 37% and 58%, respectively. The main reason for the lower efficiency of the PEMFC is that the electrochemical heat is removed by the cooling water.

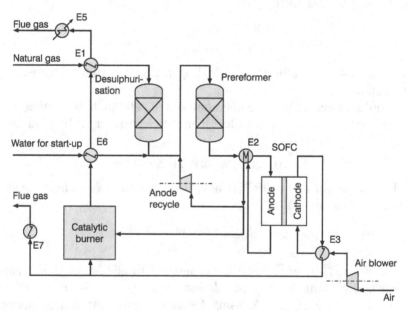

Figure 2.16 Process scheme for an SOFC plant (J.B. Hansen [62]). Reproduced with the permission of the author.

The prereformer is also used in schemes based on liquid hydrocarbons (naphtha, diesel). It may also be used for preconversion of feedstocks like coal gas, biogas and methanol resulting in a methane-rich gas. The methanation reaction provides heat for preheating and the methane allows a better temperature control of the stack by internal reforming. For coal gas it was shown [64] that the addition of a methanation step results in a relative improvement of the efficiency by about 2%.

2.4 CO rich gases

2.4.1 Town gas

Before natural gas became available in Europe in the 1970s, town gas was essential. It was originally made in coking ovens, sometimes followed by water gas shift to reduce the content of CO [77]. Later low-pressure steam reforming units with addition of air were used in different cyclic processes using butane or natural gas as feedstock (Otto-CC3P, Segas, O.N.I.A. -G.E.G.I, etc. [300]). The introduction of high-pressure steam reforming of naphtha led to the installation of a large number of big town gas units in the UK and continental Europe. These plants were based on three different principles as illustrated in Figure 2.17.

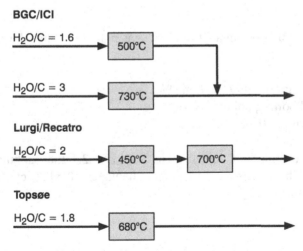

Figure 2.17 Different routes to town gas (LHV=450 MM BTU).

Two schemes (BGC [British Gas Corporation]/ICI [87] [139] and Lurgi Recatro [259]) used low-temperature adiabatic reformers (precursor for the prereforming technology, refer to Sections 1.2.5 and 5.3.2) in parallel or in series.

The Topsøe technology [493] used a single tubular reforming step operating at lower steam-to-carbon ratio and reformer exit temperature than the two other schemes and thus met the required gas composition and heating value.

2.4.2 Oxogas

CO-rich gases are used for a number of important petrochemical reactions such as oxo-synthesis converting olefins to alcohols and acetic acid. The CO content of the syngas is increased by decreasing the atomic ratio H/C in the feed gas. This is achieved at low H_2O/C, use of naphtha instead of natural gas as feedstock, use of partial oxidation, and by addition of CO_2 to the feedgas. Most of these measures increase the potential for carbon formation (refer to Chapter 5).

It is well-established that CO_2 can replace H_2O in the reforming reaction [396] [399] (refer to Section 3.5.4). CO_2 is typically added after the prereformer as shown in Figure 1.14. One example is described in Example 2.6 below.

Example 2.6
A naphtha to be used in a CO-rich syngas plant [510] has the following properties:

PNA distribution: 89/8/3 volume %
Initial/final boiling point: 36/146°C
Specific gravity: 0.67

By use of the method described in Section 1.2.1 the naphtha can be represented by a mixture: n-hexane/n-heptane: 30/57%; cyclohexane: 10% and benzene: 3%.

At the following operating conditions:

P=21 bar
T_{exit} tubular reformer 950°C
H_2O/C=1.8
CO_2/C=0.8

the following product gas (at equilibrium) is obtained (dry vol%): H_2=49.6, CO=34.8, CO_2=14.7, CH_4=1.0.

The calculation is carried out by first converting all higher hydrocarbons to CO and H_2 and then using the methods in Example 1.1 to calculate the mixture equilibrium composition.

Table 2.6 shows examples of combined steam and CO_2 reforming demonstration in a full-size monotube pilot plant (Figure 3.6) or in industry [521]. The CO production per fired fuel (duty) in the reformer is compared.

Table 2.6 CO_2 reforming of natural gas. Demonstration tests in a full-size monotube reformer. Approximate duration: 400 hours [521]. Refer to Figure 2.18.

	A	B	C	D	
Catalyst	Ni cat	Ni cat	Noble metal cat	Ni cat. S passivated	Ni cat. S passivated
	Pilot	Pilot	Industry	Pilot	Industry
H_2O/CH_4 (mol/mol)	1.38	1.64	0.93	0.70	0.95
CO_2/CH_4 (mol/mol)	0.24	0.80	0.81	2.43	0.65
P_{outlet} (bar)	23	23.5	23.5	14.5	6.3
T_{outlet} (°C)	945	945	945	945	875
Dry CH_4 leak (vol%)	4.8	2.0	4.0	0.6	
H_2/CO (vol/vol)	2.6	1.8	1.5	0.72	1.8
Relative CO prod./ fired duty	1.00	1.22	1.36	1.74	

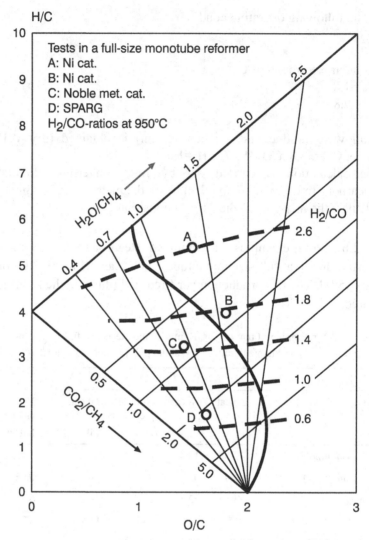

Figure 2.18 Carbon limit diagram. Thermodynamics predict carbon for conditions at the exit of the reformer. P=25.5 bar [403] [415]. The dotted lines show combinations of H_2O/CH_4 and CO_2/CH_4 resulting in the indicated H_2/CO ratios. There is thermodynamic potential for carbon in the equilibrated gas to the left of the curve (based on data for whisker carbon (250 nm), refer to Section 5.2). The points A–D refer to process conditions listed in Table 2.6. (D is shown irrespective of the lower pressure). Reproduced with the permission of Japan Petroleum Institute.

The most efficient examples are those with highest potential for carbon formation as illustrated in Figure 2.18 (refer to Section 5.2). This is handled by the use of noble metal catalysts or by the sulphur passivated reforming process (SPARG) (refer to Section 5.4).

CO_2 addition to the autothermal reformer was studied in pilot scale (100 Nm^3 NG/h). Results are summarised in Table 2.7. It is possible to achieve $H_2/CO=1$ with $CO_2/C=1.0$). If more CO_2 is added, it will pass unconverted through the ATR reactor [399].

Table 2.7 Autothermal reforming tests in a pilot plant [399].

Test	A	B	C	D	E
Feed ratio(mol/mol)					
H_2O/CH_4	0.60	0.90	0.70	0.69	0.60
CO_2/CH_4	0.	0.	0.50	0.78	1.01
O_2/CH_4	0.64	0.62	0.69	0.73	0.75
Product Gas					
Temperature (°C)	1065	1000	1018	1030	1028
Pressure (bar abs)	24.5	24.5	24.5	24.5	24.5
H_2/CO (mol/mol)	2.25	2.47	1.49	1.19	0.99
CH_4 (dry mol%)	0.42	0.90	0.32	0.11	0.11

The CO_2 conversions in various processes are compared in Table 2.8. The net CO_2 conversion is corrected for the CO_2 in the flue gas from the tubular reformer.

Table 2.8 CO_2 conversion in various processes. CO_2 reforming for $H_2/CO=1$ [399].

	Tubular reforming		SPARG	Autothermal reforming
	Conventional	Advanced		
H_2O/CH_4	2.8	2.6	0.9	0.6
CO_2/CH_4	3.4	3.0	1.6	1.0
O_2/CH_4	0	0	0	0.7
P (bar abs)	22	22	15	22
T_{exit} (°C)	850	950	935	1050
CO_2 conversion (%)	27	39	60	32
Net CO_2 conv. (%)	5	12	20	23

2.4.3 Reducing gas

Direct reduction of iron ore is an alternative to the conventional steel making in a blast furnace. It involves no melting of the ore and normally the iron ore is reduced in gas containing hydrogen and carbon monoxide. The total world production by direct reduction in 2006 was 67 million tonnes steel [283]. Most plants for direct reduction of iron ore are based on natural gas with the reducing gas being manufactured by the reforming process. A number of processes have been introduced, which are characterised by extensive integration of the reformer and the reduction furnace (Midrex, HYL, Nippon Steel, Purofer, etc.) [371]. This operation requires a nearly stoichiometric atomic ratio of oxygen to carbon in the reformed feed to ensure maximum reduction potential of the gas, which is essential for efficient utilisation of the shaft furnace. For thermodynamic and stoichiometric reasons, the consumption of hydrogen and carbon monoxide for the reduction of haematite (Fe_2O_3) is restricted to only about half of the introduced gas.

> **Note**
> The reduction of iron ore (haematite) proceeds via wüstite [495]:
>
> $Fe_2O_3 + 0.895H_2 \leftrightarrow 2.105Fe_{0.95}O + 0.895H_2O \quad -\Delta H^\circ_{727} = 184.2 \text{ kJ / mol Fe}$ (2.10)
>
> $Fe_{0.95}O + H_2 \leftrightarrow 0.95Fe + H_2O \quad -\Delta H^\circ_{727} = 10 \text{ kJ / mol Fe}$
>
> The conversion to iron is determined by the thermodynamics of the reduction of wüstite. At the temperature of industrial interest, this means a H_2/H_2O ratio of at least 2. Below this ratio, metallic iron cannot be formed.
>
> For stoichiometric reasons, the hydrogen consumption for the reduction of haematite into wüstite is only about half of the hydrogen used for reduction of the wüstite into metallic iron. As a consequence of these thermodynamic and stoichiometric restraints, only roughly 50% of the hydrogen ($1/3+1/2 \cdot 1/3=1/2$) introduced at the bottom of the shaft furnace is used for the reduction process, whereas the rest is found – diluted with steam – in the top gas [495].

In practice, the gas utilisation amounts to 30–40% and this situation makes the effective use of the unconverted gas in the effluent (top gas) from the shaft furnace decisive for the process economy. Two different principles have been applied to solve the problem (Figure 2.19). The top gas is either recycled to the shaft furnace inlet after removal of product water and carbon dioxide or the gas is recycled to the reformer inlet. For both solutions, part of the top gas is used as fuel for the reformer.

In the Midrex process (Figure 2.19a) [389] [490], the feed gas is converted into a reducing gas at approximately 950°C containing less than 5% of the oxidising components, steam and carbon dioxide. The hot gas is passed to the shaft furnace and after passing this, the cooled and dust-scrubbed top gas contains 18–20% carbon dioxide. One third of the top gas is used with natural gas as fuel in the reformer, whereas the remaining two thirds are passed to the reformer inlet after saturation with water. The hot flue gas from the reformer is utilised for preheating of feed gas and of combustion air. With this scheme, the overall consumption of natural gas amounts to less than 11 GJ per tonne of sponge iron. Similar energy consumptions are achieved in process schemes with recycle of CO_2-scrubbed top gas to the shaft furnace inlet [495] (Figure 2.19b).

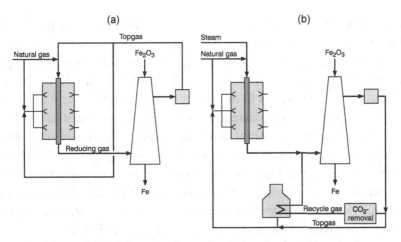

Figures 2.19a+b Process schemes for direct reduction of iron ore.

The most economical solutions avoid condensation of excess water in the reformer effluent gas. This operation requires a nearly stoichiometric atomic ratio of oxygen to carbon to attain a minimum value (<10%) of the so-called oxidation degree, η.

$$\eta = \frac{CO_2 + H_2}{CO + H_2 + CO_2 + H_2O} \cdot 100\% \tag{2.11}$$

This requires that the steam reformer is operating at a ratio (H_2O+CO_2)/CH_4 close to one, which means high risk of carbon formation (refer to Section 5.2). One solution is the use of sulphur passivated reforming allowing both schemes in Figure 2.19 to operate close to or beyond thermodynamic carbon limits (refer to Section 5.5).

There may be a risk in designing plants too highly integrated. The improved economy is paid for by higher sensitivity towards upsets. Operating problems in one part of the plant can result in serious consequences for the rest of the plant. Reformer autonomy cannot be achieved in a flow scheme such as Figure 2.19a, since any disturbance in the shaft furnace operation is transferred to the reformer. Today direct reduction plants may be based on an efficient hydrogen plant. The energy consumption may be higher, but the plant is much simpler (and cheaper).

2.5　Ammonia

For ammonia, the natural gas consumption is decisive for the production costs. The energy consumption of ammonia production has decreased over the years. A first significant step was the introduction of high-pressure steam reforming at 30–40 bar [87], which meant less work for compression of the syngas to the synthesis pressure (150–250 bar). The energy consumption was decreased by integration of the steam turbine cycle and by the application of better catalysts [494]. This resulted in an increase in the energy efficiency as shown in Figure 2.20. Still, the synthesis catalyst is not far from the promoted iron catalyst optimised by Mittasch for the Haber–Bosch synthesis [320], although several attempts [256] have been made to find a catalyst closer to the optimum activity predicted by density functional theory (DFT) analysis [255].

Today, the energy consumption for a natural gas-based plant is close to 28 GJ/t NH_3 liq. Energy consumption as low as 27.2 GJ/t NH_3 liq. has been reported [169]. Typical capacities for a new plant may be in the range of 2000–3500 MTPD.

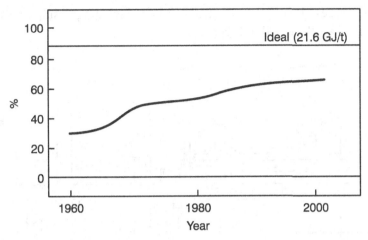

Figure 2.20 Energy efficiency for ammonia synthesis.

A typical ammonia plant is shown in Figure 2.21.

Figure 2.21 Typical natural gas-based ammonia plant. Two lines of 1000 t/d (Thal, India).

The process involves eight catalytic steps including desulphurisation of the feed as illustrated in Figure 2.22.

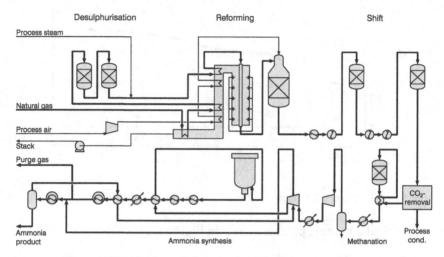

Figure 2.22 Flow sheet of typical natural gas-based ammonia plant.

Synthesis gas for ($H_2/N_2=3$) [169] is manufactured by adding nitrogen with air in a second reforming step in which oxygen reacts with non-converted methane from the primary reformer. The secondary reformer resembles the ATR reactor (Section 1.3.2), but operates at milder conditions.

Carbon monoxide is converted in high-temperature shift and low-temperature shift reactions, and after removal of CO_2 by wash and remaining CO and CO_2 by methanation, the syngas is fed to the ammonia synthesis loop. The ammonia is recovered as liquid ammonia in a chiller system driven by ammonia. The CO_2 is often utilised for reaction with ammonia to urea.

A prereformer (refer to Section 1.2.5) may be added to reduce the load on the tubular reformer. High preheat temperatures are used in such a way that the heat available in the waste heat section of the primary reformer is recovered down to a low temperature for preheat of natural gas, reformer feed, process air, and combustion air. Heat downstream of the reforming section and the shift section is recovered for production of high-pressure steam. All heat down to temperatures less than 100°C – except for the heat used in the reboiler in the CO_2-removal unit – is used for high-pressure steam production [395]. Table 2.9 shows typical operating parameters for the catalytic reactors.

Table 2.9 Natural gas for syngas-based ammonia synthesis. Typical operating parameters for catalytic reactors.

	Catalyst	P_{exit} (bar)	T_{exit} (°C)	H_2O/C	Air/C	Dry exit gas
HDS	Ni(Co)/Mo/Al$_2$O$_3$		400			
ZnO	ZnO		385			<10 ppb H_2S
Prim. reform.	Ni/support		810	3.0		12% CH_4
Sec. reformer	Ni/support	37	1000		1.4	0.4% CH_4
HTS	Fe/Cr/Al		430			3.2% CO
LTS	Cu/Zn/Al		225			0.2% CO
Methanation	Ni/support		320			<10 ppm CO + CO_2
NH$_3$ synthesis	Fe, K, Ca, Al	190	460			19.6% NH_3

The high efficiency of an ammonia plant is paid for by a large exchange of energy in the steam reformer and heat recovery units with large heat-exchange surfaces. This is evident from an exergy analysis (Section 2.1.3). Table 2.10 shows the principal main exergy losses from an ammonia plant compared with the enthalpy changes [169].

Generally, the sizes of the enthalpy and exergy are comparable to each other, but the reformer shows a large loss of exergy because of the creation of hot flue gas in the combustion process. This is compensated for by very low losses in the turbines and compressors using the high-pressure steam recovered from the waste heat.

An ammonia plant is optimised for maximum use of the waste heat, but this is linked to the application of steam turbines and based on the assumption of autonomy of the plant.

Table 2.10 Enthalpy and exergy analysis of an ammonia plant. I. Dybkjær [169].

	Enthalpy (kJ/kg NH$_3$ liq)	Exergy (kJ/kg NH$_3$ liq)
Total consumption (feed + fuel)	29.4	30.7
Losses:		
Reforming	0.4	4.9
Steam generation	0.3	2.4
Turbines/compressors	6.5	0.5
Other process steps	9.9	2.3
Energy in NH$_3$ product	17.1	20.1
Total losses	27.6	32.4
Efficiency (%)	58	66

A typical conventional ammonia plant (1500 t/d) has steam turbines with a power output of 35 MW. The syngas compressor and the ammonia compressor in the chiller system are the largest consumers. There is a flat minimum of compressor power at a synthesis pressure of approximately 150 bar [169].

The steam turbines typically operate at an efficiency of the steam cycle slightly above 30%. One might consider taking advantage of the higher efficiency (45–48%) of modern air derivative gas turbines by integrating the ammonia plant with a power plant operating with steam turbines [406]. However, there is rarely a need for more electric power at locations where cheap natural gas is available.

In view of the loss of exergy from the combustion in the reformer furnace, it is tempting to utilise the heat of the process gas from the secondary reformer as heat source for the primary reforming step in a convective reformer [395]. The amount of heat available downstream of the secondary reformer at temperatures above 450°C corresponds to about 60% of the duty required for primary reforming. If convective reforming is introduced in parallel with the tubular reformer without any changes in operating parameters on the process side, the size of the tubular reformer can be reduced to about 40%. However, the introduction of a convective reformer will cause less production of high-pressure steam. If the secondary reformer is operated with an excess of process air (about 50% above the stoichiometric amount), or if the process air is enriched to an O_2 content of about 30%, then the sensible heat in the product gas from the secondary reformer is sufficient for the primary conversion of natural gas in a convective reformer [395], meaning that the fired reformer can be eliminated. Such schemes require, however, extra installations, either for production of enriched air or for removal of excess N_2 from the synthesis gas, and they suffer of course also from the reduced efficiency of the steam production described above.

2.6 Methanol and synfuels

2.6.1 Methanol as intermediate

Methanol is a key intermediate for a number of important petrochemicals and synthesis fuels as illustrated in Figure 2.23.

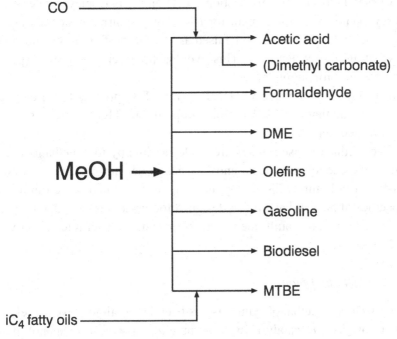

Figure 2.23 Methanol is a key intermediate.

Methanol as an energy carrier and motor fuel has been discussed for decades [144] [194] [420], lately as fuel for fuel cells. Methanol may play an increasing role for synfuels alone, via dimethylether directly or via zeolites to gasoline or as an agent for trans-esterfication of fatty oils into biodiesel.

So far, methanol has been used primarily for petrochemical purposes giving a higher $\Delta \mathcal{P}$ (refer to Section 1.1). Methanol is converted directly into formaldehyde. It is used as a reactant with CO for acetic acid and potentially for methyl formate and dimethyl carbonate [338], which might be used as a gasoline additive. Methanol is also an active alkylation agent in reactions with aromatics. It is one of the reactants for MTBE banned in many countries.

Dimethylether (DME) can easily be made from methanol by dehydration [527] [529] or in an integrated methanol/DME synthesis [170] (see below) (refer to Section 2.6.4) and it may play a role in the energy sector as a replacement for LNG for gas turbines or for LPG as domestic fuel. DME can also be used as a pollution-free diesel substitute in existing diesel engines. This would, however, require a special infrastructure for supply of DME.

Gasoline can be made via DME in the MTG process [531] or in an integrated methanol/DME/gasoline loop in the TIGAS process [492] (refer to Section 2.6.4).

The methanol-based routes are made feasible by low methanol costs. Again, the economy of scale and low gas prices are decisive factors as illustrated in Figure 1.23. Today, natural gas-based plants are running at capacities of 5000 t/d and being designed for capacities of 10000 t/d [3].

If such units are built, there may be potential for an integration with GTL plants.

2.6.2 Methanol plant

Since 1970 the methanol synthesis has been based almost exclusively on a $Cu/ZnO_2/Al_2O_3$ catalyst [65]. A natural gas-based methanol plant consists of three parts: syngas preparation, synthesis and distillation. The syngas part typically represents about 60% of the investments.

The synthesis gas for methanol should ideally have the same stoichiometry as the final product. This is expressed by the module $M=(H_2-CO_2)/(CO+CO_2)$ which should be close to 2 as shown in Table 1.6.

Note
The expression for M is derived from simple stoichiometric calculations.

$$CO + 2H_2 = CH_3OH$$
$$CO_2 + H_2 = CO + H_2O \qquad (2.12)$$
$$b c a$$

With initial amounts: a, b, and c, for CO, CO_2 and H_2, respectively. The amounts of CO and H_2 after full reverse shift are a+b and c-b, respectively, so that the final ratio H_2/CO is:

$$M = \frac{H_2 - CO_2}{CO + CO_2} = 2 \qquad (2.13)$$

M>2 means too much, whereas M<2 means too little hydrogen in the syngas. It is preferred to operate the synthesis slightly above 2 as a module slightly below 2 easily results in side products (higher alcohols). Steam reforming of natural gas results in M close to 3 resulting in a surplus of hydrogen, which is recycled to the tubular reformer as fuel.

Combined steam and CO_2 reforming can meet the correct stoichiometry. The reaction

$$0.75CH_4 + 0.25CO_2 + 0.5H_2O \leftrightarrow CO + 2H_2 \leftrightarrow CH_3OH \qquad (2.14)$$

forms the basis for a 3,000 MTPD methanol plant in Iran [240]. Liquid hydrocarbons have a lower H/C ratio than CH_4, meaning that less CO_2 is needed to meet the module M=2. The feasibility of CO_2 reforming depends strongly on the price (and pressure) of the CO_2 source. Many natural gas resources as well as biogas contain large quantities of CO_2.

Figure 2.24 shows the H_2/CO ratio and the module M obtained from lean natural gas by autothermal reforming (ATR) as a function of the steam-to-carbon ratio (S/C) and the exit temperature. Although certain adjustments are possible, direct production by ATR of a gas with a module M=2 is impossible. Such gases are best produced by a combination of steam reforming and oxygen fired autothermal reforming or, for very large capacities, by ATR followed by adjustment of the gas composition by addition of H_2 and/or removal of CO_2.

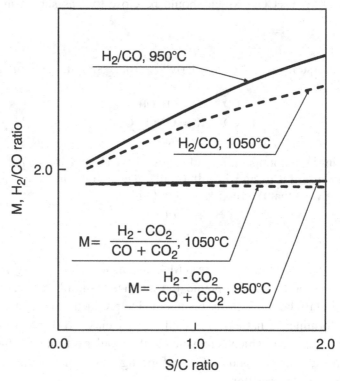

Figure 2.24 Module M and H_2/CO in raw gas from ATR. T.S. Christensen et al. [111]. Reproduced with the permission of the author and Elsevier.

For natural gas-based plants, the choice of technology depends on the scale of operation (refer to Section 1.3.5). For the present world-scale plants (3–5000 t/d), a hybrid solution, two-step reforming, is the most economic solution.

> Note
> This may be implemented in two ways [4], as shown in Figure 2.25, with the tubular reformer and the ATR either in series or in parallel. In both schemes, the conditions can be selected so the gas has a module M of about 2. For the same overall conditions, the gas compositions and temperature exit the ATR reactor should be identical. The main role for

the fired tubular reformer is to provide heat to the process streams. From thermodynamic considerations, the heat input and hence the size of the tubular reformers will be the same in the two cases independent of the flows. However, for the series scheme, it means a lower reformer exit temperature (and tube-wall temperatures). In the parallel scheme, the lower flow through the reformer means a higher exit temperature and a more expensive reformer. The mixing of the hot exit gas and cold by-pass gas may create problems.

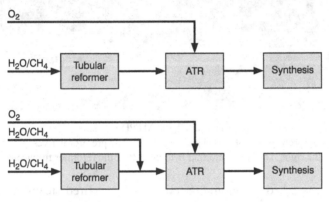

Figure 2.25 Two-step reforming.

A two-step reforming plant is shown in Figure 2.26 and typical conditions for two-step reforming methanol plants are shown in Table 2.11.

Table 2.11 Natural gas-based methanol plant. Two-step reforming (series). Typical conditions for catalytic reaction.

	Cat	P_{exit} (bar)	T_{exit} (°C)	H_2O/C	O_2/C	Exit gas, dry
HDS/ZnO	Co/Mo/ Al_2O_3/ZnO		360			
Prereformer	Ni/Support		460	1.5–2.0		70–80% CH_4
Tubular reformer	Ni/Support		700–750	1.5–2.0		25–40% CH_4
ATR reformer	Ni/Support	25–35	1000		0.35–0.5	0.6–1.0% CH_4
Synthesis	Cu/Zn/Al	75–90	240–260			M=2.05 (in)

Figure 2.26 Synthesis gas plant of a 2,500 MTPD methanol plant with two-step reforming + prereforming, Statoil, Tjeldbergodden. The prereformer vessel is shown to the right. The ATR reformer is to the left of the reformer furnace.

As discussed for ammonia (Section 2.5), the fired steam reformer in the two-step process might be replaced by a convective reformer heated by the hot syngas from the autothermal (secondary) reformer. In contrast to what is the case for ammonia plants, the heat of the ATR exit gas matches the heat required for the primary reforming [395] [479].

In the future very large (10000 t/d) methanol plants may be based on ATR alone. The ATR route is compared with the two-step reforming process in Table 2.12 [4].

Table 2.12 Comparison of two-step reforming and ATR [4].

Technology	Two-step	ATR
Single line capacity (rel.)	75	100
Oxygen consumption (rel.)	76	100
Reforming and preheat duty* (rel.)	264	100
Loop carbon efficiency (%)	95.0	90.5
Loop hydrogen efficiency (%)	81.7	92.2

* including preheat and reformer tubes

The maximum single line capacity is largest for ATR, whereas the oxygen consumption is lower for the two-step route. The ATR route yields a more reactive syngas with a higher CO/CO_2 at a lower overall steam-to-carbon ratio [4].

The maximum size of a single line natural gas-based methanol plant is limited by the volumetric flow exit the autothermal reformer [4]. Therefore, the ATR should operate at a low steam-to-carbon ratio, which results in lower flow and higher CO/CO_2 ratio. This means a more reactive syngas resulting in less recycling in the methanol synthesis. A high CO/CO_2 ratio may result in higher yields of by-products, which makes the distillation section more expensive. This is of less importance for fuel methanol.

2.6.3 Methanol via gasification

A high CO/CO_2 ratio is also characteristic of the syngas for methanol plants based on gasification of coal (petcoke, biomass, etc.) as illustrated in Table 2.13.

Table 2.13 Typical syngas characteristics of a methanol plant.

Feedstock	Technology	M	CO/CO_2
Natural gas	Steam reforming	3.0	2.8
Natural gas	Two-step reforming	2.05	3.6
Natural gas	ATR	2.00	15.5
Coal	Gasification	2.05	10.1
Biomass	Gasification	1.9	6.1

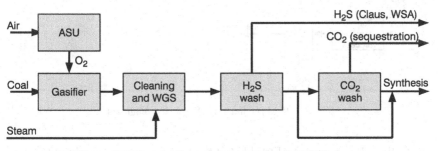

Figure 2.27 Process diagram for a coal-based MeOH plant.

The syngas is made by gasification and gas treatment adjusting to M=2 as shown in Figure 2.27. In order to split H_2S and CO_2 removal, a physical wash system (Rectisol, Selexol) is preferred (refer to Section 1.5.3).

2.6.4 Combined syntheses and co-production

Instead of using methanol as an intermediate, it can be advantageous to combine methanol synthesis with synthesis of the final product in a single process loop. One example is the methanol to gasoline process (MTG). In the MTG process, the conversion of syngas to gasoline is carried out in two separate steps with condensation and reevaporation of methanol [531] (Figure 2.28). Each process step requires a recycle stream: the methanol synthesis because of thermodynamic limitations, and the MTG process to control the temperature rise.

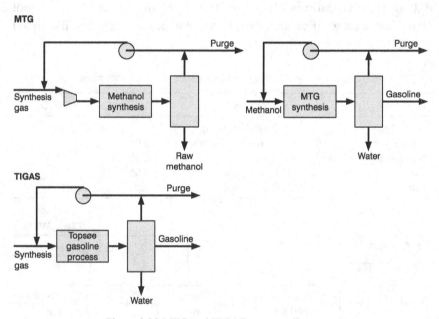

Figure 2.28 MTG and TIGAS, process diagrams.

In the TIGAS process [492] [261] the two loops are combined (Figure 2.28), resulting in a simpler and cheaper process layout. The two process steps are carried out at the same pressure. By using a combined MeOH/DME synthesis, the thermodynamics is more favourable, which can be utilised to achieve a given syngas conversion at lower pressure. This effect is more pronounced when using syngas from coal [491]. As DME is the intermediate for gasoline, the H_2/CO ratio can be pushed to 1 according to the reactions in Table 2.14:

Table 2.14 Reactions to DME.

Reaction		$-\Delta H^\circ_{298} \; \dfrac{kJ}{mol}$
R35	$3CO + 3H_2 \leftrightarrow (CH_3)_2 O + CO_2$	246
R36	$2CO + 4H_2 \leftrightarrow 2CH_3OH$	181
R37	$2CH_3OH \leftrightarrow (CH_3)_2 O + H_2O$	23

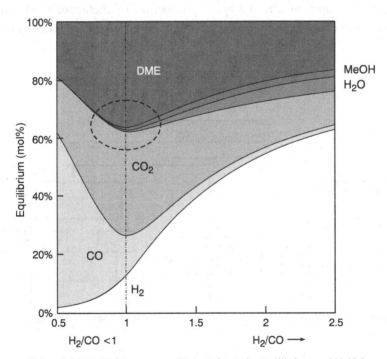

Figure 2.29 DME from syngas. Thermodynamic equilibrium and H_2/CO.

Reaction R36 in Table 2.14 requires steam addition to adjust to $H_2/CO=2$ with a resulting energy loss by condensation of steam. The adjustment of $H_2/CO=1$ for the TIGAS process can be achieved by adding steam directly to the MeOH/DME synthesis. As shown in Figure 2.29, $H_2/CO=1$ (Reaction R35 in Table 2.14) represents the optimum for DME production thermodynamically.

Similar considerations can be made for an integrated synthesis for DME production [170] and for integrated synthesis of acetic acid [406]. The cost savings in the synthesis loops should be compared with increased costs for separation of the products in a diluted product gas.

Methanol can be produced as co-product in the ammonia synthesis [239] [406]. The CO and CO_2 for the methanol synthesis is obtained by a partial bypass of the shift converters and the CO_2 wash. The methanol synthesis takes place in an adiabatic reactor or a simple once-through boiling water reactor and the unconverted syngas is passed through a methanator back to the ammonia synthesis loop.

This principle of cheap bypass production of methanol (and other products) in once-through synthesis is being considered for IGCC power plants as illustrated in Figure 2.30 [297] [318] [397] [428].

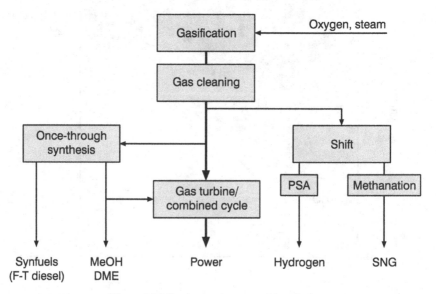

Figure 2.30 Co-production power/chemicals.

When using once-through synthesis and burning the non-converted gas in the gas turbine, it is possible to save the energy consumption used to recycle non-converted syngas in a conventional methanol plant. The key parameter is the conversion per pass which, as outlined above, can be increased with the combined MeOH/DME synthesis, by a condensing methanol synthesis [60], or by gasoline production via the TIGAS process [428]. The co-production can be balanced with the load of the power production, thus leaving the coal gasifiers operating at constant load.

The combined methanol/dimethylether synthesis pushes the equilibrium to higher once-through conversion as illustrated in Table 2.15.

Table 2.15 IGCC and once-through synthesis. (Shell gasifier) [418].

	Gas composition (vol%)				
	H_2	CO	CO_2	H_2O	Inerts
Pressure (bar g) 33	29.7	56.4	1.4	7.0	5.5
Pressure (bar g) 58	28.0	53.3	1.5	8.0	10.2

	Conversion of CO + H_2%	
	33 (bar g)	58 (bar g)
Methanol	18.6	31.6
Methanol/DME	69.0	76.0
Gasoline/TIGAS	69.0	76.0

2.6.5 Fischer–Tropsch synthesis

The manufacture of synfuels via syngas was applied in Germany during World War II with nine plants based on Fischer–Tropsch synthesis with a total capacity amounting to 700,000 tonnes per year (approximately 7000 bpd) [165]. The syngas was made by coal gasification. The original FT units in South Africa were also based on coal gasification using Lurgi dry ash gasifiers because of the high ash content of the coal. The Sasol plants mainly apply the high-temperature FT synthesis meaning that the syngas is converted to a wide spectrum of chemicals and fuels. The methane from the gasifier (refer to Table 1.17) and from the FT synthesis is converted to more syngas by autothermal reforming. The optimum syngas composition for the high-temperature FT synthesis is as for the methanol synthesis using a module M=2.

The low-temperature FT synthesis over promoted cobalt catalyst leads to the formation of paraffinic wax, which is converted into high quality diesel by mild hydrocracking as shown in the simplified process diagram in Figure 2.31 [257]. Today, there is an interest in building large-scale FT units based on conversion of natural gas to liquid products (GTL). This offers a solution for validation of "stranded natural gas", as the liquid fuels can be moved readily around the world to the markets where they are most needed [193]. As an alternative, projects involving liquified natural gas (LNG) may be more competitive if large quantities of gas are available.

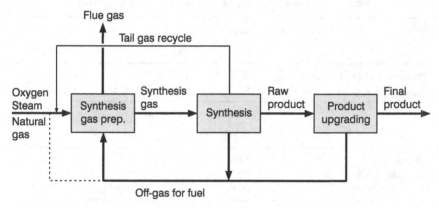

Figure 2.31 Flow scheme for natural gas- based low-temperature Fischer–Tropsch synthesis [409]. Reproduced with the permission of ACS.

Figure 2.32 shows a 34,000 bpd FT plant in Qatar, where an even larger plant (120,000 bpd) is being built by Shell [91].

Figure 2.33 shows a photo of the ATR reactor in one of the process lines of the GTL plant shown in Figure 2.32.

Figure 2.32 Oryx GTL plant in Qatar, 34,000 bpd FT diesel. Oxygen plant: Air Products; Syngas: Topsøe; FT synthesis: Sasol; Hydrocracking: Chevron. The plant is designed in two lines (17,000 bpd). Reproduced with the permission of Oryx GTL.

Figure 2.33 ATR reformer for a 17,000 bpd FT line (Oryx plant, Qatar) [332]. The reactor to the left is the prereformer. Reproduced with the permission of BC Insight.

A GTL unit based on natural gas consists of three parts as shown in Figure 2.31: syngas preparation, FT synthesis and product upgrading (hydrocracking of wax). Oxygen blown autothermal reforming (ATR) is the preferred choice for the syngas manufacture [34], which amounts to approximately 2/3 of the plant investments, with the air separation unit responsible for approximately 50% [172]. This means that the oxygen consumption (tonnes O_2/tonne FT product) is a key parameter.

The size of the syngas plant is directly related to the carbon efficiency of the FT synthesis. Non-converted syngas and light synthesis products may be recycled to the syngas unit or used as fuel as illustrated in Figure 2.31. Both situations mean a larger syngas unit per tonne of FT product. Hence, it is important that the syngas composition is tuned (adjusted) for maximum conversion per pass in the FT synthesis. The desirable composition of the syngas for the low-temperature Fischer–Tropsch corresponds to a ratio of $H_2/CO\sim2$.

CO is the only reactant for the low-temperature FT synthesis for wax and diesel and the cobalt catalyst is not active for shift reaction. It means that the syngas should have a minimum content of CO_2, which remains inert in the FT synthesis and hence represents a carbon (oxygen) loss. This is different from the high-temperature FT synthesis and the methanol synthesis for which the catalyst is active for shift reaction. CO_2 is normally returned to the syngas plant as shown in Figure 2.31.

Direct production of a gas with $H_2/CO=2$ is achieved by ATR only at very low H_2O/C ratio and high exit temperature (this may involve a risk of soot formation) as illustrated in Figure 2.24. At higher H_2O/C ratio, the value of 2 is best obtained by partial recycle of CO_2 or CO_2-rich tail gas (Figure 2.31). It may be combined with removal of H_2 from the syngas. This could be an attractive way of producing H_2 for the hydrocracking of FT wax to diesel. The POX processes result in a H_2/CO below that of ATR and hence a separate hydrogen steam reforming plant is added to meet the H_2/CO ratio [233].

A process diagram for an ATR-based syngas unit for a natural gas-based FT plant is shown in Figure 2.34. The process heat is recovered in boilers resulting in surplus steam in the plant. By replacing the boilers with a convective reformer as shown in Figure 2.35, the process heat of the exit gas from the autothermal reformer is used for steam reforming (chemical recuperation) and for preheating the feed gas to the autothermal reactor [34] [427]. This saves large fired preheaters and reduces the oxygen consumption. The convective reformer (HTER) can be in series or in parallel to the autothermal reformer [34]. The data in Figure 2.35 show that a parallel convective reformer results in higher capacity for less oxygen consumption and steam export [34] [427]. The use of this principle of "chemical recuperation" [427] may involve cooling down of the process gas through a temperature range with potential for carbon formation from CO and attack from metal dusting corrosion (refer to Section 5.2.2).

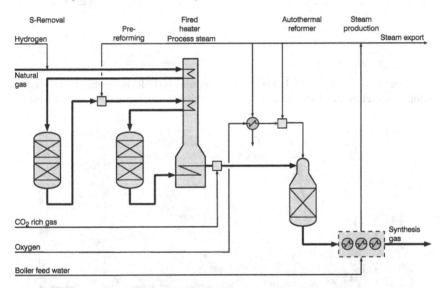

Figure 2.34 Syngas for GTL. Natural gas is desulphurised and after addition of steam, the process gas is passed through an adiabatic prereformer (converting higher hydrocarbons) and a fired preheater. Oxygen is added to the burner of the autothermal reformer and the process heat is recovered in a boiler producing steam. Bakkerud *et al.* [34]. Reproduced with the permission of Elsevier.

Figure 2.36 shows a convective reformer being installed in an FT plant in South Africa.

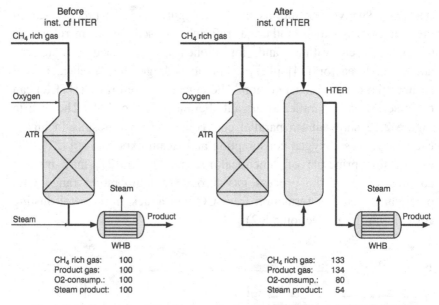

Figure 2.35 Syngas for GTL. Gas heated reforming (HTER) in combination with autothermal reforming. Bakkeud *et al.* [34]. Reproduced with the permission of Elsevier.

Figure 2.36 Installation of a convective reformer at an FT plant.

At present, FT units are built with a carbon yield to C_{5+} products of ~75% [477] corresponding to a CO_2 emission of 1.1 tonnes/tonne C_5+ product and a thermal efficiency of approximately 65% [193]. The diesel product is sulphur-free and has a high cetane number (70+). This makes the GTL plant economically feasible at oil prices above 25–40 USD/b depending on location and gas price. There may be room for improvement of the thermal efficiency, for instance by adding gas heated reforming for heat recovery as illustrated in Figure 2.35.

The methanol synthesis has a higher thermal efficiency (approximately 72%) [420] than the FT synthesis. CO_2 is a reactant in the synthesis which gives room for a high carbon efficiency and less constraints on the syngas unit.

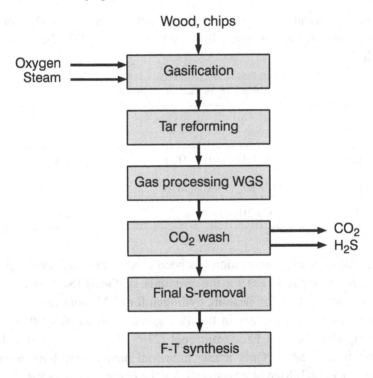

Figure 2.37 Biomass to Fischer–Tropsch synthesis. Adapted from K. Salso, Andritz. [437].

It may be argued that energy efficiency is of less importance when natural gas is cheap, but high energy efficiency means small feed pretreat units and reduced requirements for utilities and hence less investments. Moreover, high efficiency means less CO_2 production. Therefore, it is important that the syngas composition is tuned (or adjusted for maximum conversion per pass) in the FT synthesis.

There is an increasing interest in coupling gasification of biomass to the FT synthesis. One example is the process diagram shown in Figure 2.37. The tar is removed by reforming before gas cleaning, water gas shift and CO_2/H_2S wash in a coal-based plant.

2.6.6 SNG

Synthetic natural gas or substitute natural gas (SNG) can be manufactured by the methanation reaction [219] [423]. Reactions are seen in Table 2.16.

Table 2.16 Methanation reactions.

Reaction		$-\Delta H^o_{298} \; \frac{kJ}{mol}$
R38	$CO + 3H_2 \leftrightarrow CH_4 + H_2O$	206
R39	$CO_2 + 4H_2 \leftrightarrow CH_4 + 2H_2O$	165
R40	$2CO + 2H_2 \leftrightarrow CH_4 + CO_2$	247

For many years, methanation has been used as the final clean-up step in preparing synthesis gas for the ammonia synthesis (Section 2.5) and lately as one solution for cleaning hydrogen for PEM fuel cells.

During the energy crisis in the 1970s, methanation of synthesis gas from naphtha [84] and later from coal [219] [448] was considered for manufacture of SNG. Only a few industrial plants were built because energy prices stabilised at a relatively low level during the 1980s. Today, coal conversion is in focus in the light of the limited oil resources and the wish for security of energy supply. At locations with diminishing resources of natural gas, the manufacture of SNG from coal or biomass may become feasible [504].

The methanation reactions are the reverse of the reactions for steam reforming. Nickel and other Group VIII metals are active. Both CO_2 and CO are converted, meaning that the syngas should have a module M=3 (refer to Table 1.6).

The heat of Reaction R38 in Table 2.16 is high and amounts to 20% of the heating value of the syngas ($H_2/CO=3$) and much larger temperature increases are observed in SNG processes than during methanation for clean-up purposes. Consequently, the heat recovery is a key element of an industrial process. It is a challenge to achieve equilibration of the exothermic process at low temperature, while the heat is recovered at high temperature to provide useful energy (superheated steam). In thermodynamic terms, this means minimising the exergy loss.

A syngas of $H_2/CO=3$ at 30 bar would result in an adiabatic temperature increase to 923°C with an inlet temperature of 300°C [219]. However, the high temperatures will require expensive construction materials for reactors and heat exchangers followed by a series of reactors for final methanation. Moreover, at typical conditions, there is thermodynamic potential for carbon formation at high temperatures.

Another solution is to dilute the syngas by recycling product gas, thus reducing the temperature increase to what can be handled by a methanation catalyst [219] [423]. The higher the temperature increase, the lower the recycle ratio and sizes of equipment and compressor.

This formed the basis for a high-temperature methanation process (TREMP) [423], which was studied extensively in the ADAM-EVA project around 1980. The reaction heat was recovered as superheated steam. Figure 2.38 shows a process diagram of typical temperature profiles. The temperature increased from 300 to approximately 650°C over the first reactor followed by two clean-up reactors.

The operation means catalyst sintering at high temperatures and exposure to high partial pressures of carbon monoxides at low temperatures. It is evident that the catalyst must maintain activity at low temperature after having been exposed to high temperatures.

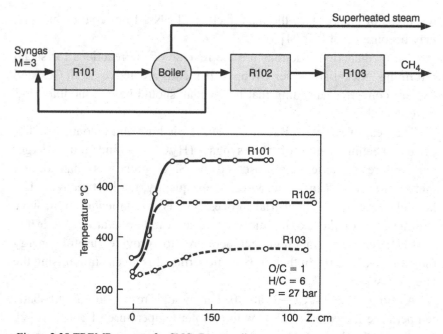

Figure 2.38 TREMP process for SNG. Process diagram with three methanation reactors and corresponding temperature profiles [219].

The high heat of reaction means that high conversions are achieved even at pressures around 10 bar in contrast to ammonia and methanol synthesis. Figure 2.39 shows the isothermal conversion to methane, DME and methanol at pressures equal to 10 and 50 bar. In all cases the conversion decreases strongly with temperature, but at usual reactor outlet temperatures the conversion to methane is still large. This may be an advantage for once-through conversion in co-production schemes for IGCC plants (refer to Section 2.6.4) and for low-pressure gasifiers for biomass.

As for the combined MeOH/DME synthesis, it is possible to carry out the methanation reaction at $H_2/CO=1$ according to Reaction R40 in Table 2.16. This is not possible over a nickel (or other Group VIII metals) because of a big potential for carbon formation (refer to Chapter 5).

However, it has been demonstrated on sulphide catalysts [236] [467], which are related to sour shift catalysts (refer to Section 1.5.2).

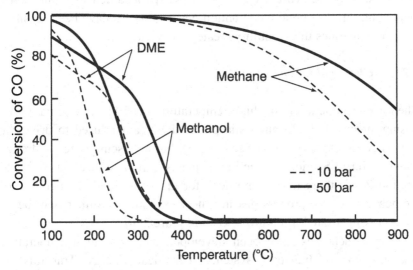

Figure 2.39 Equilibrium conversion to methanol, DME and methane as a function of temperature and discrete values of pressures: 10 and 50 bar. Feed: methanation: $H_2/CO=3$, methanol and methanol/DME: $H_2/CO=2$.

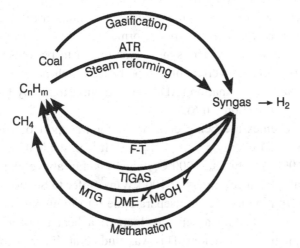

Figure 2.40 The synfuel cycle [426]. Reproduced with the permission of Elsevier.

The reforming and gasification reactions described in Chapter 1 and the synfuel syntheses described in Section 2.6.5 are interlinked in what can be termed the synfuel cycle [413] [415] illustrated in Figure 2.40. Depending on the economic situation, various routes have been feasible. This demonstrates the flexibility created by syngas.

2.7 Chemical recuperation

Chemical recuperation of high-temperature heat is an alternative to raising steam [427]. In ammonia plants the steam is used to drive the syngas compressor (refer to Section 2.5), but as discussed for hydrogen plants (refer to Section 2.2) and GTL plants at remote locations (refer to Section 2.6.5) there may be no need for the steam, and recuperation of the heat of the hot process gas in a heat exchange reformer may be an option (as illustrated in Figure 2.35 and Figure 2.36).

Process schemes have been developed to utilise high-temperature heat for energy transport systems (Chemical Energy Transmission Systems – CETS). The steam reforming reaction fulfils the criteria of reversibility, high heat of reaction, and temperature level of the forward and backward reaction (methanation, refer to Section 2.6.6). This was the basis for the German ADAM-EVA system [189] [285], in which the hot helium (950°C, 40 bar) from a high-temperature nuclear reactor was used as heating source for a convective reformer (EVA) converting methane (natural gas) into syngas. The syngas was meant to be transported over long distances and converted into superheated high-pressure steam by high-temperature methanation (ADAM) as demonstrated by the TREMP process (refer to Section 2.6.6).

Similar schemes are being considered for transport of energy from concentrated solar energy plants [44] [155]. It is possible to achieve solar fluxes at 1000 kW/m^2 at 1000°C, which is far above the fluxes in conventional tubular reformers of 100 kW/m^2 (refer to Sections 1.2.2 and 3.2.2). One solution has been to separate the reforming reaction from the solar receiver by having a separate loop of a heat transfer media (as helium for the nuclear reactor). This was studied at first in a sodium heat pipe reformer [363]. However, reactors for direct absorption of the solar heat have since been developed [154] [234] [524]. The use of CO_2 reforming may represent an advantage [368] by avoiding a steam cycle and condensation of steam in the pipelines during load variations. CO_2 reforming may be carried out with noble metal catalysts [368] (refer to Sections 2.4.2 and 3.5.4), but in a closed CO_2-reforming/methanation loop the addition of steam may be required to eliminate carbon formation in the methanation step (refer to Chapter 5), which results in reduced efficiency of the loop [185]. To overcome some of the potential difficulties when applying CO_2 or steam reforming loops for the very transient operation of a solar plant, other attempts have dealt with using ammonia decomposition/synthesis [306]. However, the heat of reaction is smaller and there appears to be an inherent loss of energy via the condensation and re-evaporation of ammonia.

Schemes have been developed for chemical recuperation of the exhaust heat from gas turbines for large-scale power production [318] [402]. The combustion temperature for gas turbines is normally controlled by operating with surplus air, but dilution represents an exergy loss or a decrease in the maximum achievable efficiency. One way of solving the problem [402] is to use the exhaust heat from the gas turbine to convert the methane feed by the steam reforming reaction in a heat exchange reformer to a fuel gas with less heating value per volume, thus decreasing the demand for dilution. Still, at a turbine inlet temperature of 1350°C the benefits of chemical recuperation remains marginal, because only 18% of the natural gas is converted into syngas due to thermodynamic constraints. At future turbine inlet temperatures of 1500°C, the higher exhaust temperature from the second turbine creates more useful heat for heat exchange reforming and it means that more than 1/3 of the natural gas can be converted, resulting in a similar gain in efficiency. A similar increase in efficiency can, however, be achieved by steam injection without reforming reaction.

The thermodynamic constraint is eliminated when integrating a gas turbine cycle with a high-temperature fuel cell (SOFC) with internal reforming as illustrated in Figure 2.41 [415] [454].

The exhaust gas from an SOFC fuel cell provides a hydrogen containing clean fuel gas mixture for the combustion chamber of the gas turbine allowing a more efficient control of the maximum temperature than is achievable by adding the surplus air. Such schemes are claimed to have electric efficiencies close to 70% [454].

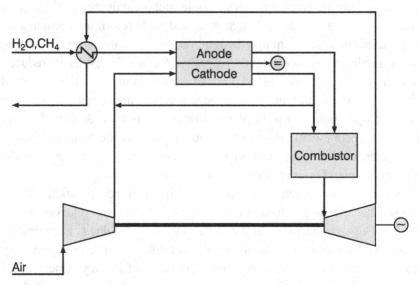

Figure 2.41 SOFC and gas turbine cycle [415]. Reproduced with the permission of Elsevier.

As a thermochemical cycle chemical recuperation is a heat engine converting high-temperature heat into chemical energy. Like a steam cycle resulting in mechanical energy the efficiency depends on the operating temperature and the irreversibilities of the individual process steps [156].

Part II
Steam Reforming Technology

Topsøe monotube reformer pilot plant Houston

3 Technology of Steam Reforming

3.1 Early developments

The steam reforming reaction was introduced to industry more than 75 years ago, although the reaction had not been subject to many scientific studies [381] [389]. Early work in the 19th century included reactions of hydrocarbons and steam over calcium oxide [487] and a cyclic process using nickel [543].

Laboratory studies on the reforming reaction were published in the 1920s [327] showing that product gases close to thermodynamic equilibrium were obtained over nickel catalysts. Fischer and Tropsch [192] observed no significant differences when steam was replaced by carbon dioxide [18]. It is surprising that Sabatier did not report any studies on steam reforming as he studied numerous reactions on nickel catalysts. Among these was the methanation reaction reported as early as 1902 [432].

The reforming process was adopted mainly in the US where natural gas was easily available as feedstock, whereas reformers in Europe were initially reduced to operate on propane and LPG.

The first patent on a tubular reformer using supported nickel catalysts was obtained by BASF in 1912 [348] [544]. A later patent (1928) dealt with a heated tubular reactor [546]. A license was given to Standard Oil, New Jersey, and the first industrial reformer was started in Baton Rouge in 1930 [93]. Six years later, the first ICI reformer was commissioned at Billingham [203].

The industrial breakthrough of the steam reforming technology came in 1962, when ICI succeeded in starting two tubular reformers operating at pressure (15 bar) and using naphtha as feedstock [87]. The first reformer designed by Topsøe was started in 1956 and the first naphtha reformer in 1965, followed by a hydrogen plant in 1966 operating at 42 bar [381] [413].

The introduction of high-pressure reforming meant that the energy consumption of the ammonia synthesis could be decreased significantly (refer to Section 2.5) As a result, a number of natural gas-based ammonia plants were built in the US, primarily by M.W. Kellogg in the mid-1960s [224] [312].

Figure 3.1 MWK ammonia plant, Texas City (1960s) [312]. Reproduced with the permission of PenWell.

At the same time in Europe, steam reforming of naphtha became the key technology in the build-up of the UK town gas. Apart from the tubular reforming technology provided by ICI [87] and Topsøe [493], low-temperature steam reforming of naphtha in an adiabatic reactor was used to produce methane-rich gas. This process was pioneered by British Gas, originally using methanol as fuel (refer to Section 1.4.1) [126] [139] [150]. It formed the basis for later developments for SNG (refer to Section 2.6.6) and prereforming (refer to Section 1.2.5) [126] [139].

The Topsøe reforming technology was pioneered in a full-size monotube pilot plant during the 1960s as shown in Figure 3.2. The plant was running on naphtha.

Figure 3.2 Pilot Plant. Frederikssund, Denmark (around 1966).

The steam reforming process became the preferred technology for synthesis gas for methanol and other petrochemicals, oxo-alcohols, acetic acid, etc. [88], [425].

The steam reforming process may appear straightforward from an overall consideration as the product composition is determined by simple thermodynamics, but in reality it is a complex coupling of catalysis, heat transfer and mechanical design.

The progress in steam reforming technology has resulted in less costly plants, in part because of better materials for reformer tubes, better control of carbon limits, and better catalysts and process concepts with high feedstock flexibility. This progress has been accompanied by a better understanding of the reaction mechanism, the mechanisms of carbon formation and sulphur poisoning, and the reasons for tube failure.

3.2 Steam reforming reactors

Steam reforming reactors are conveniently divided into:

- direct fired steam reformers, where the reactor is a furnace with burners supplying the endothermic heat of reaction directly to the catalyst in heated tubes as in a tubular reformer;
- indirectly heated reformers, where the catalyst is placed in an adiabatic bed. The reaction heat is either supplied by preheating the feed as for an adiabatic prereformer or the heat is supplied by internal combustion in a burner as in an autothermal reformer;
- convective reformers with internal heat exchange. A convective reformer may also be fired directly; and
- steam reformers with mechanical integration of catalyst and reactor as in micro-scale reactors.

3.2.1 Role of catalyst

In all reactor types, the desired conversion takes place on the catalyst. Side-reactions, such as carbon formation (refer to Chapter 5), are eliminated by use of the proper catalyst types, controlled catalyst bed inlet temperatures and a given H_2O/C ratio depending on the type of feed. The role of the catalyst is in all cases to achieve equilibrium conversion and maintain low pressure drop and in tubular and convective reformers to keep low tube metal temperatures, too.

The space velocity Nm^3 C_1/m^3 catalyst/h has conventionally been used to determine the necessary catalyst volume for a given amount of feed, but this parameter gives no information about pressure drop, heat transfer or risk of unwanted side-reactions, so advanced rigorous models are necessary to evaluate a design (refer to Section 3.3).

The catalyst must also ensure equal distribution of flow between the tubes in a tubular reformer and across the catalyst bed in an adiabatic reformer. Mal-distribution may cause local overheating or channelling and hence shorter life of the reactor. It means that the pressure drop should be carefully monitored during catalyst loading. The steam reformer is also one of the first reactors in a syngas process and it is exposed to impurities remaining in the feed and the steam, in particular sulphur (refer to Chapters 4 and 5).

The catalyst should be mechanically stable under all process conditions as well as conditions during start-up and shutdown of the reformer. In particular, resistance to conditions during upsets may become critical. Breakdown of catalyst pellets may cause partial or total blockage of some tubes resulting in "hot spots", "hot bands" or totally hot tubes [389] in tubular or convective reformers. The formation of carbon may result in the same problems.

The shape of the catalyst pellet should be optimised to achieve maximum activity with minimum increase in pressure drop. The pressure drop depends strongly on the void fraction of the packed bed and decreases with increasing particle size (refer to Section 3.3.4). Similarly the effective catalyst activity per volume is roughly proportional to the surface-to-volume ratio of the catalyst pellet due to the low catalyst effectiveness factor (typically below 0.1 for a tubular reformer) (refer to Section 3.4).

Activity $\alpha \dfrac{1-\varepsilon}{d_p}$

$\Delta P \; \alpha \dfrac{1-\varepsilon}{\varepsilon^3} \cdot \dfrac{1}{d_p}$

Figure 3.3 The catalyst shape is a compromise between catalyst activity and pressure drop. The particles shown have an external diameter of 16 mm [425]. Reproduced with the permission of Wiley.

In order to fulfil the above-mentioned requirements to the catalyst, the particles are usually made of ceramic material of cylindrical shape with one or more internal holes as seen in Figure 3.3. Solutions based on the use of ceramic foam as catalyst support have also been considered [365] together with other shapes such as monoliths and even catalysed hardware [397]. The use of such catalyst formulations are often developed together with new reactor concepts taking advantage of a higher effectiveness factor than achievable in conventional steam reformers (refer to Section 3.3.8).

Simplified equations to estimate relative catalyst activity and bed pressure drop are shown in Figure 3.3 below:

Example 3.1
Compare the relative catalyst activity and pressure drop of catalyst fillings with cylinders of size 10x10 mm, and rings of size 16/8x10 mm. The outer diameters are 10 and 16 mm, respectively, and the particle height is 10 mm in both cases. The inner diameter of the ring is 8 mm. The bed void fraction is calculated as a sum of two parts. The first part is the void fraction of the catalyst bed when considering the particles as full cylinders. This part, ε_{cyl} is assumed to be 40%. The second part is the contribution from the holes so that the actual bed void fraction of the ring is calculated from:

$$\varepsilon = \varepsilon_{cyl} + (1 - \varepsilon_{cyl}) \left(\frac{d_{inner}}{d_{outer}} \right)^2$$

In both cases we need the volume equivalent particle diameter, d_p, which is the diameter of a sphere having the same volume as the actual particle (see Equation 3.29).

Application of the formulas in Figure 3.3 gives the following results:

	Cylinder	Ring
Outer diameter (mm)	10	16
Height (mm)	10	10
Inner hole diameter (mm)	0	8
Equivalent volume particle diameter, d_p (mm)	11.4	14.2
Void fraction	0.40	0.55
Relative pressure drop	0.82	0.19
Relative catalyst activity	0.0053	0.0032

It is seen that the ring represents the optimum, since the loss in activity is of minor importance due to the low effectiveness factor, compared with the large decrease in pressure drop.

In indirectly heated reformers operating conditions may vary considerably, but the requirements to the catalyst are similar to those in the tubular reformer. The catalyst activity must be larger for low-temperature adiabatic prereformers, where catalyst effectiveness factors will be outside the strongly diffusion-controlled regime. This implies that a simplified activity expression as in Figure 3.3 is not valid for a prereformer. In an autothermal reformer the requirements to the catalyst includes high temperature stability properties and a good distribution over the catalyst bed of the partly converted gas from the burner. In micro-scale reactors catalyst stability is a major issue.

3.2.2 The tubular reformer

In the tubular reformer, the catalyst is placed in a number of high-alloy reforming tubes placed in a furnace. The outer diameter of the tubes ranges typically from 100 to 200 mm and the tube length from 10 to 13 m (Figure 1.13). Typical inlet temperatures to the catalyst bed are 450–650°C, and product gas leaves the reformer at 700–950°C depending on the application. The average heat flux from the burners may be as large as 150 kW/m^2 based on m^2 of inner tube surface.

Tubular reformers operate at high mass velocities corresponding to Reynold numbers larger than 5000. Today such reformers are built for capacities up to 300,000 Nm^3/h of H$_2$ or syngas.

Example 3.2
Estimate the number of tubes in the hydrogen reformer from Example 1.3. The average heat flux is 90 kW/m^2, the inner tube diameter is 100 mm and the tube length 13 m. Calculate the space velocity and the actual inlet superficial velocity assuming ideal gas.
The total heat transferred is 108 MW (see Example 1.3), so the required heat transfer area is 1200 m^2. Each tube has an internal surface area equal to 4.08 m^2, so the number of tubes is estimated as 294.

The total catalyst volume will be 30 m^3 and since the feed flow of CH_4 is 500 mol/s, the space velocity is 1345 $Nm^3 C_1/m^3$ catalyst/h. The gas velocity based on empty tube at the inlet conditions is around 1.5 m/s and the gas mass velocity is 13 kg/m^2/s.

The tubular reforming furnace is a radiant box including burners and a convection section to recover the waste heat of the flue gases leaving the radiant section. Industrial-size sidewall-fired reformers are seen in Figure 1.12 and Figure 1.13. The thermal efficiency of the tubular reformer and waste heat recovery section approaches 95%, although only 50–60% of the heat is transferred to the process through the tubes. The remaining part is recovered from the flue gas. This heat is used for preheating of the reformer feed, combustion air, and for steam production.

Burners may be located on the walls, in the top, on the bottom or a combination of these [389] as illustrated in Figure 3.4 below:

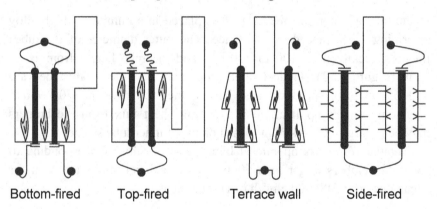

Bottom-fired   Top-fired   Terrace wall   Side-fired

Figure 3.4 Reformer furnace configurations [389]. Reproduced with the permission of Springer.

In top- and bottom-fired box type furnaces, tubes are arranged in a number of parallel rows with the burners between the tube rows either in the top or bottom. The flue gas is taken out in the bottom of the top-fired and in the top of the bottom-fired reformer. These types have fewer burners.

Sidewall-fired and terrace wall-fired furnaces are long furnace chambers with the tubes arranged along the centre line and a matrix of burners located on the two opposing furnace walls. The flue gas leaves the furnace in the top for transfer to the waste heat channel. The advantage of these furnace types is that the process gas and flue gas outlet temperatures can be controlled individually by varying firing intensity on bottom and top burner rows, respectively. Although the total number of burners is larger than for the top- and bottom-fired types, each row of burners can be controlled so it is not necessary to control burners individually.

Schematic temperature and heat flux profiles for a top-fired and a sidewall-fired reformer for identical process outlet conditions are seen in Figure 3.5 below. The top-fired furnace has a high heat flux at the inlet, whereas the sidewall-fired furnace has a more equally distributed heat flux profile. The top-fired furnace has an almost flat tube temperature profile, whereas in a sidewall-fired furnace the tube-wall temperatures increase down the reformer. The terrace-wall fired reformer has profiles similar to the sidewall-fired reformer, whereas the bottom-fired reformer has a larger heat flux in the lower part of the reformer.

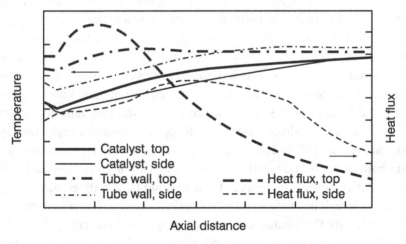

Figure 3.5 Schematic temperature and heat flux profiles in a top-fired and sidewall-fired furnace. Top: Top-fired; Side: Sidewall-fired.

The combustion of natural gas with air is accompanied by formation of nitrogen oxides. This becomes significant above 1400–1500°C, which is easily surpassed in normal diffusion flames. The problem is solved by use of low NOx burners and further reduction can be obtained by cleaning the flue gas using selective catalytic reduction (SCR) with ammonia [510]. Low NOx burners have longer flames in order to limit the maximum temperatures, but this has a significant impact on the heat flux and temperature profiles in the reformer, since it tends to make the flux profiles wider. The burner type must be taken into consideration in the design.

The reformer tubes are exposed to high temperatures and pressures and are made of high alloy steels (for example HP-BST, a 25/35 Cr/Ni Nb Ti alloy). The tubes are supported in the top or bottom with mechanical flexibility in the opposite end to absorb thermal expansions. The high alloy tubes are expensive and account for a major part of the total cost. The reliability of the tubes is important since tube failure can result in long down-periods for retubing and hence loss of production. The mechanical tube design depends not only on the maximum metal temperature, but also on the maximum temperature gradient across the tube wall and the number of scheduled (and unscheduled) start-ups and shutdowns. The latter is in particular important where the steam reformer is considered as a utility, such as in a refinery. The method to calculate the tube-wall thickness is based on mechanical creep-rupture strength as a function of temperature, age and number of cycles. Such data must be obtained from the tube manufacturer. At high operating temperature the tube metal creeps under stress. The creep rate, which is almost constant over time, increases almost linearly with applied stress and exponentially with temperature [389]. The creep eventually results in rupture along the grain boundaries of the metal. This means that for a given tube its life is very sensitive to small changes in the maximum tube-wall temperature.

The constraints in designing tubular and convective reformers are hence primarily the mechanical material properties. The catalyst volume may then be considered a derived property. However, high catalyst activity is essential to ensure low tube-wall temperature. A higher catalyst activity can be utilised to absorb the same reaction heat at a lower temperature [389] (refer to Example 3.4).

3.2.3 Scale-up of steam reforming technology

In addition to construction material properties, scale-up of reforming technology to industrial size requires reaction engineering knowledge concerning catalyst, heat transfer, pressure drop and flow under the often severe conditions in a steam reformer. Some of these properties can be found in laboratory and bench-scale experiments and some can be found by conventional chemical engineering tools and CFD modelling, but still a direct scale-up to an industrial plant of a new steam reforming reactor type may be too risky and an intermediate verification will be necessary. For this purpose a pilot plant can be used to perform measurements and evaluations to ensure a safe scale-up to an industrial unit.

For heated and convective reformers a scale-up requires at least one full-size tube due to the interaction between complicated heat transfer, catalyst and in particular mechanical design. For an adiabatic reactor scale-up can in principle be carried out using the reactor diameter, but the basic experiments must be truly adiabatic and at industrial mass velocities, so also in this case a pilot plant can be used (refer to Section 3.5.1). For other equipment items, such as burners, scale-up will usually require a combination of flow experiments, CFD modelling and full-scale tests.

Figure 3.6 shows a full size mono-tube tubular reformer pilot plant. Convective reformers of various configurations have been tested at the same pilot site (refer to Figure 3.11).

The furnace – with the exhaust chimney on the top – has radiant wall burners in six levels on two opposite walls to supply the heat of reaction to a single full-size tube located in the centre. Such a configuration is appropriate since it allows simulation of almost any outer tube-wall temperature profile by variation of the burner firing pattern.

Figure 3.6 Topsøe reforming pilot plant in Houston, Texas.

Although carefully designed experiments have been carried out in laboratory, bench-scale and pilot plants, the final verification also requires feedback from an industrial plant, although inevitably the number of measurements here is significantly smaller. It is, however, important that the same tools and principles for measurements and simulation are used in the scale-up process to provide the necessary knowledge for verification.

3.2.4 Plant measurements

In addition to internal reactor measurements, plant measurements must include flows, pressures and temperatures around the reactors, and gas analyses. Flows, temperatures and pressures can be followed dynamically from the control system, whereas complete gas analyses are only available off-line. Tools to investigate a process dynamically are important for plant operation and on-line optimisation [82] [92] [309], whereas model evaluation and validation is performed off-line using steady-state measurements. Collecting and retrieving the necessary measurements (including tube-wall temperatures) for a rigorous evaluation of a syngas plant with a tubular reformer may take several days.

Before the validation of steady-state models, it must be realised that all experimental measurements have errors.

Example 3.3
In syngas plants only dry gas analyses are available and as a consequence, the concentration of water in the outlet gas from a steam reformer must either be determined by measuring the steam-to-dry gas ratio or it must be calculated using redundant measurements. Usually the last method is preferred due to large experimental uncertainties in the steam/dry gas measurement, where just 3% deviation in the measurement will correspond to about 10°C in temperature approach to the methane reforming reaction.

Consistent values for the process variables around the reactors must be known and for this purpose data reconciliation [72] [118] [351] is used to find a set of process variables, $\underline{y}$ satisfying the basic mass and heat balances $\underline{f}$ not only around each individual process unit, but also around the plant:

$$\underline{f}(\underline{y}) = 0 \quad (3.1)$$

In data reconciliation a mathematical method varies a set of independent variables $\underline{x}$ till the total sum of squares over all plant measurements, SQ, attains a minimum:

$$SQ = \sum_i \left(\frac{y_{rec,i} - y_{meas,i}}{\sigma_i} \right)^2 \quad (3.2)$$

Here, y_{meas} is the measured value, y_{rec} the reconciled value, and σ is the standard deviation for the measurement. The latter can be obtained by statistical methods for each instrument, but gross errors may be present in the measurements, implying that methods must be available to judge if a deviation results from a gross error or a random error [228] [536].

The independent variables $\underline{x}$ include a subset of the process and key reaction characterisation variables, such as conversions and temperature approaches. Relative catalyst activities and heat transfer coefficients can also be used as independent variables in data reconciliation. The choice is hence between:

A: Use the same equilibrium type reactor models as in the process simulation using conversion and temperature approaches to characterise the reactions.

B: Use a rigorous reactor model, such as the two-dimensional model described in Section 3.3.

In A the specific reactor parameters must then be found in a subsequent step. From a mathematical point of view, however, it would be more correct to include the rigorous reactor model in B directly, thus eliminating step A so that consistent process data and reactor parameters are found simultaneously. However, due to experimental uncertainties in the measurements − even from a very well-controlled pilot plant − apparent model limitations, limited amount of data, and/or model deficiencies, such an approach will seldom be able to give a reliable representation of all measured properties of the reactor and will result in wrong parameters. It is therefore preferable to execute the reconciliation as a two-step procedure, where the conversion and equilibrium reactor models of type A are used in the first step followed by the rigorous reactor model B in the second step to find the reactor parameters, but using the reconciled measurements from the first calculation. This is a permissible split for the syngas processes where conversions and temperature approaches are sufficient to characterise the reactions.

The model equations in method A are identical to those used in process simulation (refer to Section 2.1), and the independent variables are selected as feed flows and compositions, outlet temperatures, and conversions or temperature approaches to equilibrium. The solution requires a robust method to solve the minimisation problem, such as the successive quadratic programming method [72] [173] with bounds on the independent variables.

Usually a proper statistical discrimination between different reactor models cannot be based on experiments alone, since too few measurements are available. The discrimination is then done by "proposing" a model and testing both the reliability of the model and whether the found parameters are within acceptable physical boundaries.

3.2.5 Reformer temperature measurements

Temperature measurements at high temperatures need special considerations, in particular when the purpose is to determine heat transfer parameters.

Reliable catalyst temperatures inside the bed can be measured by thermocouples loaded individually to avoid disturbing the flow pattern as would be the case if a central thermopocket is used. Radial temperatures can be measured using thermocouples placed on a bracket pointing directly against the flow direction to avoid disturbance from false heat conduction.

In fired reformers, outside tube-wall temperatures must be measured carefully [19] [131]. Apart from their impact on tube life, they are key variables in the evaluation of heat transfer coefficients. Measurements may be carried out using commercial IR cameras with different wavelengths, but in a pilot plant such measurements are best carried out using Pt/Rh thermocouples attached on the cold side of the tubes. In an industrial plant thermocouples must be embedded, but a gold cup pyrometer [131] can be used locally to obtain accurate measurements.

Note
A tool to measure tube-wall temperatures is the IR camera. It interprets the radiation emitted from a tube surface as coming from a black body and converts it to a temperature. Provided the tube is clean, it typically has an emissivity in the order of 0.8–0.9 and it will hence reflect some of the radiation coming from the warmer furnace walls and flames. The instrument will consequently interpret a measured temperature higher than it actually is. The IR camera has filters so that only radiation at a certain wavelength μ_L is measured and the equation to correct the

measured temperature $T_{pyrometer}$ to the true temperature T_{tube} can then be written as:

$$\frac{1}{e^{\frac{14388}{\mu_L \cdot T_{pyrometer}}} - 1} = \frac{\varepsilon_t}{e^{\frac{14388}{\mu_L \cdot T_{tube}}} - 1} + \frac{F_{PT}(1-\varepsilon_t)}{e^{\frac{14388}{\mu_L \cdot T_{furnace}}} - 1} \quad (3.3)$$

F_{PT} is a view factor, which can here be set to unity, since the measured temperature is the one on the front. The furnace temperature varies significantly on the furnace wall, so an average value recorded by the IR camera is used. In a pilot plant the correction is small if the measurements are made on the cold side of the tube, whereas in an industrial reformer the measured temperatures are those on the hot side along the furnace and larger corrections result.

The correction is largest in the top since the temperature difference between furnace wall (or flame) and catalyst tube wall is largest here, whereas it decreases towards the bottom of the tube.

For a further description of IR cameras, reference is given to commercial literature.

The tubes are always fired from two sides. This implies that the cold sides away from the flames are partly in shadow from the other tubes when seen from the burners and the front side towards the burners are warmer. This results in an uneven heat flux distribution around the tube periphery, which must be taken into consideration in the design.

From a radiation point of view, the heat flux distribution is a function of the tube spacing given as the ratio between the centre-to-centre distance C_t and the tube diameter d_t, so that the ratio of maximum heat flux at the front face to the average heat flux – used on the catalyst side – is as expressed by Hottel [244]:

$$\frac{q_{max}}{q_{av}} = \varepsilon_t \left(\frac{\pi \cdot d_t}{2 \cdot C_t \cdot F_{PT}} + \frac{1-\varepsilon_t}{\varepsilon_t} \right) \quad (3.4)$$

The view factor F_{PT} can be expressed by the ratio C_t/d_t, but a correction for multiple radiation and in particular circumferential heat conductance in the tube wall must also be taken into consideration for a full-size tubular reformer [244]. For a mono-tube pilot plant, the tube spacing has no meaning, so a value fitted to proper measurements on the hot and cold sides must be used to evaluate the reactor models.

Furnace gas temperatures are measured using an exhaust pyrometer (refer to Figure 3.16), where a small amount of flue gas is sucked through a pipe with a protected thermocouple.

3.3 Modelling of steam reforming reactors

Reactor modelling is an essential step in a scale-up of new steam reforming reactors. Fixed-bed modelling has always been a challenge to chemical engineers and it is one of the first applications where computers have been used extensively for the tedious property calculation in the numerical solution. An early example by Kjaer is shown in [274], and later in [275]. Modelling of steam reformers is widely used and among the early ones are Hyman [250] from 1968 and an early Topsøe publication [493].

The tools for fixed-bed reactor modelling are well described in textbooks on reaction engineering, e.g. by Aris [24], and Froment and Bischoff [199]. The mathematical model consists of a set of coupled ordinary or partial differential equations, which – at least in cylindrical geometry – can be solved with a standard numerical technique along the length of the reactor. Such conventional modelling of catalytic reactors has reached maturity and reactor modelling now incorporates fluid dynamics to investigate flow patterns in industrial reactors and their impact on performance. The use of micro-scale integrated non-cylindrical reactors and catalyst shapes or reactors with a flame-controlled reaction are all future challenges in reaction engineering, where computational fluid dynamics (CFD) must be used to solve the problems [159]. A key concern, as described by Dudukovic [166], is, however, that many kinetic and transport correlations are not only old (often more than 50 years), but have also been developed and correlated for the conventional models. Scale-up thus requires a critical evaluation of all transport correlations, before they can be applied in CFD models. In addition CFD models are time-consuming and with a requirement of maybe several hundred reactor simulations per day, the conventional reaction engineering models are still widely used, although their shortcomings are obvious.

The required correlations are:

- Kinetics of the reactions and their dependence on operating conditions;
- Mass transport of components and heat transport to and from the catalyst surface and through the pore system;
- Transport of heat to and from the catalytic reactor; and
- Pressure drop.

Applied and fundamental research can give adequate descriptions of these phenomena, but usually scale-up requires experimental work due to the shortcomings of the correlations. Steam reforming reactor experiments at pilot scale are costly, so parameter identification is usually combined with other key points such as durability of a specific catalyst system or construction material.

Conventional modelling of the catalyst side distinguishes the significance of the catalyst particle itself for the reactor performance and how heat transfer is taken into consideration. The following modelling concepts are generally accepted [199]:

Homogeneous: gas and catalyst in the bed is a continuum so that temperature, pressure, and concentrations do not distinguish whether they are inside the catalyst or in the gas phase. The kinetic rate expressions are so-called "effective" or "pellet kinetic" rates, where the catalyst effectiveness factors have been included in the reaction rate expressions.

Heterogeneous: intra and external gradients are included and catalyst effectiveness factors are calculated continuously during the integration. The kinetic rate expressions are intrinsic rates. This is definitely the preferred model, but rigorous modelling of transport processes is tedious for all catalyst sizes and shapes in syngas processes.

One-dimensional: only the variation of the average temperature in the catalyst bed in the flow direction is considered.

Two-dimensional: also gradients in the radial direction are considered. The model is called two-dimensional since rotational cylindrical geometry is assumed; if this is not the case, rigorous CFD modelling is necessary.

In a tubular or convective steam reformer with heat transfer to the catalyst bed, radial temperature gradients of temperature and conversion will develop in addition to the axial gradients so the two-dimensional model must be used. This model represents the heat transfer mode and aims at modelling the heat transfer in the catalyst bed firstly as heat transfer through the catalyst bed and secondly as heat transfer through an additional (artificial) film resistance between the catalyst at the wall and the wall itself. Although the sizes of the radial gradients in a reformer are small compared with the gradients at the wall, they must be known when predicting risk of carbon formation, usually originating on the catalyst closest to the tube wall (refer to Section 5.2.4 and Figure 3.8).

Internal combustion must be modelled in a separate step due to the different heat transfer modes as considered in Sections 3.3.5 and 3.3.6. Similarly modelling of the internal combustion in an ATR must also be carried out in a separate step.

In an adiabatic reformer no heat transfer takes place and the one-dimensional model is sufficient, but the two-dimensional model can also be used in this case since it simplifies to the one-dimensional model.

The heterogeneous model is preferred in both cases, but due to the low effectiveness factors, the homogeneous model may be used in practice since results will be almost identical.

Axial dispersion is usually not taken into consideration when simulating industrial syngas reactors, mainly due to the large mass velocities, but even if steep gradients are present, axial dispersion is only relevant in a small section of the industrial reactor, so the error introduced by neglecting axial dispersion is small. In dynamic modelling and in laboratory reactors axial dispersion should be included in the evaluation.

Other model considerations are necessary for simulating bench-scale, laboratory reactors and micro-scale reactors with integrated catalyst and reactor. Flow regimes are at different gas velocities and catalyst particle sizes are smaller than in an industrial reactor or the catalyst is integrated with the wall. In this case a rigorous CFD model is required, which includes entry phenomena and heat transfer to confining walls along the flow direction and heat conduction through construction materials and thermocouples. Laboratory and bench-scale reactors will be discussed in Section 3.5.1 and micro-scale reactor concepts in Section 3.3.8.

Traditionally a one-dimensional homogeneous model has been used for the steam reforming catalyst bed, such as in [250] [381] [493], whereas a one-dimensional heterogeneous model has been used in [143] [181] [201] [525]. A two-dimensional homogeneous model has been used in [389] [395], whereas the corresponding heterogeneous model has been used in [142] [342] [415], however, without significant difference between that and the one-dimensional model. Only when carbon formation is considered, the two-dimensional model is necessary, refer to Section 5.2.4. Below is described the two-dimensional heterogeneous reactor model, since all the other models may be derived from it.

3.3.1 Two-dimensional reactor model

The two-dimensional model must be able to simulate the industrial reactor types described in Chapter 1. A generalised configuration is shown in Figure 3.7, where the catalyst bed is confined between outer and inner channels. A centre tube may also be present.

The conversion, temperature, and pressure drop in the reactor are calculated by simultaneous integration of the balances for mass, heat, and momentum as formulated by Froment and Bischoff [199]:

$$\frac{\partial}{\partial z}(u \cdot C_A) = \frac{D_{r,eff}}{r} \cdot \frac{\partial}{\partial r}\left(r \cdot \frac{\partial C_A}{\partial r}\right) + R_A \cdot \rho_{bulk} \quad (3.5)$$

$$\rho \cdot C_p \cdot u \cdot \frac{\partial T}{\partial z} = \frac{\lambda_{r,eff}}{r} \cdot \frac{\partial}{\partial r}\left(r \cdot \frac{\partial T}{\partial r}\right) + \sum_j R_j \cdot \rho_{bulk} \cdot (-\Delta H_j) \quad (3.6)$$

$$\frac{dP}{dz} = -\left[\frac{\Delta P}{\Delta z}\right]_{cat-bed} \quad P = P_{inlet} \quad \text{at } z = 0 \quad (3.7)$$

The mass balance is for a formation of component A and j is the reaction index.

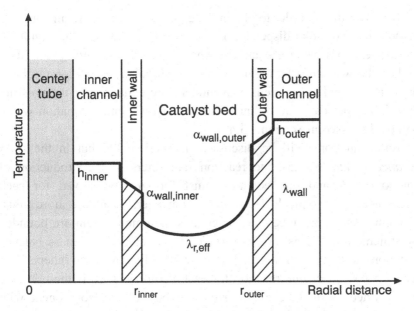

Figure 3.7 Two-dimensional model – basic heat transfer features.

In the derivation, the effective radial diffusivity, $D_{r,eff}$ is assumed identical for all components, and both $D_{r,eff}$, and the effective thermal conductivity in the bed, $\lambda_{r,eff}$ are assumed independent of radial distance. The gas flow is assumed to be plug flow. That is not correct since a velocity profile develops, the form of which depends on the catalyst shape, the ratio between tube and particle diameter, and the radial variation of the void fraction. The local gas velocity is large close to the wall, then decreases about one particle diameter from the wall and levels off towards the centre as shown in [136] [294] [514]. Simulation of such a velocity profile requires CFD [484], but since the shape of the velocity profile does not change fundamentally for a given catalyst type, the plug-flow assumption is acceptable. Lerou and Froment [294] correlated experimental data using both a plug flow and a velocity profile and could correlate the radial effective thermal conductivity using almost the same equation. The wall coefficient is more difficult to correlate due to the large variation in velocity profile close to the wall. Heat transfer formulas (refer to Section 3.3.2) found using the plug-flow assumption should consequently be used with great caution in CFD calculations.

It is seen that in order to obtain the equations for the one-dimensional model, the two radial dispersion terms must be removed. This results in the proper mass balance equation, whereas the flux boundary condition at the tube wall must be included in the heat balance equation to obtain the conventional one-dimensional model. The equation for the pressure drop is in both cases an ordinary first-order differential equation which will be discussed in Section 3.3.4.

Rate equations will be discussed in Section 3.5, but in the *mass balance equation* (3.5), the reaction rate refers to the production of component A and one equation must formally be solved for each component. A typical feed may, however, contain more than 25 components and since the conversions of some of them are bounded by stoichiometry, it is more convenient to transform the mass balance equations so that only one equation must be solved for each independent reaction instead of one reaction for each component. R_j is the effective rate defined as the rate in mol/s/kg cat formed of a component with stoichiometric coefficient +1 and appearing in reaction j only. The reaction rate is based on mass of catalyst, whereas Equation (3.5) is per m³ bed. This is the reason for including the bulk catalyst density ρ_{bulk} calculated from the void fraction ε and the catalyst pellet density as: $\rho_{bulk} = (1-\varepsilon)\rho_p$.

The relation between the formation rate of component A and the general rates is found using the stoichiometric coefficients for component A as:

$$R_A = \sum_j \nu_{A,j} \cdot R_j \quad (3.8)$$

In the heterogeneous model, the concept of the effectiveness factor is used. It is defined as:

$$\eta_j = \frac{R_{int,average,j}}{R_{int,bulk,j}} \quad (3.9)$$

where R_{int} is the intrinsic rate of reaction. The effective rate based on bulk properties to be used in (3.5) is then:

$$R_j = \eta_j \cdot R_{int,bulk,j} \quad (3.10)$$

since the intrinsic rate is evaluated at the bulk conditions. To include the conversions of all components in the reactions, it is convenient to introduce the fractional conversion, ξ_j in reaction j, as the amount in mol/s formed via reaction j per unit feed mixture of a component with stoichiometric coefficient +1 appearing in reaction j only. ξ_j thus corresponds to the general rate R_j and it has the same definition as used for solving chemical equilibrium (refer to Example 1.1).

If the fractional conversions are known, the mole fraction of component i, y_i, can be determined by use of the stoichiometric coefficients v_{ij} from the following equation:

$$y_i = \frac{y_{0i} + \sum_j v_{ij} \cdot \xi_j}{1 + \sum_j \xi_j \sum_k v_{kj}} \qquad (3.11)$$

where y_{0i} is the mole fraction at the inlet and k is a component index.

Concentration can be converted to mole fraction from the gas law and after combining the resulting equation with Equation (3.11) for a single hypothetical component A appearing in reaction j only, the result can be combined with Equation (3.5). The mass velocity $G = u \cdot \rho$, which is independent of conversion, can be inserted to obtain the final equation for reaction j:

$$\frac{\partial \xi_j}{\partial z} = \frac{d_p}{Pe_M} \cdot \frac{1}{r} \cdot \frac{\partial}{\partial r}(r \cdot \frac{\partial \xi_j}{\partial r}) + \frac{M_{w,0}}{G} \cdot R_j \cdot \rho_{bulk} \qquad (3.12)$$

where the so-called Peclet number for radial mass transfer is:

$$Pe_M = \frac{d_p \cdot G}{\rho \cdot D_{r,eff}} \qquad (3.13)$$

The boundary and initial conditions for the mass balance equations are:

$$\frac{\partial \xi_j}{\partial r} = 0 \text{ at } r = r_{inner} \text{ and } r = r_{outer} \qquad (3.14)$$
$$\xi_j = 0 \text{ at } z = 0$$

This is the final form of the mass balance equation for the catalyst bed.

Pe_M is evaluated in analogy with heat transfer through the bed, since the same molecular mechanism is used for the transport due to the highly turbulent flow and the parabolic temperature profile.

The *heat balance equation* (3.6) can be rewritten as:

$$\frac{\partial T}{\partial z} = \frac{d_p}{Pe_H} \cdot \frac{1}{r} \cdot \frac{\partial}{\partial r}\left(r \cdot \frac{\partial T}{\partial r}\right) + \frac{1}{G \cdot C_p} \cdot \sum_j R_j \cdot \rho_{bulk} \cdot (-\Delta H_j) \quad (3.15)$$

where the summation is over all independent reactions, and the effective radial heat transfer is described by the radial Peclet number for heat transfer:

$$Pe_H = \frac{d_p \cdot G \cdot C_p}{\lambda_{r,eff}} \quad (3.16)$$

The boundary condition reflects the basic assumptions in the two-dimensional model for a cylindrical tube, where the heat is transferred from an outer wall to the catalyst bed:

$$\lambda_{r,eff} \cdot \frac{\partial T}{\partial r}\bigg|_{r=r_{outer}} = \alpha_{wall,outer} \cdot \left(T_{wall,outer} - T_{cat,outer}\right) \quad (3.17)$$

The temperature difference on the outer side is across the (hypothetical) film between the wall, T_{wall}, and the outermost layer of the catalyst bed, T_{cat}, (refer to Figure 3.7) with the heat transfer coefficient given as $\alpha_{wall,outer}$. The equation thus reflects the observation that the effective thermal conductivity decreases strongly close to the wall, but instead of letting it be a function of the radial distance, it is interpreted as a parameter $\alpha_{wall,outer}$, as discussed in Section 3.3.2.

In the general case with two surrounding channels to the catalyst bed, the boundary conditions for the heat transfer from the two channels with temperatures T_{inner} and T_{outer} are:

$$-\lambda_{r,eff} \cdot \frac{\partial T}{\partial r}\bigg|_{r=r_{inner}} = \pm U_{inner} \cdot \left(T_{inner} - T_{cat,inner}\right) = \pm q_{inner}$$
$$\lambda_{r,eff} \cdot \frac{\partial T}{\partial r}\bigg|_{r=r_{outer}} = \pm U_{outer} \cdot \left(T_{outer} - T_{cat,outer}\right) = \pm q_{outer} \quad (3.18)$$

The actual sign for the heat flux is determined by the flow direction in the channel relative to the flow direction in the catalyst bed. If a channel is absent, the corresponding gradient is zero.

Technology of Steam Reforming

In convective reformers, the flow in the heat exchange channel(s) may be co-current or counter-current to the flow in the catalyst bed. In the counter-current case, iteration is performed on the outlet temperature and pressure of the channel until balance at the outlet of the catalyst bed.

The heat transfer coefficient U is calculated as the sum of the resistances from the catalyst bed to the corresponding channel including the wall (assuming they are all referred to the same heat transfer area):

$$\frac{1}{U_{outer}} = \frac{1}{\alpha_{wall,outer}} + \frac{1}{h_{outer}} + \frac{r_{outer} \ln\left(\frac{r_{outer}}{r_{inner}}\right)}{\lambda_{wall}} \quad (3.19)$$

The heat flux boundary condition at the wall of the catalyst bed is used when heat is exchanged with an external channel, where the heat transfer is described using a flux profile. This is the case for an adiabatic catalyst bed as in a prereformer or an autothermal reformer where the heat flux is zero so that the heat transfer equations simplify to a one-dimensional model. The heat flux boundary condition is also used when the tubular reformer is coupled with a furnace type model (refer to Section 3.3.6).

The initial condition for the temperature is simply:

$$T = T_{inlet} \quad \text{at} \quad z = 0 \quad (3.20)$$

For a tube with no inner channels the boundary conditions for conversion and temperature simplifies to the gradients being zero at the centre. In the general case with heat exchange to inside and outside channels, this feature is not valid, and the following dimensionless variable is introduced:

$$x = \frac{r - r_{inner}}{r_{outer} - r_{inner}} \quad (3.21)$$

If inner and/or outer channels are present, one heat transfer and one pressure drop equation are set up for each of such channels, since the mass balance is trivial. For the outer channel, the two equations are:

$$F_{outer} C_p \frac{dT_{outer}}{dz} = U_{outer} S_{outer} \cdot (T_{outer} - T_{cat,outer})$$

$$\frac{dP_{outer}}{dz} = -\left[\frac{\Delta P}{\Delta z}\right]_{outer} \quad (3.22)$$

Note

The final two-dimensional mathematical model thus consists of one partial parabolic differential mass balance equation (3.12) with boundary and initial conditions in (3.14) for each of the j reactions and one partial parabolic differential heat transfer equation (3.15) with boundary conditions in (3.17), (3.18) and initial conditions in (3.20). Simultaneously the pressure drop ordinary differential equation (3.7) and the differential equations for the temperature and pressure in each of the surrounding channels in (3.22) must be integrated. Catalyst effectiveness factors in the catalyst bed must be available in all axial and radial integration points using the methods in Section 3.4.

The equations for the catalyst bed combined with the boundary conditions can be solved using modern methods such as those available in CFD. A fast method uses discretisation in the radial direction by the orthogonal collocation method by Villadsen and Michelsen, where the dependent variable is only considered in the so-called collocation points [512]. This method reduces each partial differential equation to a corresponding number of coupled ordinary first-order differential equations with the axial distance as independent variable. The selected number of collocation points depends on the stiffness of the actual problem. Only with very steep gradients are more than five collocation points necessary to predict the extrapolated properties close to the reactor wall with sufficient accuracy to predict the risk of carbon formation.

Combustion in a channel is represented by a heat flux profile, and iterations must be carried out until convergence of the heat flux and temperature profiles by use of the boundary condition at the wall, Equation (3.18). If the heat flux is given in one reactor part, the resulting tube-wall temperature may be used in the other reactor part to calculate a new heat flux profile. The solution method is successive iteration with controlled step sizes.

3.3.2 Heat transfer in the two-dimensional model

The heat transfer parameters in the two-dimensional model are the effective radial conductivity $\lambda_{r,eff}$ and the wall heat transfer coefficient α_{wall}. Reviews of the correlations using the plug-flow assumption can be found in [159] [199] [274] [288].

Correlations of the effective radial conductivity usually have the following form:

$$\frac{\lambda_{r,eff}}{\lambda_g} = \frac{\lambda_{r,stag}}{\lambda_g} + \frac{Re\,Pr}{Pe_{turb}}$$

$$Re = \frac{G \cdot d_p}{\mu} \quad Pr = \frac{C_p \cdot \mu}{\lambda_g}$$

(3.23)

The first term is the stagnant term, which depends on particle and gas thermal conductivity, particle to particle heat transfer and radiation, as shown in the literature [43] [199]. At the high mass velocities in tubular reforming, the turbulent term is dominating, so the stagnant part cannot be evaluated from industrial measurements, whereas it is important for bench-scale experiments. Pe_{turb} is the limiting turbulent value of the radial Peclet number. Expressions depend on the tube-to-particle diameter ratio [199]. The following equation is seen in [43]:

$$Pe_{turb} = 8\left\{2 - \left[1 - \frac{2d_p}{d_t}\right]^2\right\}$$

(3.24)

The tube-to-particle diameter ratio is usually larger than 5 to assure an appropriate packing at industrial conditions. Values around 3.3 have been used in the pilot plant in Figure 3.6, but such small values require very careful loading. The length parameter in the Re number in Equation (3.23) depends on particle shapes and has also been measured for rings, where a length effect has been found [42].

The wall coefficient is usually correlated as a Nusselt number equation, i.e.

$$Nu = Nu_{stag} + Nu_{turb} + Nu_{rad}$$

$$Nu = \frac{\alpha_{wall} \cdot d_p}{\lambda_g}$$

(3.25)

$$Nu_{turb} = A \cdot Re^a \, Pr^b$$

Both the stagnant and the particle to wall radiation parts are small compared with the turbulent part at industrial syngas conditions even at the high temperatures. For the turbulent part different values of the parameters A, a, and b can be found in the literature [199] [274], depending on particle type, size, and tube-to-particle diameter ratio.

In bench-scale units detailed measurements can be used to determine the two terms individually, but if only a central thermowell is used, the overall heat transfer coefficient must be fitted to the one-dimensional model. The relation between the two models can be found by comparing the two models assuming that the temperature profile is parabolic in fully developed turbulent flow and that the main resistance is at the wall [296] (refer to Figure 3.8):

$$\frac{1}{U_{\text{one-dimensional}}} = \frac{1}{\alpha_{\text{wall,outer}}} + \frac{d_p}{8 \cdot \lambda_{r,\text{eff}}} \quad (3.26)$$

The constant "8" in the last term changes with the relative values of the two resistances, but it is evident that the two parameters are cross-correlated as expressed by the Biot number:

$$Bi = \frac{\alpha_{\text{wall,outer}} \cdot d_t}{\lambda_{r,\text{eff}} \cdot 2} \quad (3.27)$$

Theoretically it may be better to correlate the data using the Biot number [157] as a function of Reynolds number, but at the high Reynolds numbers in industrial reformers the two formulations become similar.

The power of the Prandtl number, b, is usually 0.33, and in general of no concern in the literature, but it should be noted that it may change significantly in the reactor from inlet to outlet. This is not only caused by the temperature, but also by changes in gas composition, since steam and hydrogen have different thermal conductivities.

The power of the Reynolds number at high superficial velocities has values between 0.5 and 1 in the literature. Many references use 0.75. This has also been found in the pilot plant in Figure 3.6, but the power decreases at very high loads [119]. The reason is probably a combined impact of the superficial mass velocity on the radial effective conductivity and the heat transfer coefficient at the tube wall emphasising the need for a fundamental description using CFD. Results shown by Dixon *et al.* in [160] [484] for steam reforming in narrow tubes are promising.

In tubular reformers the catalyst particles are usually cylindrical and the tube-to-particle diameter is larger than 5–10. Smaller particles increase pressure drop, whereas the use of larger particles requires considerable skill during loading. To limit tube and operating costs, this ratio is decreasing. Dixon [158] measured heat transfer for spheres, cylinders and rings of various sizes with ratios less than 4 and found that the wall heat transfer decreased significantly for ratios below 2, whereas above 2 the impact was weaker for cylinders and rings compared with spheres.

Example 3.4
By use of the expression for the one-dimensional heat transfer coefficient in Equation (3.26) it is now possible to establish the heat transfer balance for the one-dimensional model for a tubular reformer tube considering only steam reforming of methane. The equation is established from the two-dimensional model in Equation (3.15), the boundary condition in Equation (3.17) and the derived heat transfer coefficient for the one-dimensional model in Equation (3.26) as:

$$G \cdot C_p \frac{dT}{dz} = \frac{4}{d_t} U_{one-dimensional}(T_{wall,outer} - T_{cat}) + R_{ref} \cdot \rho_{bulk} \cdot (-\Delta H_{ref})$$

The left-hand side and the heat transfer coefficient are primarily determined by the operating conditions in the furnace and the tube. This implies that the difference between the outer tube wall and the catalyst temperatures is inverse proportional to the catalyst activity. An active catalyst is consequently necessary to limit the outer tube-wall temperature.

3.3.3 Heat transfer parameters in syngas units

Absolute values of the heat transfer parameters that can be used for scale-up are difficult to determine in bench-scale units due to the very high gas velocities and heat fluxes in industrial units. Attempts to determine them in pilot plants operating at industrial conditions but without reaction are also highly uncertain due to small driving forces. The steam reforming reaction is, however, strongly endothermic and limited by chemical equilibrium. This implies that for a new catalyst, the reaction will be close to chemical equilibrium in the major part of the tube, so variation in catalyst activities will only have a small impact on the temperature profile (refer to Section 3.3.7). If, however, heat transfer is increased, a corresponding amount of heat will be absorbed resulting in a higher equilibrium conversion and a higher catalyst temperature. The heat transfer coefficients can hence, as a good approximation, be considered independent of kinetics, when evaluating heat transfer data from measurements including reaction, as long as the catalyst is highly active. This is in contrast to an exothermic reaction, where catalyst activity and heat transfer are strongly coupled and works in opposite directions so an independent determination of the two parameters in the same experiment will be highly uncertain.

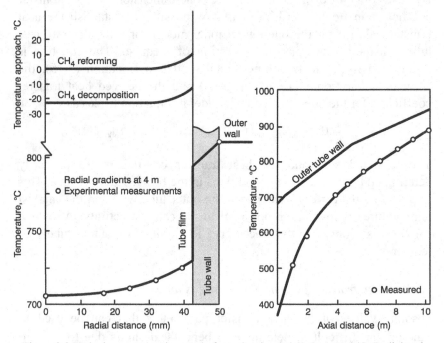

Figure 3.8 Reducing gas experiments (refer to Section 2.4.3). Low H_2O/C. Radial and axial temperature measurements. Measurements as points – calculated values as lines [393]. Reproduced with the permission of Elsevier.

Recorded temperature profiles from the monotube pilot plant in Figure 3.6 are shown in Figure 3.8 [393]. The simulation uses the two-dimensional homogeneous model using the outer tube-wall temperature profile as a representation of the firing. The heat transfer parameters are correlated using the methods in the previous chapter.

Figure 3.8 shows that extrapolation of the measured radial temperature profile to the wall is straightforward, but in reality it may be discussed if the catalyst particle having contact with the wall has a temperature between the inside catalyst bed and the actual wall temperature. This has, however, not been observed when the catalyst is fully active [119]. If, however, catalyst deactivation takes place, the temperature of the outermost catalyst layer and hence the outer tube-wall tempearure will of course increase rapidly.

It is also seen in Figure 3.8 that an error equal to 10% in the wall heat transfer coefficient corresponds to more than 10°C on the outer tube-wall temperature and this has a significant impact on the mechanical design.

The actual catalyst shape influences the heat transfer coefficient through its actual size and shape. Void may enter some correlations, e.g. in [296] using equations for pressure drop (refer to Section 3.3.4), but they do not distinguish between external and internal void. CFD simulation in [160] shows a minor impact of internal void on the radial temperature profiles.

Below results from experiments in the monotube pilot plant in Figure 3.6 are discussed. The process was operating at ordinary syngas conditions for an ammonia plant. The load varied between half and full load, and catalyst particles with different hole fractions but identical outer dimensions were used [119]. The particles have the same length as diameter and relatively small hole fractions. The particle sizes are shown in Table 3.1 below:

Table 3.1 Particle sizes in heat transfer test in a tube with an inner diameter of 84 mm. Mass flow: 7–14 kg/m²/s Average heat flux: 46–93 kW/m².

	Cylinder	Ring	7-hole development type
Outer diameter (mm)	16.1	16.7	16.2
Height (mm)	15.6	16.5	16.2
Inner hole diameter (mm)	0.0	7.8	3.0
Equivalent volume particle diameter, d_p (mm)	18.3	17.6	16.9

The measured axial temperature profiles for the three different shapes are similar to those in Figure 3.8, but little difference between the recorded temperature profiles on the outer tube wall and in the catalyst bed could be seen between those three particles having almost identical outer dimensions. Detailed evaluation further revealed that the calculated heat transfer coefficients at the same operating conditions differed by only a few per cent in spite of a significant impact on the pressure drop.

Similar results have also been obtained for different commercial and development cylindrical type catalysts with one or more holes in a 70 mm tube, where the tube-to-particle diameter ratio ranged from 3.4 to 5, but at identical operating conditions [120]. The average heat flux was as large as 160 kW/m^2. It should be added that the pressure drop varied with more than a factor 2 between the individual experiments. The small overall particle size dependence on the heat transfer coefficient in Equation (3.25) where the d_p is raised to the power a-1 appears thus to be correct.

Nearly all measurements in the literature to determine heat transfer parameters are without reaction as discussed in [11] [520]. The latter reference concludes that fundamentally there is no reason for any impact of the reaction itself, but the catalyst has, of course, a major impact on the temperature within the tube and on the tube wall, since a highly active catalyst will give significantly lower temperatures compared with a catalyst with low activity (refer to Example 3.4). But this does not prove that the reaction itself has an impact on the heat transfer rate, and it is usually not included in the modelling.

Since a major fraction of the heat added in a tubular reformer is used for reaction, heat fluxes are significantly higher in the tubular reformer with reaction compared with bench-scale measurements without reaction. Parameter identification in measurements without reaction is highly uncertain in a tubular reformer due to the small temperature differences, but experiments with reactions under very different operating conditions can be used to check if the reaction has any impact on the heat transfer rate.

Table 3.2 below shows basic parameters for two runs, A and B, carried out in the pilot in Figure 3.6 with almost identical superficial mass velocity and operating conditions. The major difference was the H_2O/C and the average heat flux [120]. If the reaction had an impact, larger heat transfer coefficients would be expected for operating conditions with a larger reaction duty as seen in run B.

Table 3.2 Impact of reaction on heat transfer coefficients. Comparison of data sets at different operating conditions. Pilot plant in Figure 3.6.

	Run A	Run B
Superficial mass flow (kg/m^2/s)	19.8	20.9
Average heat flux (kW/ m^2)	109.	133.
Relative flow of CH_4 (%)	100.	150.
H_2O/C	2.45	1.46
Conversion of CH_4 (%)	71	58
Reaction/heating (%)	77/23	78/22

The corresponding temperature profiles recorded are seen in Figure 3.9.

It is seen that Run B with the high reaction duty and the low H_2O/C requires higher tube-wall temperatures to transfer the 22% larger duty. The distribution between reaction and heating of the transferred duty is almost identical in the two runs, so the reaction duty is also 22% larger in Run B.

The superficial mass velocity is slightly higher in Run B compared with Run A, but Reynolds numbers are almost identical due to differences in viscosity. After taking the differences in transport properties due to the smaller water concentration in Run B into consideration, the calculated difference between correction factors to the heat transfer coefficient is less than 1% to simulate the temperature profiles in the catalyst bed properly. It is hence concluded that the conventional Nu-Re number formulation in Equation (3.25) using the superficial mass velocity as key parameter is sufficiently accurate and that the reaction itself only has a negligible impact on the size of the heat transfer coefficient. This observation is in agreement with [520].

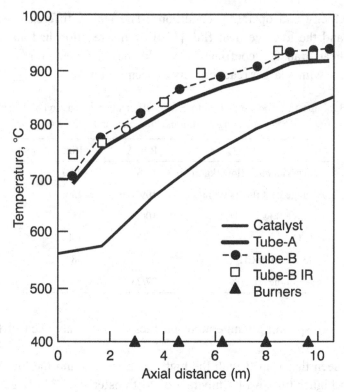

Figure 3.9 Experimental temperature profiles for runs A and B in Table 3.2. Catalyst temperature profiles are similar in the two runs.

3.3.4 Pressure drop

To calculate pressure drop in a fixed bed, the Leva equation or the Ergun equation [199] is usually used. The two equations are shown below.

The functional form of Equation (3.28) of both equations has been derived by assuming that the void volume can be converted to an equivalent tube. Both equations are valid in the laminar and turbulent regions. The Leva equation depends on the values of the friction factor f_m and N, which for turbulent flows are 0.7 and 2, respectively, and the first term in the Ergun equation diminishes in the turbulent region. The two functional terms are consequently almost identical in the turbulent region – Re>1000 – except for a constant. This correction constant must, however, always be determined by experimental measurements since the pressure drop depends strongly on the surface smoothness of the particle.

Leva:

$$-\frac{\Delta P}{\Delta z} = \frac{2 \cdot f_m \cdot G^2}{\rho} \cdot \frac{(1-\varepsilon)^{(3-N)}}{\varepsilon^3 \cdot d_p \cdot \psi^{(3-N)}}$$

Turbulent regime:

$f_m \approx 0.7$
$N \approx 2$ (3.28)

Ergun:

$$-\frac{\Delta P}{\Delta z} = \frac{150 \cdot \mu \cdot G}{\rho \cdot \psi^2 \cdot d_p^2} \cdot \frac{(1-\varepsilon)^2}{\varepsilon^3} + \frac{1.75 \cdot G^2}{\rho \cdot \psi \cdot d_p} \cdot \frac{1-\varepsilon}{\varepsilon^3}$$

The particle size enters the equations through the volume equivalent particle diameter, d_p, and so-called surface shape factor, ψ, defined as

$$d_{eq} = \sqrt[3]{\frac{6}{\pi} V_p} \quad \psi = \frac{\pi d_p^2}{S_x} \quad d_{eq} = d_p \cdot \psi \quad (3.29)$$

where V_p is the actual particle volume and S_x its surface area [440]. The equivalent particle diameter, d_{eq}, is identical to the one used in calculation of the catalyst effectiveness factor.

The shape factor is valid for a full cylinder and for similar particles. For particles with one or more holes, the usual shape factor concept predicts too large pressure drops [119]. It appears that the flow through the inner holes is not just a function of void for all particle shapes.

A key variable in Equation (3.28) is its dependence of the void fraction, ε, and the dependence shown above is the background for the simplified relative pressure drop equation in Figure 3.3.

Flow patterns in a narrow fixed bed are complex, so experimental measurements at industrial superficial gas velocities are necessary to determine pressure drops for new catalyst shapes. Such experiments need not include a reaction.

3.3.5 Convective reformers

In a tubular reformer the product gas exits the reformer at temperatures around 700–950°C and the flue gas exits at temperatures around 900–1100°C. This implies that significant internal temperature differences in a tubular reformer are found, and that heat recovery is important to limit the amount of lost thermodynamic work in the process (refer to Figure 2.2 and Section 2.1.3).

In a convective reformer, however, flue gas and product gas are further cooled by heat exchange with the catalyst bed through inner or outer channels as shown schematically in Figure 3.7. The internal temperature differences are decreased, thus limiting the lost work, but due to the smaller temperature differences higher metal temperatures may be found so the design constraint has focus on finding appropriate construction materials allowing convective heat exchange at high temperatures.

Cooling of the product gas can be carried out by "adding" a bayonet tube in the catalyst bed (refer to the Topsøe TBR reformer mentioned in Section 2.2.3) [20]. If the fuel is natural gas combusted with a stoichiometric amount of air, the adiabatic flame temperature is above 2000°C. The object of a convective reformer will then be to recover the heat from this stream until the outlet temperature is as close as possible to the process inlet temperature. In a fired tubular reformer, heat is mainly transferred by radiation allowing an almost stoichiometric combustion. Gas radiation emissivity decreases strongly below 1000°C, so the low-temperature heat must be transferred by convection. In practice this transition between radiation and convection is the major design challenge for convective reformers, as convection gives higher material temperatures due to the improved heat transfer.

The convective reformer is compact, but has larger heat transfer areas compared with a tubular reformer due to the smaller heat transfer coefficients in the cold end. A fair comparison should thus include the heat exchange areas in the waste heat section in the tubular reformer, but still the fired tubular reformer remains the most economical solution for large-scale reforming. The economy of scale is more advantageous for a tubular reformer, whereas a convective reformer has an upper size for each module so many modules are needed for large plant sizes.

In practice a fired convective reformer is feasible only in processes where fuels with small heating values are available or when heating can be provided from other sources such as He from a nuclear reactor in the ADAM-EVA process (refer to Section 2.7) [464]. Fuels with small heating values are present as off-gases from product purification, e.g. the off-gas from a PSA unit in a hydrogen plant (refer to Section 2.2.2) or the anode exit stream in a fuel cell plant (refer to Section 2.3.1). In particular the emerging interest in large fuel cell plants (>1MW) has resulted in many new convective steam reformer designs.

Other applications of convective reforming are associated with recovery of the process gas heat in ammonia and methanol plants [395] [479] and lately also in ATR-based syngas units for GTL plants as discussed in Chapter 2. Design and simulation of such a reformer is shown in [519], where the two-dimensional heterogeneous model has been applied using the kinetics in [525]. A homogeneous model has also been used, and almost identical temperature profiles have been found.

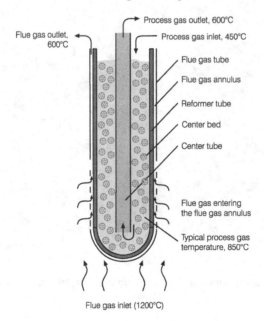

Figure 3.10 A Topsøe convective reformer. Broman et al. [89].

The principle in the Topsøe convective reformer from the plant shown in Figure 2.7 is shown in Figure 3.10 above.

The reformer in Figure 3.10 is applied in a hydrogen plant where the flue gas has an adiabatic inlet temperature close to 1200°C. It is cooled by counter-current heat transfer with the catalyst bed in a system where the flue gas in the bottom is added gradually. The product gas is cooled in a bayonet tube.

Another convective reformer is shown in Figures 3.11 and 3.12. It was developed for use in low-pressure systems where a diluted fuel is available as off-gas from a fuel cell [475]. The reformer has two parallel catalyst beds. The burner is located in the centre and the flue gas is added between the two beds where the catalytic reaction can absorb the heat to limit wall temperatures. By use of this mechanical configuration, the flue gas has co-current flow in the second bed and counter-current flow in the first bed to extract as much heat as possible. Results are shown in [114] [502].

Figure 3.11 Topsøe convective steam reforming pilot, Houston, Texas. Capacity 1000 Nm^3 H$_2$/h.

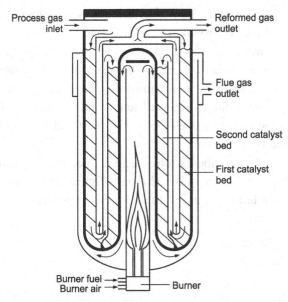

Figure 3.12 A convective reformer concept. Udengaard et al. [502]. A burner in the centre tube and flue gas are cooled by two concentric catalyst beds with internal heat exchange to cool the product gas.

3.3.6 Tubular reformer furnace chamber

The purpose of the furnace in the tubular reformer is to provide the heat of reaction as energy efficient as possible. The furnace is often the largest single equipment item in a syngas process and it contributes a major part both of the exergy losses and the total cost.

The model for the tubular reformer furnace must provide:

- Outer tube-wall temperatures;
- Catalyst tube heat flux profiles; and
- Flue gas exit temperature.

All these properties depend on the furnace configuration (refer to Figure 3.4), burners, fuel type and how the firing is carried out. The major part of the energy is transferred by radiation, so the furnace acts as the "source", whereas the "sink" is a tube side reactor model such as the two-dimensional model shown in Section 3.3.1.

A proper optimisation of a steam reformer must always be based on a furnace model, since the delivered heat flux profile is bounded by the furnace configuration and the flexibility of the burners. Seen from an exergy point of view outer tube-wall temperature and heat flux profiles may decrease the exergy losses [335], but it should be checked if they can be provided by a furnace.

Figure 3.13 below shows on the left the sidewall-fired tubular reformer seen from the end wall. The right figure shows ignited burners on the wall to the left and catalyst tubes on the right side.

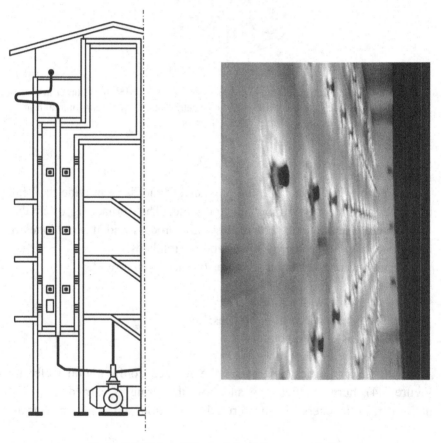

Figure 3.13 Sidewall-fired tubular reformer – end view.

The "source" model for the furnace is a radiation type model using the radiation behaviour in the furnace as defined in Hottel and Sarofim [243]. Earlier examples are a single gas – single grey zone model [305], and a single gas – speckled surface furnace model [463] for the sidewall-fired furnace. Both models assume a well-stirred furnace and constant properties on the flame side. These models use a single gas temperature and cannot predict the impact of a change in firing pattern. Similar considerations can be used for the terrace wall-fired reformer, whereas the top-fired reformers have large gradients so modelling requires the flame to be taken into consideration.

A rigorous furnace model requires CFD and is intimately connected to the functioning of the individual burners. Originally burners were of a premix type with no preheat of combustion air, but in order to increase firing efficiency, combustion air preheat is now common, but this increases NOx formation. New burner design decreases NOx formation by delayed mixing of fuel and combustion air streams to limit maximum temperatures in the flame. Modelling of such burners requires full CFD, but for the daily work it may be sufficient to use a model with lumped parameters for given burner types to determine flame lengths and combustion heat release. Rao *et al.* [358] and Plehiers and Froment [352] have considered the burners as point sources of radiative heat transfer in a sidewall-fired furnace. A model for a top-fired furnace is shown in [325].

In the *multi-zone radiation model* [243] the furnace is divided into a number of gas, burner, furnace and tube wall zones and the radiation exchange between them is established. The flue gas flow is specified as net plug flow. In the top-fired furnace there is a net down-flow, whereas the sidewall-fired and the terrace wall-fired reformer furnaces both have a net upward plug flow of flue gas. In both cases a significant recirculation of combustion gases takes place, so it is allowed to assume that a zone size with homogeneous properties similar to a stirred-tank reactor can be selected. The method requires evaluation of a large number of radiation direct exchange areas, but to simplify the model, the following is assumed:

- The furnace is symmetrical around the tube plane;
- The temperature along the length of a tube row is constant at a certain axial distance from the top so that the exchange areas can be summed up to one area;
- The tubes are modelled as a grey plane with an average emissivity depending on spacing and actual tube emissivity;
- The combustion gas is mixed completely with the flue gas in the burner gas zones;
- The flue gas can be modelled as a one-clear + three-grey gas radiation model with parameters depending on the ratio H_2O/CO_2; and
- The burner combustion profile and flame length must be decided. This requires CFD to create simplified models.

The furnace is divided into zones. By use of the methods in [243] [362], the *direct exchange areas* are calculated using simplified geometrical approaches or they may be calculated using a Monte-Carlo technique [352]. The direct exchange areas are then converted to *total exchange areas* and finally to *directed flux areas*, which are the ones used in the balance equations by the methods in [243].

Each gas volume zone, i, has a gas temperature $T_{g,i}$ and each surface zone has a surface temperature $T_{s,i}$.

The balance for a surface zone, i, expresses that the incoming radiation flux from all other surface, k, and gas zones, l, minus the outgoing radiation flux plus the heat flux delivered by convection from the adjacent gas zone minus the heat flux through the zone surface is equal to zero. A surface equation is then:

$$\sum_k \overrightarrow{S_k S_i} \cdot \sigma \cdot T_{s,k}^4 + \sum_l \overrightarrow{G_l S_i} \cdot \sigma \cdot T_{g,l}^4 - \left\{ \sum_k \overleftarrow{S_k S_i} + \sum_l \overleftarrow{G_l S_i} \right\} \cdot \sigma \cdot T_{s,i}^4 + \quad (3.30)$$

$$U_i \cdot A_{s,i} \cdot (T_{g,i} - T_{s,i}) - q_{s,i} \cdot A_{s,i} = 0$$

Note that the heat is only transferred by convection from the adjacent gas zone to either the tube using U_t or the furnace wall using U_f. The heat flux q is either going to the tube or is a heat loss through the furnace wall.

For volume zone, i, the balance equation expresses that the incoming radiation flux from the other volume, l, and surface zones, k, minus the adiation flux to the other gas and surface zones plus the net heat release is equal to zero. The net heat release in the zone includes the heat of combustion plus heating of the flue gas and minus heat delivered by convection to the adjacent surface zones. A volume equation is then:

$$\sum_l \overrightarrow{G_l G_i} \cdot \sigma \cdot T_{g,l}^4 + \sum_k \overrightarrow{S_k G_i} \cdot \sigma \cdot T_{s,k}^4 - \left\{ \sum_l \overleftarrow{G_l G_i} + \sum_k \overleftarrow{S_k G_i} \right\} \cdot \sigma \cdot T_{g,i}^4 -$$
$$U_{t,i} \cdot A_{t,i} \cdot (T_{g,i} - T_{t,i}) - U_{f,i} \cdot A_{f,i} \cdot (T_{g,i} - T_{s,i}) + Q_{comb,i} = 0.$$
(3.31)

The convection heat transfer from the gas volume has of course only contributions from the two neighbouring tube and furnace wall zones. The heat release in a volume zone Q_{comb} is from the combustion. For a sidewall-fired reformer it can be calculated considering the burner as a point source [352] [358] or simply as a duty heat release in the nearby gas volume assuming a well-stirred combustion. For a top-fired furnace an appropriate duty heat release model must be used.

The advantage of the formulation described above in the zone model is that once the models for the flue gas and the furnace geometry have been selected, the cumbersome evaluation of the direct exchange areas need only be carried out once for a particular geometry.

The final zone model results in a number of coupled non-linear equations as shown above. The combination with the catalyst tube model is carried out by converging the heat flux profiles in the two models.

Verification of the furnace model requires data from industrial furnaces, but such units can by nature only operate at pre-determined operating conditions. Good agreements are shown in [352] [525].

The simplified furnace model described above has also been verified against detailed measurements on the pilot reformer in Figure 3.6. Figure 3.14 below shows a comparison between measured and calculated outlet catalyst temperatures and flue gas temperatures for a number of different operating conditions [120] and it is seen that good agreement is obtained.

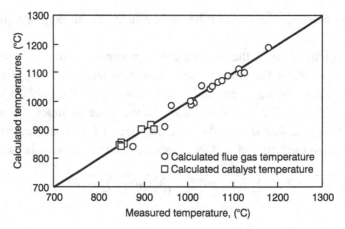

Figure 3.14 Verification of furnace model – significant variation of operating conditions. Loads from 50-200%. Measurements as points – trend line is shown.

For the sidewall-fired furnace changing the firing profile may be necessary. A comparison between the measured and calculated flue gas temperatures when closing burners from the top in the same runs is seen in Figure 3.15 [120].

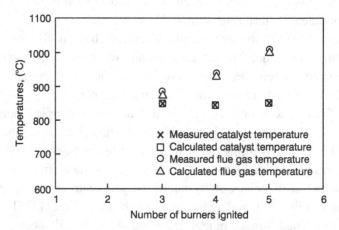

Figure 3.15 Verification of furnace model – reducing burner loads. S/C=2.5, T_{out}=850°C, q_{av}=109 kW/m². Burners counted from bottom of furnace. Measurements and calculated values as points.

Recording of measurements is seen in Figure 3.16.

Figure 3.16 Verification of modelling. One of the authors (LJC) measures flue gas temperatures using an exhaust pyrometer.

In spite of good results using a simplified burner model it is evident that CFD is the proper tool to model a furnace, but a simplified realistic model is necessary in the daily work. A comparison between a full CFD model and the simplified burner model is shown in [330], where CFD has been used to calculate exact flow and temperature fields in the sidewall-fired furnace using appropriate burner models. It is evident that modelling of flue gas exit flows in the top of a sidewall-fired furnace and also the flow distribution in a top-fired furnace along the walls requires CFD.

3.3.7 Tubular reforming limits of operation

The *limit of operation* of a tubular steam reformer is often requested. One such limit is the maximum metal temperatures on the catalyst tubes (refer to Section 3.2.2), but it must also be known how this limit is influenced by variations in the catalyst activity and heat transfer.

This can easily be calculated by extrapolating the models shown above, but a check has also been carried out on the pilot in Figure 3.6.

Three individual runs were performed, where the firing profile was changed as dramatically as possible. A survey of the runs is shown in Table 3.3 [120].

Table 3.3 Determination of the operating limit by changing the firing profile. Load of CH_4, 120%, H_2O/C=3.3, T_{out}=820°C, q_{av}=132 kW/m².

Run A	Run B	Run C
Equal firing in 5 rows	No firing in bottom row. Equal firing in the other 4 rows.	Only firing in the 2 bottom rows.

The temperature profile in Run A corresponds to an ordinary sidewall-fired furnace. The temperature profiles in Run B allow establishment of reforming equilibrium at the outlet due to the reduced firing in the bottom. The profile in Run C simulates maximum heat flux at the very bottom similar to a bottom-fired reformer. It is evident that such changes can only be performed in a pilot plant.

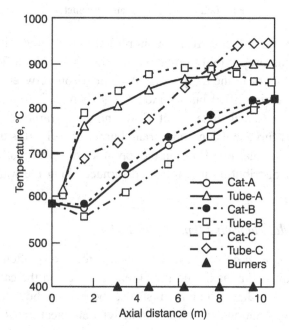

Figure 3.17 Operating limit. Temperature profiles for runs A, B and C in Table 3.3. Axial distance from top of tube. Measurements as points. Lines are connecting points.

The recorded temperature profiles are shown in Figure 3.17 and it is seen that the difference between the tube-wall temperatures in the bottom was up to 100°C.

The simulations can be used to calculate the approach to chemical equilibrium for the reaction in the tube. Figure 3.18 shows the measured temperature approach to equilibrium at the very bottom of the reformer as a function of the calculated local heat flux.

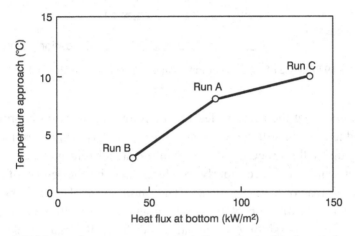

Figure 3.18 Measured temperature approach to the methane reforming reaction at the outlet of the reformer as a function of local heat flux for the three runs, A, B, and C.

It is seen that the relationship is almost linear as will be further discussed in the example in Section 3.5.1.

It is evident that a high activity catalyst is able to absorb even very high local heat fluxes and maintain a condition close to chemical equilibrium.

3.3.8 Micro-scale steam reforming reactors

Micro-scale plate-type reformers and multi-channel reformers are being developed for compact units for small-scale operation [188] [282] [319] [488] – for instance for the use in cars. A better catalyst utilisation (larger catalyst effectiveness factor) can be achieved by catalysing the heat transfer surfaces (catalysed hardware) and by leaving the tubular constraint [166] (refer to Section 1.2.4). Some designs involve a reforming catalyst on one side of the metal wall and a combustion catalyst on the other side [252] [476] [488] [511] as illustrated in Figure 3.19 below.

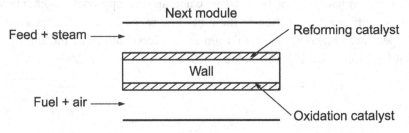

Figure 3.19 Principle of methane reforming micro-scale reactor with coupling to the combustion side.

This implies that the heat needed for the reforming reaction is provided by conduction, but still the heat transfer over the gas film determines the heating up of the process gas. The multi-channel reformer designs are compact with estimated hourly productivities in the order of 1000 $Nm^3\ H_2/m^3$ reactor/h [319] [488]. The multi-channel reformers may result in fast heating up of the feed stream within milliseconds [313], hence with a possibility to avoid coke from thermal cracking in preheaters and over deactivated catalyst.

High heat fluxes (100 kW/m^2) can be obtained in spite of the mainly laminar flow in the multi-channel systems, where the Nusselt number is constant. This is mainly due to the short distance for heat transfer and temperature driving forces around 100°C as described by Haynes [223].

The micro-channel reformers used show high heat transfer rates and the design achieves a high productivity per volume [295]. However, there is little economy of scale and the feasibility is limited to small- to medium-size capacities.

Rigorous modelling must take the selected geometry into consideration and this usually requires CFD. It must, in addition to heat transfer across the wall and film mass and heat transfer between the gases and catalysts, also include axial heat conduction in the metals due to steep temperature profiles. In addition transient behaviour and interaction between the steep temperature profiles must be understood for a proper design, especially when a reasonable catalyst deactivation is assumed. Scale-up is formally simple since it includes adding modules on the top of each other, but a scale-up must in addition to proper manifolding and gas distribution also consider the possibility of leaks, so in practice full-scale experiments will be necessary, in particular to investigate the long-term behaviour [166].

3.4 Modelling of the catalyst particle

The heterogeneous reactor model requires catalyst effectiveness factors. They are found by solving the coupled mass and heat transfer model for a catalyst particle. For simple rate expressions and particle geometries, such as spherical and long cylinders, analytical solutions are available in Froment and Bischoff [199] and Aris [24]. The effectiveness factor can in this case be calculated from the Thiele modulus, ϕ, which is proportional to the characteristic particle diameter, d_{eq}. For large values of the Thiele modulus, the effectiveness factor becomes reverse proportional to the Thiele modulus and hence the equivalent particle diameter as shown in Figure 3.3. For industrial catalyst particles − such as those shown in Figure 3.3 − it is formally necessary to use 3-dimensional CFD, as done by Dixon et al. in the steam reforming model for narrow tubes [484]. For the catalyst particles with holes CFD was used by Alberton et al. [14] using the kinetic expressions by Xu and Froment [525]. They found that the effectiveness factor varied linearly with surface-to-volume ratio or the reverse d_{eq} as seen in Equation (3.29) confirming the limiting value of the effectiveness factor as the reverse ordinary Thiele modulus, but significant variations depending on kinetics and operating temperatures were found. In [310] similar considerations could also be simplified to a linear model.

Below is shown the derivation using spherical geometry based on [113]. The spherical geometry is seldom used in steam reforming, but conversion to other particle shapes can be carried out using the equivalent particle diameter as described by Aris [24].

3.4.1 Catalyst particle model

The basic steady-state diffusion *mass transport* balance equation for formation of component A in a particle in spherical geometry is:

$$D_{eff,A} \left\{ \frac{d^2 C_A}{dr^2} + \frac{2}{r} \cdot \frac{dC_A}{dr} \right\} + R_{int,A} \cdot \rho_p = 0 \qquad (3.32)$$

It is assumed that the effective diffusion coefficient is independent of the radial distance in the pellet. This is discussed further in the next chapter.

The boundary conditions in the centre and at the surface are:

$$\frac{dC_A}{dr} = 0 \quad \text{at} \quad r = 0$$
$$C_A = C_{surf,A} \quad \text{at } r = r_p \qquad (3.33)$$

where $C_{surf,A}$ is the surface concentration of component A.

The differential equation describing *heat transport* in a spherical catalyst particle is:

$$\lambda_p \left\{ \frac{d^2 T}{dr^2} + \frac{2}{r} \cdot \frac{dT}{dr} \right\} + \sum_j R_{int,j} (-\Delta H_j) \cdot \rho_p = 0 \qquad (3.34)$$

The summation is over all independent reactions. The boundary conditions are:

$$\frac{dT}{dr} = 0 \quad \text{at } r = 0$$
$$T = T_{surf} \quad \text{at} \quad r = r_p \qquad (3.35)$$

where T_{surf} is the surface temperature of the particle.

In addition to the particle equations for the inside of the particle, it is necessary to describe the equations for the surface conditions in order to combine them with the bulk conditions.

The basic equations for mass transfer of a component A and heat transfer through the gas film are:

$$N_{surf,A} = k_{surf,A} \cdot (y_{surf,A} - y_{bulk,A})$$
$$q_{surf} = h_{surf} \cdot (T_{surf} - T_{bulk}) \qquad (3.36)$$

Correlations for the catalyst surface mass transfer coefficient k_{surf} and the heat transfer coefficient h_{surf} are seen in [199] [440]. They are usually only valid for spheres or full cylinders, but extrapolations to other particle shapes can be carried out by use of the equivalent diameter and the shape factor definition in Equation (3.29). Using these simplified surface transfer equations in CFD calculations should be checked carefully.

The last equation needed is the definition of the catalyst effectiveness factor for reaction j:

$$\eta_j = \frac{\int_0^{r_p} 4 \cdot \pi \cdot r^2 \cdot R_{int,j} \cdot \rho_p \cdot dr}{\frac{4}{3} \cdot \pi \cdot r_p^3 \cdot R_{int,bulk,j} \cdot \rho_p} \tag{3.37}$$

A mixture is thus described by one mass balance for each component as shown in Equation (3.32) and one heat balance as shown in Equation (3.34).

Similar to the two-dimensional reactor model, the particle model can also be reduced to the solution of one differential equation for each of the independent reactions [113]. This is carried out by use of the key component concept. A key component is an active component in each of the independent chemical reactions not participating in other reactions. Usually the limiting reactant for a reaction is chosen as a key component.

The reduction of the number of differential equations in the solution of the catalyst effectiveness factor problem is performed using the stoichiometric coefficients $v_{i,j}$. The rate of production of component A from all R reactions is given in Equation (3.8) and a similar equation can be written for the production of key component k:

$$R_{int\,key,k} = \sum_j v_{k,j} \cdot R_{int,j} \tag{3.38}$$

If $R_{int,key,k}$ is known, $R_{int,j}$ can be calculated from a similar equation:

$$R_{int,j} = \sum_k v_{j,k}^{-1} \cdot R_{int,key,k} \tag{3.39}$$

where v^{-1} is defined as the inverted stoichiometric matrix for the key components. The dimension of the matrix is the number of independent reactions and it is calculated from a sub-matrix of $v_{i,j}$ containing the stoichiometric coefficients of the key components. In order to use this concept, the matrix must not be singular, which implies that the effectiveness factor can only be solved for independent reactions.

The rate of production of each component to be used in (3.32) can now be calculated from the rate of productions of the key components as:

$$R_{int,i} = \sum_j v_{i,j} \cdot \sum_k v_{j,k}^{-1} \cdot R_{int,key,k} \qquad (3.40)$$

This equation also gives a relationship between the mass balance equations for each component (3.32) and the heat balance in (3.34) allowing an analytical solution after insertion of the boundary conditions (3.33) and (3.35) [113]. This allows calculations of temperatures and concentrations in the interior of a catalyst particle from the key component concentrations in the particle and the temperature and concentration at the surface. The next step is to combine this solution with the surface heat and mass flux equations (3.36) and the equation defining the effectiveness factor in Equation (3.37) by use of the surface mass flux relations for the key components [113]. The final equation is:

$$\frac{k_{surf,k}}{C_{tot}}(C_{surf,k} - C_{bulk,k}) = \frac{1}{r_p^2} \int_0^{r_p} r^2 \cdot R_{int,k} \cdot \rho_p \cdot dr \qquad (3.41)$$

This equation is used as an integral boundary condition at $r=r_p$, where the surface concentration $C_{surf,k}$ is the unknown function.

Note
The final model hence includes one ordinary second-order differential equation (3.32) with an integral boundary condition at the surface (3.41) for each of the reactions. The numerical solution of the catalyst effectiveness factor can be carried out using the orthogonal collocation method by Villadsen and Michelsen [512]. The steam reforming reaction takes place mainly in the outer shell of the catalyst particle, since large particles are used to limit pressure drop. In this case it is advantageous to divide the catalyst pellet into two sections, an inner section and an outer section, divided by a spline point and with an appropriate coupling between the two sections. The spline collocation method has been used by [525] and [181]. A description of the method, spline collocation, can be found in [512].

3.4.2 Effective diffusion coefficients

Effective diffusion coefficients in catalyst particles are calculated as functions of bulk gas diffusion coefficients, pore volume distribution specified as particle porosity, ε_p, as a function of pore radius and the so-called tortuosity factor, τ, which describes the actual "road" a molecule must travel. The use of different effective diffusion models is discussed in the literature [199] [436] and performance of measurements in [221]. Below is shown the basic parallel pore model, where the effective diffusion coefficient, D_{eff}, is calculated from the particle porosity, the tortuosity factor, and the diffusion coefficient in the bulk and the Knudsen diffusion coefficient, D_{bulk} and D_K [199] [389] [440] as:

$$D_{eff} = \frac{\varepsilon_p}{\tau} \cdot \frac{1}{\frac{1}{D_{bulk}} + \frac{1}{D_K}} \qquad (3.42)$$

The contribution from each pore segment is considered to be parallel so the individual contributions are simply added.

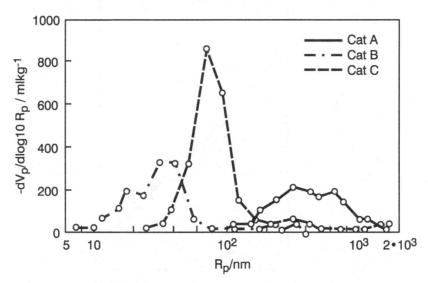

Figure 3.20 Pore volume distribution of typical reforming catalyst [389]. Reproduced with the permission of Springer.

D_{bulk} decreases proportionally with pressure, whereas D_K is independent of pressure and proportional to the pore radius, R_p. Pore volume distributions for typical reforming catalysts are shown in Figure 3.20 [389]. Catalyst A represents a typical high-temperature ceramic catalyst with low surface area, whereas higher surface areas have been obtained in catalysts B and C.

The effective diffusion coefficients of molecules in catalyst pores at the high pressures of industrial reformers are dominated by the bulk diffusion coefficient, whereas the Knudsen diffusion has significant influence at atmospheric pressure. This is illustrated in Figure 3.21. Consequently, low-pressure laboratory tests of large catalyst particles can be misleading for evaluation of the activity of reforming catalysts and they must aim at determination of the intrinsic activity with relatively fine catalyst powders.

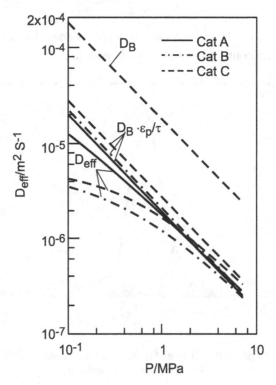

Figure 3.21 Effective diffusion coefficients of typical reforming catalyst [389]. Reproduced with the permission of Springer.

3.4.3 Simulation of a hydrogen plant reformer

Simulation of tubular steam reformers and a comparison with industrial data are shown in many references, such as [250]. In most cases the simulations are based on measured outer tube-wall temperatures. In [181] a basic furnace model is used, whereas in [525] a radiation model similar to the one in Section 3.3.6 is used. In both cases catalyst effectiveness factor profiles are shown. Similar simulations using the combined two-dimensional fixed-bed reactor, and the furnace and catalyst particle models described in the previous chapters are shown below using the operating conditions and geometry for the simple steam reforming furnace in the hydrogen plant, Examples 1.3, 2.1 and 3.2. Similar to [181] and [525], the intrinsic kinetic expressions used are the Xu and Froment expressions [525] from Section 3.5.2, but with the parameters from [541].

The resulting temperature profile on the outer tube wall, the average catalyst temperature and the concentration of CH_4 down the tube are shown in Figure 3.22.

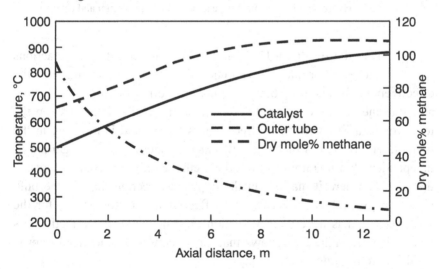

Figure 3.22 Hydrogen plant. Temperature and CH_4 concentration profiles.

The temperature approaches to methane reforming are shown in Figure 3.23 as well as the catalyst effectiveness factors for the reforming and shift reactions.

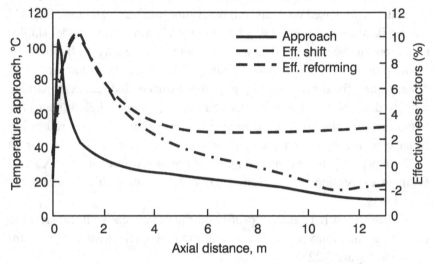

Figure 3.23 Hydrogen plant. Temperature approach to equilibrium and catalyst effectiveness factors.

Similar to the results in [181] and [525] it is seen that the reactions are close to equilibrium in the major part of the tube and that the effectiveness factors for both reactions are correspondingly low. Note also that the effectiveness factor for the shift reaction is positive in the top, crosses a line with zero effectiveness factor, and is negative in the bottom. The reason is that the shift reaction should not be regarded as independent, but it simply follows the methane reforming reaction inside the particle where it may go in the opposite direction than in the bulk phase. In [525] it is shown that the effectiveness factor profile for the shift reaction has a vertical deflection tangent where the effectiveness factor changes sign. This shows that the effectiveness factor concept is valid for the limiting reaction only.

3.5 Reaction kinetics

3.5.1 Industrial rates and the scale-down problem

Most of the catalyst volume of the tubular reformer will operate close to equilibrium (refer to Figure 3.23), and the primary driving force for reaction is the change of methane concentration at equilibrium with the increasing catalyst temperature. This will be illustrated in the example below.

Example 3.5
In the catalyst particle, the steam reforming reaction approaches chemical equilibrium, where all reactions become first order in their limiting component. This implies that a simplified intrinsic rate equation may be written as the following function of the distance to equilibrium [389]:

$$R_{int,CH_4} = k_{int}\left(p_{CH_4} - p_{CH_4,equil}\right)$$

It is convenient to insert the ratio β between the reaction quotient, Q_r, and equilibrium constant as an approximation for the ratio between the equilibrium and actual CH_4 concentrations:

$$\beta = \frac{Q_r}{K_{eq}} = \frac{p_{CO} \cdot p_{H_2}^3}{p_{CH_4} \cdot p_{H_2O} \cdot K_{eq}} \approx \frac{p_{CH_4,equil}}{p_{CH_4}}$$

At equilibrium β is exactly equal to unity. The rate equation can then be written as:

$$R_{int,CH_4} = k_{int} \cdot p_{CH_4}(1-\beta)$$

β is also related to the temperature approach to equilibrium. This can be seen when approximating the equilibrium constant by $\ln(K_{eq})$=A-B/T_{eq} and similarly $\ln(Q_r)$≈A-B/T (B is positive). The resulting equations can be written as:

$$1-\beta = 1 - \frac{Q_r}{K_{eq}} \approx T - T_{eq} = \Delta T_{ref}$$

If Equation (3.5) is simplified to the one-dimensional model by neglecting the radial concentration gradient and a constant gas density is assumed, the mass balance equation for the one-dimensional model can be written after insertion of the intrinsic rate equation for CH$_4$ as:

$$-u\frac{dp_{CH_4}}{dz} = \eta_{CH_4} \cdot R_{int,CH_4} \cdot \rho_{bulk} \approx \Delta T_{ref}$$

The equation can be further simplified after multiplication with dT/dT and assuming that the temperature gradient of the methane concentration is equivalent to the equilibrium concentration temperature gradient. The equation can now be rewritten as:

$$\Delta T_{ref} \approx -u\frac{dp_{CH_4}}{dT}\frac{dT}{dz} \cong -u\frac{dp_{CH_4,equil}}{dT}\frac{dT}{dz}$$

This implies that increasing load or temperature gradient will result in a larger approach to equilibrium. This was discussed in Section 3.3.7, Figure 3.18. In an adiabatic reactor (prereformer), dT/dz approaches zero as the reactor approaches equilibrium, meaning a small reaction rate in most of the reactor.

A low effectiveness factor — see Figure 3.23 — implies that the effectiveness factor and thus the effective rate for the steam reforming of methane is inversely proportional to the Thiele modulus [199] and hence the equivalent particle diameter assuming that the particle is isotherm. For a first-order equilibrium rate expression, a general effectiveness factor can be evaluated as shown in [199] [389]. For a large equilibrium constant, this equation can be simplified to:

$$R_{CH_4} \approx \frac{6}{d_{eq}}\sqrt{D_{eff} \cdot k_{int,vol}} \cdot p_{CH_4}(1-\beta) \qquad (3.43)$$

The rate constant must be measured in gas volume per particle volume for a proper conversion, so the conversion from the mass particle based k_{int} requires the molar volume of the gas. In addition it is also assumed that the particle is isothermal so in practice conversion of intrinsic rate to effective rate requires fitting of the rate constant. Most important is, however, that the effective energy of activation is half the intrinsic value and that the rate is inversely proportional to the equivalent particle diameter. At reactor conditions the catalyst bulk density must be used for a comparison of the activity of different catalyst particles as shown in Figure 3.3.

If such an effective rate equation is used in the hydrogen reformer example in Section 3.4.3 instead of the intrinsic one from Section 3.5.2, no apparent difference between the temperature and concentration profiles in Figure 3.22 can be observed, which confirms that the effectiveness factor in a high-temperature tubular reformer is low. At lower temperatures, such as those relevant for a prereformer, the simplifications in the conversion are not correct and intrinsic kinetics must be used.

As for many industrial reactions, the high reaction rate means strong transport restrictions and the reaction is also characterised by a high, negative heat of reaction. Consequently, the reforming process is subject to significant mass and heat transport restrictions having the potential of making it difficult to investigate the reaction fundamentally at laboratory scale; laboratory results can thus easily be misleading.

The scale-down to laboratory reactors involves a change of a number of reaction conditions [415] [393]. Industrial reactors operate at high superficial gas velocity and Reynolds numbers (Re=1000–10000), but laboratory reactors typically operate at much lower Reynolds numbers, implying that the resistance to convective heat and mass transfer around each particle is significantly larger. It is evident that neither heat fluxes nor mass flows can be reproduced in laboratory bench-scale units.

In laboratory experiments, the resistance to transport in the gas phase near the catalyst particle may be so large that the gas at the external surface of the catalyst pellet will be very close to equilibrium. Consequently, in investigations of reforming of higher hydrocarbons, there will hardly be any higher hydrocarbons left at the catalyst surface. This can easily lead to fallacious results, for instance, in thermogravimetric analysis (TGA) experiments with low gas velocities [390]. It is essential to know the catalyst temperature or to be able to estimate it. Even in investigations of small catalyst particles designed for measurements of intrinsic rates, temperature gradients can be substantial, as shown in Figure 3.24 [403] [415]. Evidently it is difficult to achieve gradientless reactor performance at temperatures above 600°C with normal industrial catalysts. The temperature gradients may easily lead to the measurement of apparent activation energies that are lower than the actual values.

At the high gas velocities required for measuring intrinsic rates at low conversions away from equilibrium, the lack of back-diffusion of product gases may lead to oxidation of the nickel catalyst, as methane behaves as an inert in the Ni/NiO equilibrium (refer to Section 4.1). The problem is solved by addition of hydrogen to the feed. Axial dispersion also plays a significant role in determining the stability of carbide catalysts (refer to Section 4.1).

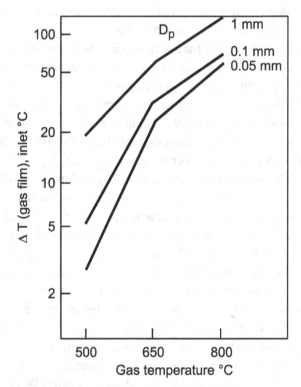

Figure 3.24 Temperature gradients in steam reforming gradient-less micro-reactors [403] [415]. $H_2O/CH_4=4$; $H_2O/H_2=10$, P=1 bar abs. The calculations were made for a heterogeneous one-dimensional reactor model for various catalyst particle diameters, d_p. Xu and Froment kinetics [525] were used. ΔT_{film} represents the temperature drop from the bulk gas phase to the surface of the catalyst particle.

An alternative to kinetic measurements in gradientless reactors is to measure pellet kinetics by using industrial-size pellets in internal recycle reactors, which at high recycle rates approach the character of a perfectly mixed continuously stirred tank reactor (CSTR). By varying the feed rate, it is possible to determine the kinetics at the high conversions encountered in large-scale reformers. However, the mass velocity in the reactor is still one order of magnitude lower than that in an industrial reformer [393]. Therefore, temperature gradients caused by the reforming reaction cannot be eliminated, and it is necessary to measure the catalyst pellet temperature in addition to the gas temperature. For this purpose, a thin thermocouple may be fitted in a drilled hole in one of the pellets. Figure 3.25 shows the basis for measurements [393] as calculated by the method in Section 3.4.

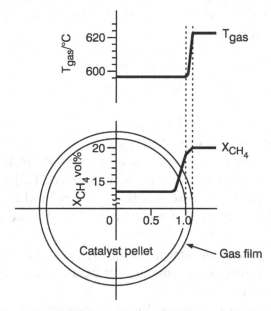

Figure 3.25 Steam reforming of methane. Berty CSTR reactor [393]. Temperature and concentration gradient. Reproduced with the permission of Elsevier.

3.5.2 Intrinsic kinetics. Steam reforming of methane

Various approaches have been applied to establish intrinsic kinetics of steam reforming of hydrocarbons [415].

The first kinetic studies of the steam reforming reactions were strongly influenced by diffusion restrictions [13] and it was not before the Temkin group [58] studied the reaction on nickel foils in the 1960s that more precise information was achieved [58] [271].

Early work on the kinetics of the steam reforming of methane [59] was based on the assumption that the methane adsorption was rate-determining, in agreement with the general assumption of a first-order dependence on methane concentration. Later work by Khomenko *et al.* [271] avoided the discussion of a rate-determining step; instead, the researchers inserted the quasi steady-state approximation in terms of the Temkin identity [75], and the following rate expression was obtained for the temperature range of 470–700°C:

$$R_{int,CH_4} = \frac{k \cdot p_{CH_4} \cdot p_{H_2O}(1-\beta)}{f(p_{H_2O}, p_{H_2})\left[1 + K_w \frac{p_{H_2O}}{p_{H_2}}\right]} \qquad (3.44)$$

where $f(p_{H_2O}, p_{H_2})$ is a polynomial in $f(p_{H_2O}, p_{H_2})$. Equation (3.44) contains five temperature-dependent constants.

Although attempts have been made to establish kinetics for steam reforming on the basis of a micro-kinetic approach [23] and lately on first principles [264], most work is based on empirical kinetics, which has been sufficient to develop highly sophisticated models for tubular reformers.

There is a general agreement [58] [379] [516] that the reaction is first-order with respect to methane and that the activation energy is in the range of 100–120 kJ/mol.

The comprehensive work by Xu and Froment [525] established a complex Langmuir–Hinshelwood expression, using a classic approach, on the basis of 280 measurements made with a Ni/MgAl$_2$O$_4$ catalyst, but it is restricted to a narrow range of parameters: temperatures of 500–575°C, pressures of 3–15 bar and molar H$_2$O/CH$_4$ ratios of 3–5. A number of rate equations were established on the basis of a mechanism involving 13 steps, assuming one of the steps to be rate-determining.

It was shown that CO_2 is produced not only by the shift reaction, but also by steam reforming. Hence, rates for the three reactions were found to give the best agreement with the measurement:

$$1: CH_4 + H_2O \leftrightarrow CO + 3H_2$$

$$R_{int,1} = \frac{k_{int,1}}{p_{H_2}^{2.5} \cdot Z^2} \cdot \left(p_{CH_4} \cdot p_{H_2O} - \frac{p_{H_2}^3 \cdot p_{CO}}{K_{eq,1}} \right) \tag{3.45a}$$

$$2: CO + H_2O \leftrightarrow CO_2 + H_2$$

$$R_{int,2} = \frac{k_{int,2}}{p_{H_2} \cdot Z^2} \cdot \left(p_{CO} \cdot p_{H_2O} - \frac{p_{CO_2} \cdot p_{H_2}}{K_{eq,2}} \right) \tag{3.45b}$$

$$3: CH_4 + 2H_2O \leftrightarrow CO_2 + 4H_2$$

$$R_{int,3} = \frac{k_{int,3}}{p_{H_2}^{3.5} \cdot Z^2} \cdot \left(p_{CH_4} \cdot p_{H_2O}^2 - \frac{p_{H_2}^4 \cdot p_{CO_2}}{K_{eq,3}} \right) \tag{3.45c}$$

The denominator is:

$$Z = 1 + K_{a,CO} \cdot p_{CO} + K_{a,H_2} \cdot p_{H_2} + K_{a,CH_4} \cdot p_{CH_4} + K_{a,H_2O} \cdot \frac{p_{H_2O}}{p_{H_2}} \tag{3.45d}$$

Because the three reactions are not independent as required by the model for the effectiveness factor in Section 3.4.1, it is necessary to combine the three rates into two, one for the conversion of CH_4 and one for the production of CO_2.

$$\begin{aligned} R_{int,CH_4} &= R_{int,1} + R_{int,3} \\ R_{int,CO_2} &= R_{int,2} + R_{int,3} \end{aligned} \tag{3.46}$$

These two expressions include five temperature-dependent constants. They show a small negative reaction order of the overall pressure, in agreement with the data of Figure 3.26.

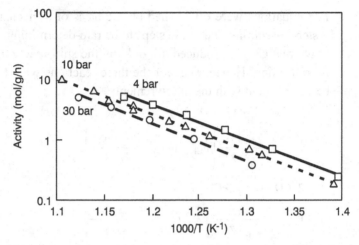

Figure 3.26 Steam reforming of CH_4 [415]. Plug-flow reactor $H_2O/CH_4=4$, $H_2O/H_2=10$, and 0.2 g of $Ni/MgAl_2O_4$ catalyst (with particle diameters in the range of 0.16-0.3 mm); space velocity=1.7 10^6 (vol. total feedgas/vol. cat. bed/h). Reproduced with the permission of Elsevier.

The negative reaction order was confirmed by measurements with industrial-size catalyst pellets in an internal recycle reactor (CSTR, Berty reactor), as shown in Figure 3.27.

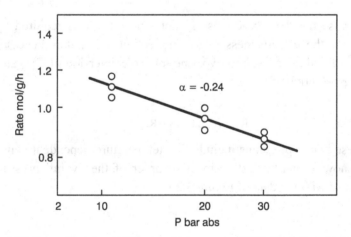

Figure 3.27 Steam reforming of CH_4 at various pressures [415]. Data were obtained in a Berty reactor; $T_{gas}=600°C$, $H_2O/CH_4=3$, $H_2O/H_2=5$, mass of catalyst=20 g, catalyst pellet: cylinder with 7 holes (diam./height/holes: 16/8/7x3 mm), space velocity=10^6 vol total feed/vol cat.bed/h. Reproduced with the permission of Elsevier.

Table 3.4 Reforming kinetics. Parameters for Equation (3.45). Rate constants and activation energies [415].

Parameter in rate equation	Xu and Froment [525]		Avetisov et al. [541]	
	A (mol/g/h)	E (kJ/mol)	A (mol/g/h)	E (kJ/mol)
k_1	$4.225 \cdot 10^{15}$	-240.1	$1.97 \cdot 10^{16}$	-248.9
k_2	$1.995 \cdot 10^{6}$	-67.13	$2.43 \cdot 10^{5}$	-54.7
k_3	$1.020 \cdot 10^{15}$	-243.1	$3.99 \cdot 10^{18}$	-278.5
K_a,CO	$8.23 \cdot 10^{-5}$	70.65	$3.35 \cdot 10^{-4}$	65.5
K_a,H_2	$6.12 \cdot 10^{-9}$	82.90	$2.06 \cdot 10^{-9}$	58.5
K_a,CH_4	$6.65 \cdot 10^{-4}$	38.28	$6.74 \cdot 10^{-3}$	34.1
K_a,H_2O	$1.77 \cdot 10^{5}$	-88.68	$9.48 \cdot 10^{4}$	-74.9

$Ni/MgAl_2O_4$ catalyst, Ni surface area estimated to be 3 m^2/g.

An alternative analysis of the data by Xu and Froment was carried out by Avetisov et al. [28] [541] involving more data and resulting in better agreement with calculated and measured values of partial pressures of CO. The kinetics constants are compared in Table 3.4 [415].

The model by Xu and Froment includes an unlikely negative heat of adsorption of steam which implies an increase in the coverage of oxygen atoms with temperature. Furthermore, the model includes a significant surface coverage by methane. However, as discussed in Chapter 6, surface science studies show that methane dissociation does not proceed via an adsorbed precursor state. This inconsistency may be a result of the narrow range of the H_2O/CH_4 ratio and the temperature applied in the investigation by Xu and Froment and, further, because the mechanism of the steam reforming reactions apparently cannot be represented by a single rate-determining step over a wide range of conditions. An alternative approach is to assume that the reaction is controlled by two steps. This was supported in early work [379] [389] and has been supported by recent fundamental studies [264] (refer to Chapter 6).

Example 3.6
When using Boudart's approach for two-step kinetics [73] and the following simplified sequence with two irreversible steps [389]

$$CH_4 + n* \rightarrow CH_x -*_n + \frac{4-x}{2} H_2$$

$$CH_x -*_n + OH-* \rightarrow CO + \frac{x+1}{2} H_2 + (n+1)* \quad (3.47)$$

$$H_2O + * \leftrightarrow OH-* + \tfrac{1}{2}H_2$$

$$H_2 + 2* \leftrightarrow 2H-*$$

the following expression can be derived, assuming OH-* as the most abundant reaction intermediate (mari):

$$R_{int,CH_4} = \frac{k \cdot p_{CH_4}}{\left[1 + K_H \cdot \sqrt{p_{H_2}} + K_w \dfrac{p_{H_2O}}{p_{H_w}}\right]^n} \quad (3.48)$$

The support is probably involved in the reforming reaction by influencing the activation of steam. This phenomenon is reflected by the terms $K_{a,H2O}$ or K_w or α_{H2O} in power-law kinetics [379] [389] and by the overall pressure dependency of the reforming rate, as shown in Figure 3.27.

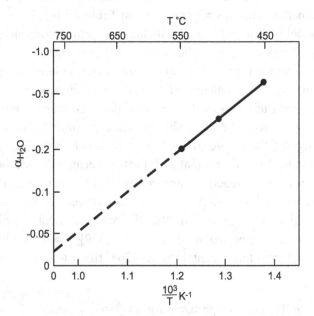

Figure 3.28 Steam reforming of ethane. Ni, MgO catalyst, temperature dependence of reaction order for water α_{H_2O} in a power-law kinetic expression [381].

Studies of steam reforming of ethane (1 bar, 500°C) [379] [381] showed that the reaction order with respect to steam varies with composition of the catalyst with presence of alkali resulting in large negative reaction orders. The retarding effect of water decreases with temperature as illustrated in Figure 3.28, thus avoiding the inconsistency with expressions (3.45).

The work by Wei and Iglesia [516] confirmed the first-order kinetics with respect to methane (Figure 3.29), but with no impact on the rate of other reactants.

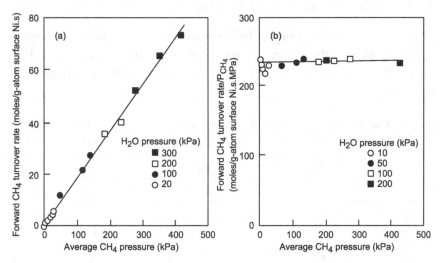

Figure 3.29 Reaction kinetics. Steam reforming of methane. Ni cat. 600°C. Wei *et al.* [516]. Reproduced with the permission of the authors and Elsevier.

The work gives strong evidence that activation of methane is the rate-determining step. Isotope exchange studies on Ni/MgO catalysts of 600°C with CH_4/CD_4 showed no formation of mixed isotopes, which would have been the case if the methane adsorption step was at equilibrium. Furthermore, Wei and Iglesia found no impact on the rate of other reactants.

Similar results were obtained on a number of group VIII metals (Ni, Pt,Pd, Ru,Rh, Ir) [517]. The activity of the metals is discussed in Chapter 6.

210 Concepts in Syngas Manufacture

The results of Wei and Iglesia are in contrast to an overall pressure dependency being slightly negative as described above. The discrepancy might be related to the fact that the studies by Wei and Iglesia were carried out at high temperatures (580–730°C). From Figure 3.29 it may be deducted that the negative reaction order α_{H2O} increases from -0.6 at 500°C to an almost insignificant value above 600°C. A similar change to first-order kinetics at temperatures above 600°C was also observed by Avetisov et al. [28].

3.5.3 Steam reforming of higher hydrocarbons

The steam reforming of higher hydrocarbons shows reaction orders with respect to hydrocarbons less than one [389] [497] as shown in Table 3.5. This reflects a stronger adsorption of the higher hydrocarbons than of methane.

Table 3.5 Steam reforming of higher hydrocarbons. Ni/MgO catalyst. 1 bar abs, 500°C. Intrinsic rates. Power-law kinetic expression in partial pressure (atm.abs) [389] [497].

	Specific rate mol/m^{2}Ni/h	E kJ/mol	Kinetic order:		
			C_nH_m	H_2O	H_2
methane	0.061	110	1.	-	-
ethane	0.12	76	0.54	-0.33	0.2
n-butane	0.138	78	-	-	-
n-heptane	0.16	68	0.2	-0.2	0.4

*) 500°C, 1 bar abs, $H_2O/C_nH_m=4$, $H_2O/H_2=10$.

The conversion of higher hydrocarbons on nickel takes place by irreversible adsorption with only C_1 components leaving the surface [389]. The first reaction step is probably related to hydrogenolysis:

$$C_2H_6 + H_2 \leftrightarrow 2CH_4 \quad -\Delta H^0_{298} = 65 kJ/mol \quad (3.49)$$

The hydrogenolysis at low temperatures of hydrocarbons higher than ethane shows that nickel attacks selectively the ends of the chains [308] [389] by successive α-scission, in contrast to, for instance, platinum. This means that methane is the main product on nickel.

Hydrocarbons show different reactivities for the steam reforming reaction. Literature [346] [389] reports different sequences of reactivities depending on temperature and the active metal.

Results in Table 3.5 show that methane has a lower reactivity than the higher hydrocarbons. The apparent activation energy is higher for methane than for ethane and n-butane, the two latter showing nearly identical activation energies. On a molar basis and at a given steam-to-carbon ratio, butane is less reactive than ethane, whereas the reactivities remain similar on a carbon atom basis.

Other reforming studies [379] on pure hydrocarbons at 500°C and 30 bar (representative of the inlet of a tubular reformer) showed that apart from benzene, most hydrocarbons present in normal naphthas are more reactive than methane. The molar reactivity decreased with increasing molecular weight for a given steam-to-carbon ratio, confirming the trend in Table 3.5. Accordingly, the reactivity of full range naphtha is less than that of light naphtha. This is illustrated in Figure 3.30 [386] [389], which also shows significant effect of benzene on the reactivity.

The limitations in converting heavy feedstocks in tubular reforming appear to be related to the desulphurisation step rather than to boiling-point temperature range. When sufficiently desulphurised, light gas oil and diesel could be completely converted into C_1 components. In practice, pore condensation in the desulphurisation catalyst dictates the feedstock limitation.

Analysis of higher hydrocarbons in the reactor effluents during tests on pure hydrocarbons showed compositions very close to the one of the feedstocks [381]. Benzene represented an exception to this picture.

For adiabatic prereformers it was demonstrated [347] that the axial temperature profile could be calculated simply from the fractional conversion of the higher hydrocarbons (naphtha), assuming all other components at equilibrium.

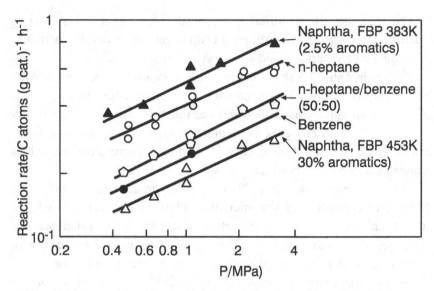

Figure 3.30 Steam reforming of various hydrocarbon feedstocks, Ni/MgO catalyst, 4.5x4.5 mm cylinders. $H_2O/C=4$. [386].

3.5.4 CO_2 reforming

There is general agreement that CO_2 reforming on nickel can be described by the same kinetics as steam reforming [59] [81] [396] [516] [538].

The change in mechanism in operation with carbon dioxide instead of steam would have little practical impact on reforming, because steam will be present not far from the inlet, but also in the centre of the catalyst particle as a consequence of the low effectiveness factors of catalysts in industrial reformers. CO_2 reforming results in lower atomic ratio of H/C which means a higher risk of carbon formation (refer to Section 5.2.4).

It was found that for other metals than nickel, CO_2 reforming at 500°C was slower than steam reforming [28] [396]. This was explained by different heats of adsorption of CO [396]. This was not observed in the studies by Wei and Iglesia [516] [517] – probably due to higher temperatures applied.

There are few studies of CO_2 reforming of higher hydrocarbons [357] [472].

4 Catalyst Properties and Activity

4.1 Catalyst structure and stability

4.1.1 Reactions with the support

The industrial catalyst is exposed to severe conditions in a tubular reformer involving steam partial pressures close to 30 bar and temperatures well above 800°C. The support should be able to withstand these conditions without loosing strength. Furthermore, it should not contain volatile components. Some of the support reactions are listed in Table 4.1. The conditions in prereformers are less demanding.

Table 4.1 Support reactions.

Reaction	
R41	$SiO_2 + 2H_2 \leftrightarrow SiH_4(gas) + 2H_2O$
R42	$SiO_2 + 2H_2O \leftrightarrow Si(OH)_4(gas)$
R43	$Al_2O_3 + H_2O \leftrightarrow 2AlO(OH)$
R44	$Al_2O_3 + KOH \leftrightarrow 2KAlO_2 + H_2O$
R45	$MgO + H_2O \leftrightarrow Mg(OH)_2$

High area supports such as γ-alumina, chromia, etc. can be used for catalysts for low-temperature adiabatic reforming, but these supports suffer from substantial sintering and weakening at temperatures above 500°C. The deterioration is strongly accelerated by the high steam partial pressure and stability tests at atmospheric pressure can therefore be misleading. Stabilisation methods applied in, for instance, auto-exhaust catalysts may become ineffective.

Catalysts stabilised with a cement may show shrinkage and decrease in strength after exposure to high temperatures. Therefore, there has been a trend towards a greater use of ceramics-based catalysts.

Because silica is volatile ($Si(OH)_4$ from Reaction R42 in Table 4.1) at high temperatures in high-pressure steam, it is now excluded from catalysts for steam reforming [85] [389], unless it is combined with alkali. For the same reason, silica-free materials are applied for the brick-lined exit gas collector and in autothermal reformers. Silica would be slowly removed from the catalyst (or brickwork) and deposited in boilers, heat exchangers and catalytic reactors downstream of the reformer.

Alkali used as promoter to eliminate carbon formation may escape slowly from the catalyst. The alkali loss is enhanced by high temperature but may to some extent be controlled by addition of acidic components (refer to Section 5.3.3). The volatised alkali may deposit in colder parts of the plant where the resulting hydroxyl ions will strongly promote stress corrosion in stainless steel. Moreover, alkali may react with some catalyst support materials such as alumina-forming the weak β-alumina (refer to Reaction R44 in Table 4.1), resulting in a decrease in the mechanical strength.

While resistant to high temperature, catalysts based on magnesia are sensitive to steaming at low temperatures because of the risk of hydration (refer to Reaction R45 in Table 4.1) The reaction may result in breakdown of the catalyst because it involves an expansion of the molecular volume. The equilibrium constant for the reaction is plotted in Figure 4.1 [381]. It is seen that at the pressures typical of tubular reformers, hydration cannot take place at temperatures above approximately 350°C. Kinetic studies of the hydration reaction have shown that magnesia reacts via liquid phase reaction in which water condenses in the pores. Therefore, the rate depends on the relative humidity of the atmosphere, and in practice hydration is a problem only when the magnesia-based catalyst is exposed to liquid water or is operated close to the condensation temperature [381]. Hydration is eliminated for carriers where magnesia has reacted with alumina to form magnesium aluminium spinel (MgAl$_2$O$_4$). The spinel support is robust and can withstand high operating temperatures.

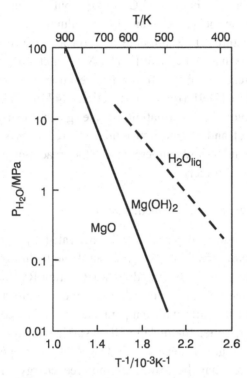

Figure 4.1 Equilibrium steam pressure for Mg(OH)$_2$=MgO+H$_2$O [381].

Typical reactions between nickel and support are listed in Table 4.2.

Table 4.2 Nickel reactions with support.

Reaction		$-\Delta H^o_{298} \quad \dfrac{kJ}{mol}$
R46	NiO + Al$_2$O$_3$ ↔ NiAl$_2$O$_4$	5.6 for α-Al$_2$O$_3$
R47	xNiO + (1−x)MgO ↔ Ni$_x$Mg$_{1-x}$O	
R48	NiO + H$_2$ ↔ Ni + H$_2$O	1.2

At steaming conditions at high temperature formation of the blue nickel aluminium spinel (Reaction R46 in Table 4.2) may start at temperatures above about 700°C [389], but a less well-defined interaction between nickel oxide and η- or γ-alumina is apparent already at lower temperatures. It is possible to form a "surface spinel" below 600°C, which may hardly be identified by X-ray methods alone.

A similar trend is observed for the reaction between nickel and magnesium oxide [389] (refer to Reaction R47 in Table 4.2). High temperatures favour the formation of the green, nearly ideal, solid solution of nickel and magnesium oxide, whereas less well-crystalised structures are found at low temperatures. Other reactions may lead to the formation of mixed spinels.

4.1.2 Activation and nickel surface area

The catalyst in industrial plants can be activated by various reducing agents such as hydrogen, ammonia, methanol and hydrocarbons added to steam [389]. The reaction with hydrogen (Reaction R48 in Table 4.2) is nearly thermoneutral and accordingly, the equilibrium constant, $K_p = p_{H_2O} / p_{H_2}$ varies little with temperature. As indicated by the upper curve in Figure 4.2, metallic nickel will be stable with approximately 0.3 and 0.6 vol% hydrogen in steam at 400°C and 800°C, respectively.

In practice, K_p may be lower, as the free energy of nickel oxide decreases due to interaction with the support material as illustrated in Figure 4.2.

As an example, the free energy of nickel oxide dissolved in magnesia can be expressed by:

$$G = G^o + RT \ln(x_{NiO}) \tag{4.1}$$

where x_{NiO} is the mole fraction of nickel oxide in the ideal solid solution. For a given H_2O/H_2 ratio, there will be a certain mole fraction of nickel dissolved in magnesium oxide.

For practical purposes, support interaction and crystal size effects mean that reduced catalysts will be oxidised when exposed to steam containing small amounts of H_2 or to mixtures of steam and reducing agents resulting in an equivalent oxidation potential. The actual equilibrium ratio depends on the structure of the individual catalysts.

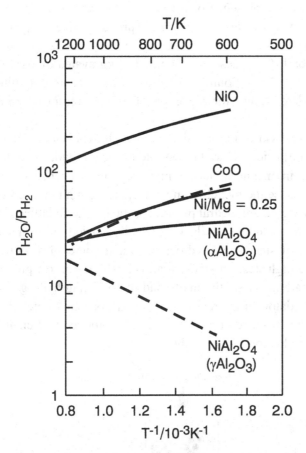

Figure 4.2 Equilibrium constants for catalyst reduction [389]. Reproduced with the permission of Springer.

The reduction of pure nickel oxide by hydrogen starts at temperatures of 200–250°C. Supported catalysts require higher temperatures to show a reasonable reduction rate. This may be ascribed to interaction with the support as indicated above. The addition of small amounts of platinum, palladium or copper to the catalyst may enhance the activation rate, probably by providing sites for the dissociation of hydrogen [389].

When the formation of nickel aluminium spinel has taken place, temperatures above 800°C [389] may be required for complete reduction. Even without the presence of a spinel phase, alumina-supported nickel catalysts may show less reducibility, probably due to penetration of aluminium ions in the nickel oxide surface layers during the impregnation [364]. These efforts may be aggravated by oxide additions (La_2O_3, MgO) [366]. Counter diffusion of nickel and aluminium appear to be rate-determining for reduction of nickel present in spinel phases [389].

Magnesia dissolved in the nickel oxide phase, even in small amounts, drastically influences the reduction rate of nickel oxide. The diffusion of nickel ions in magnesia has a high activation energy (180 kJ/mol), probably because the nickel ion must diffuse from a preferred octahedral position through a tetrahedral position in the magnesia lattice [389].

Activation in industrial plants is usually carried out by means of the feed stream of steam and hydrocarbons and sometimes with hydrogen addition at a high steam-to-carbon ratio (5–10) and at low pressure [389]. The hydrocarbons crack thermally and the resulting hydrogen or carbon acts as an initiator for the reduction process. As soon as metallic nickel is available, the steam reforming process will produce sufficient hydrogen for a quick reduction of the catalyst.

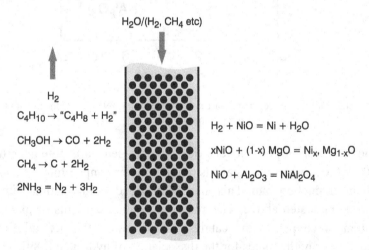

Figure 4.3 Activation of tubular reformer. Back-diffusion of hydrogen.

The temperature at which the activation is "ignited" depends on the reactivity of the hydrocarbon and on the state of the catalyst. Often the activation starts in the hotter part of the tube and the activation zone moves backwards to the colder inlet of the tube. The mechanism probably involves back diffusion of hydrogen from one activated pellet to the neighbouring inactivated pellet as illustrated in Figure 4.3.

Due to the thermo-stability of methane, the reduction is more easily initiated when the natural gas contains higher hydrocarbons. This is in particular the case if the catalyst is in a state where, for thermodynamic reasons, a low H_2O/H_2 is required (refer to Figure 4.3).

The addition of small amounts of hydrogen to the feed stream (a few percent on steam basis) may have a tremendous effect on starting the activation in the inlet part of the tube. This can be done by recycling some product hydrogen or by addition of small amounts of components easily dissociated into hydrogen (methanol, ammonia, etc.). Another possibility is installation of prereduced catalyst at the inlet of the tube.

Even at normal operation, the high stability of methane may result in H_2O/H_2 ratios over the inlet part of the catalyst bed being higher than required to maintain the catalyst in a reduced state. Consequently, some steam reformers operate with a non-activated inlet layer. This might be accepted in older plants designed for mild conditions, but in modern reformers operating at high heat flux it may become critical by causing higher tube-wall temperatures and higher risk of carbon formation (refer to Chapter 5). In such plants a small recycle of product hydrogen will help to ensure maximum activity of the catalyst. This is normally fulfilled by the hydrogen recycle stream to the HDS unit.

In laboratory units for steam reforming of methane, operating at differential conditions, hydrogen must be present in the feed stream to avoid oxidation of the catalyst as discussed in Section 3.5.1.

4.2   Nickel surface area

4.2.1   Measurement of nickel surface area

The activity of a nickel catalyst is related to the nickel surface area. It is important to refer reaction rates to unit surface area in terms of the specific activity or turn-over frequency (TOF) in order to compare catalysts and to analyse catalytic phenomena. Several methods have been used to determine nickel surface areas, [415] including chemisorption, X-ray diffraction (XRD), transmission electron microscopy (TEM) and magnetic measurements [167].

XRD provides a quick and easy method to determine the volume-averaged nickel particle size, but generally smaller nickel particle sizes are indicated by XRD than by other methods. One reason is that nickel particles are polycrystalline. The presence of twin planes in the TEM images (Figure 4.4) of nickel particles further supports this conclusion.

TEM provides size distributions of the nickel particles and information about the structure of the catalyst that can hardly be obtained by any other technique, but only a very small part of the sample is analysed.

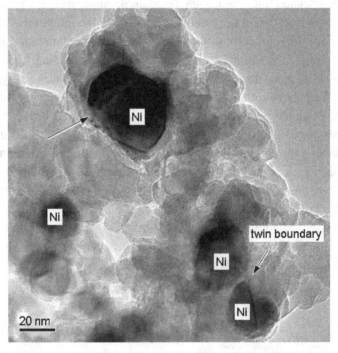

Figure 4.4 TEM image of a Ni/MgAl$_2$O$_4$ spinel catalyst. The arrows show the presence of internal faults in sintered nickel particles [415] [449]. Reproduced with the permission of Elsevier.

The most widely used method for measurements of the nickel surface area is chemisorption. This method requires [381]:

- A well-defined "saturation" layer (which need not be a "monolayer" with a simple one-to-one relation between chemisorbed molecules or atoms and metal surface atoms).
- A correlation between the adsorption capacity and the metal surface, expressed as the adsorbed amount per unit surface area of the metal or as the mean area of an adsorption site.

Chemisorption of various gases has been proposed for determination of the surface area of nickel [389]. Oxygen could involve a severe reconstruction of the nickel surface and the evaluation of results obtained with carbon monoxide could be impeded by the different ways in which carbon monoxide may chemisorb on nickel.

Chemisorption of hydrogen is a widely used method for determination of the surface area of nickel in supported catalysts. The H_2/Ni is complicated because of different states of chemisorbed hydrogen. At low temperatures (-196°C), a fast uptake results in a saturation layer with H/Ni less than one [46] [381]. At higher temperatures, the chemisorption becomes activated with hysteresis phenomena in isobars — and with some of the hydrogen present in sub-surface sites [389].

Conditions for obtaining a saturation layer with a H/Ni ratio of 1.1 [38] were defined by comparing the chemisorption of hydrogen at room temperature to BET measurements of the surface area of nickel powder, when using a mean nickel-size area of $6.77 \, 10^{-2}$ nm^2. Good correspondence was also observed between nickel crystallite size determined by hydrogen chemisorption at room temperature and that determined by XRD and TEM [403] [415].

Note
Procedure for hydrogen capacity [389]
After reduction, the sample is evacuated at 400°C for 1–2 hours until approximately 10^{-3} Pa. H_2 uptake is measured at room temperature using

45 minutes for equilibration at P_{H2} in the range 0–0.1 bar. The hydrogen capacity is determined by extrapolation to zero P_{H2}.

Another method for determination of the nickel area is based on chemisorption of hydrogen sulphide [389]. Hydrogen sulphide is the stable sulphur compound at conditions for tubular reforming. It is the most severe poison for the reaction (refer to Section 5.4). The adsorption of hydrogen sulphide on nickel is rapid even below room temperature [376] [381]. At temperatures of industrial interest hydrogen sulphide is chemisorbed dissociatively on nickel.

$$Ni + H_2S \leftrightarrow Ni - S + H_2 \tag{4.2}$$

where Ni represents an ensemble of nickel atoms on the surface. Stable saturation uptakes of sulphur are observed at ratios of H_2S/H_2 from $10\ 10^{-6}$ up to close to $1000\ 10^{-6}$ above which formation of bulk nickel sulphide takes place [372]. This means that the saturation layer is stable at H_2S/H_2 ratios several magnitudes below the ratios required for formation of bulk sulphide. Adsorption isotherms are discussed in Section 5.4.

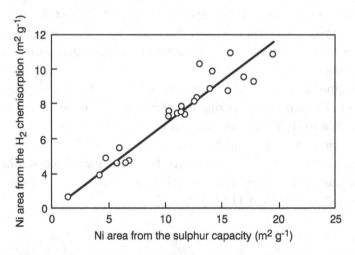

Figure 4.5 Nickel surface areas determined by chemisorption of hydrogen and hydrogen sulphide, respectively (using procedures described with test) [415]. Reproduced with the permission of Elsevier.

A similar behaviour has been observed for other systems including the chemisorption of hydrogen sulphide on other metals (Ag, Fe, Cu, Co, Ru, Pt) [523].

The surface layer has the structure of a two-dimensional sulphide growing as islands with well-defined structure. This was well described by early LEED and Auger experiments [174] [336] [344] and it was shown [376] that the monolayer of sulphur (S/Ni=0.5) expressed as the sulphur capacity (440 ppmS/m^{-2} Ni) correlated with the surface area determined by hydrogen chemisorption (Figure 4.5). Today, the structure of the surface sulphide can be seen by scanning tunnelling microscopy (STM) [51] [430] as shown in Figure 4.6.

$Ni_{surface} + H_2S \leftrightarrow S-Ni_{surface} + H_2$

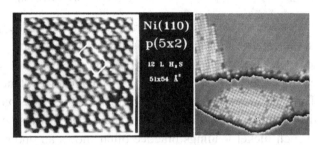

Figure 4.6 Chemisorption of sulphur on nickel surface structures observed by STM Besenbacher et al. [51] [425]. Reproduced with the permission of the authors.

Note
Procedure for sulphur capacity s_0 [389]
After reduction, the sample is exposed to a flow of H$_2$S/H$_2$=10–20 10^{-6} at 500–550°C. After equilibration (1 to 2 days as estimated from flow, sample size, particle size and expected s_0), s_0 is determined by chemical analysis. The method cannot be applied for catalysts containing Ca or Ba and only with difficulty for pyrophoric catalysts.

The nickel area is calculated by using s_0=440 wt ppm equivalent to 1 m^2 g^{-1}, s_0 being the sulphur capacity of the catalyst (µg S per g Ni).
A mean nickel particle diameter can be estimated from

$$d_{Ni} = 3 \cdot 10^3 \frac{x_{Ni}}{s_0} \quad (4.3)$$

d_{Ni} is given in nm, and x_{Ni} is the weight % nickel in the reduced state. The corresponding dispersion (%) of nickel can be calculated from

$$D_i = 0.034 \frac{s_0}{x_{Ni}} = \frac{101}{d_{Ni}} \% \tag{4.4}$$

The turnover frequency (molecules site^{-1} s^{-1}) may be calculated from

$$TOF = 4.78 \cdot 10^3 \frac{R_{int}}{s_0} = 10.86 \cdot R_{int,surf} \tag{4.5}$$

in which R_{int} and $R_{int,surf}$ are intrinsic rates mol hydrocarbon/g cat/h and mol hydrocarbon/m^2 Ni/h, respectively.

4.2.2 Nickel surface area and catalyst preparation

The nickel surface area is generally increased with higher nickel contents in the catalyst [389], but the dispersion or utilisation of the nickel tends to decrease with increasing nickel content. Accordingly, many commercial catalysts are optimised at nickel contents of 10–15 wt%. Through special preparation methods, it is possible to obtain high dispersions even at high nickel contents, but because of sintering effects at high nickel loadings practice often shows an optimum in nickel area [389] depending on nickel content.

The activation procedure influences the size of the resulting nickel surface. The highest nickel area is obtained by using dry hydrogen during heating up as well as during the reduction. The presence of steam during heating up and reduction may result in significantly smaller nickel areas [381] [389]. This effect can be explained by the assumption that steam oxidises the smallest nuclei of nickel or prevents their formation. Consequently, the number of nuclei and the resulting number of crystals is decreased, which means a smaller area.

4.2.3 Sintering

Most of the tubular reformer is operating above the Tammann temperature of nickel ($T_{Tammann}=½·T_{melt}=581°C$) and the catalyst is thus exposed to sintering, resulting in growth of the nickel particles and a decrease in nickel surface area and in activity. However, the sintering phenomenon is more complex and may take place below the Tammann temperature [452] as well.

Sintering is influenced by many parameters, including temperature, chemical environment, catalyst composition and structure, and support morphology. The most important parameters are the temperature and the atmosphere in contact with the catalyst.

Recent studies using *in situ* high-resolution transmission electron microscopy (HRTEM) and measurements of specific metal surface areas [452] have given new insight into the understanding of sintering of nickel catalysts. One mechanism considers particle migration and coalescence, where particles move over the carrier and collide. This mechanism is dominating at low temperatures. At high temperatures, particle coarsening proceeds via transport between metal particles of metal atoms or small agglomerates over the carrier (atom migration, Ostwald ripening). The time dependencies of crystal growth are $t^{1/7}$ and $t^{1/3}$ for particle migration and coalescence and atom migration, respectively. The cross-over temperature depends on process conditions (H_2O/H_2, etc.) and catalyst composition. The presence of water greatly accelerates sintering. This supports the model [452] that the movement is associated with diffusion of Ni_2-OH complexes on the nickel surface. The support can affect the sintering in various ways. It has been proposed that the pore structure, the morphology, and phase transitions of the support determine the final particle size of the nickel.

Experiments showed [425] [449] that the particle size after prolonged sintering depends little on nickel loading. In practice, it appears that a maximum particle size stabilises at a given temperature as the movement of these particles becomes very slow [415].

Investigation of nickel catalyst samples after ten years of operation in a commercial steam reformer [453] showed an increase in nickel crystal size with temperature, whereas the turnover frequency (specific activity) was almost unchanged, i.e. comparable with that of unused catalyst.

To describe the sintering of metal particles, the time dependency of the two mechanisms are $t^{1/7}$ and $t^{1/3}$ for coalescence and atom migration, respectively, as expressed by:

$$\frac{d_{Ni}}{d_{Ni}^o} = (1+kt)^n \quad (4.6)$$

Catalyst samples were exposed to a synthesis gas equilibrated at 600°C and 30 bar abs for more than 8000 hours in a pilot plant [423]. Samples were taken during the run. The relative nickel particle diameters were fitted by Equation (4.6) resulting in a value for n=0.31 indicating that at 600°C sintering proceeds via the atom migration sintering.

As shown in Figure 4.7 [452], the sintering rate is significantly higher above 600°C and the impact of H_2O/H_2 is higher, indicating a change in mechanism from coalescence to Ostwald ripening. This is also indicated by observations using *in situ* HRTEM [452].

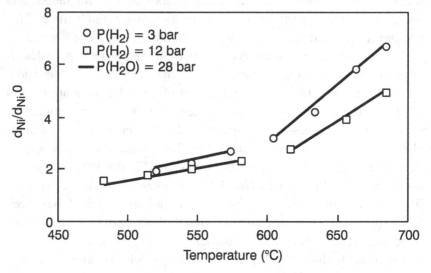

Figure 4.7 Sintering of $Ni/MgAl_2O_4$ catalyst, 700 hours. Sehested [452]. Reproduced with the permission of Elsevier.

Results of sintering tests with a high-temperature catalyst [381] [389] are shown in Figure 4.8. At 550°C, the sintering was insignificant. Even the non-sintered catalyst had large nickel particles. This observation, rather than the rule of Tammann, may explain the lack of sintering at 550°C. Values of n in Equation (4.6) can be obtained from the data in Figure 4.8, the values being 7.7 and 6.9 at temperatures of 700 and

800°C, respectively. This value is close to what is expected for spherical particle migration and coalescence [415].

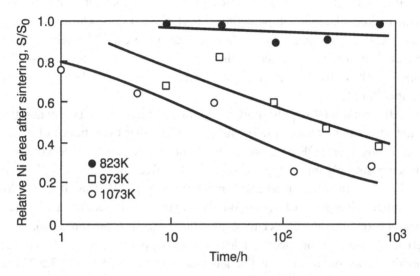

Figure 4.8 Sintering of nickel surface on ceramic reforming catalyst [381].

4.3 Catalyst activity

4.3.1 Group VIII metals

The Group VIII metals are active for steam reforming. The turn-over frequency, TOF, is typically 2–5 s^{-1} at 500°C for steam reforming of methane on nickel ($H_2O/CH_4=4$. $H_2O/H_2=10$, 1 bar abs [389]).

Alloying nickel with groups 1B metals (Cu,Ag) may cause a drastic decrease in the activity [49] [389].

Sintering of the nickel crystals results in no significant change in TOF. However, for small nickel particles ($d_{Ni}<10$ nm) there is a significant increase in TOF with increased dispersion as discussed in Chapter 6.

Cobalt shows a lower activity than nickel [389] – probably attributable to the process conditions, with the H_2O/H_2 value being close to that causing oxidation of the metal. Iron is active for steam reforming, but only at strongly reducing conditions as observed in the hot part of a direct reduction shaft furnace (refer to Section 2.4.3).

Rhodium and ruthenium show TOF values about ten times higher than nickel, platinum or palladium [264] [272] [379]. Only one study [517] claimed platinum to be the most active metal. The sequence of metal activities is discussed further in Chapter 6. Rhodium is also the best catalyst for selective steam dealkylation of toluene at low temperature [213].

All Group VIII metals show a significant decrease in TOF when the catalyst contains alkali [389] [401]. It means that even traces of alkali metal may blur TOF measurements if not recognised. The deactivation depends strongly on the type of support, the effect of alkali being less on acidic supports which bind alkali more strongly [389]. The alkali results in larger adsorption of steam on the catalyst reflected by a negative reaction order with respect to steam (large K_w in Equation 3.48). This effect is most pronounced on less acidic supports. This all hints that alkali works on nickel via the gas phase (refer to Section 3.5.2 and Section 5.3.2).

As discussed earlier (Chapter 3), the effectiveness factor of the catalyst in the tubular reformer is small. Even the low activity of alkali-promoted catalysts usually results in satisfactory performance in industry. However, as outlined in Example 3.4, high catalyst activity means that the same performance can be attained at a lower tube-wall temperature and hence leaves room for longer tube life or operation at higher heat fluxes (i.e. smaller reformer).

4.3.2   Non-metal catalysts

Difficulties in desulphurising heavy feedstocks have led to attempts to use non-metallic catalyst for steam reforming. Molybdenum carbide and tungsten carbide are catalysts for steam and CO_2 reforming and catalytic partial oxidation [122] [450]. However, in synthesis gas the carbides are stable only at elevated pressures (approximately 9 bar) and they are transformed into the oxides at ambient pressure.

Molybdenum carbide will hardly be stable in a plug-flow reactor [450], because the carbide will be oxidised at the inlet. The molybdenum

carbide will only be stable at high conversions (i.e. high concentration of CO) or in case of significant back-diffusion (as observed in "pancake" reactors (refer to Figure 1.20) and in CSTR reactors).
The rates obtained on carbide catalysts are significantly lower than those obtained on Group VIII metal as illustrated in Figure 4.9.

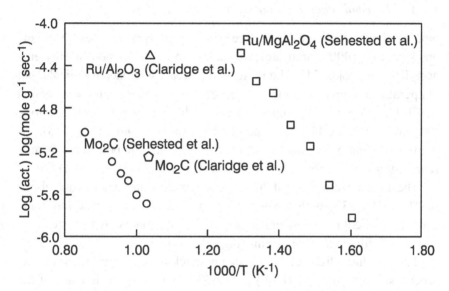

Figure 4.9 Reforming with carbide catalysts. Claridge *et al.* [122], Sehested *et al.* [450]. Reproduced with the permission of Elsevier.

Other attempts deal with the use of ceria as catalyst for internal reforming on SOFC anodes. Although CeO_2 has an activity of almost two orders of magnitude less than nickel [321] [323], the low activity might be sufficient for the reforming process [299] in a high-temperature fuel cell operating at 800°C, where the balance between the rate of reforming and the rate of the electrochemical reaction is critical [401].

Note
With an activation energy of 105 kJ/mol, the ratio of rates at 800°C and 500°C is about $\ln(k_1/k_2)=105 \cdot 10^3/R \cdot (1/773-1/1073)=4.57$, or $k_1/k_2=96$.

A calcium aluminate catalyst was tested in the Toyo Process [489], but ratios at 850°C were still a fraction of what is obtained on nickel at 500°C. The Toyo process is close to "catalytic steam cracking" discussed in the following.

4.3.3 Thermal reactions – catalytic steam cracking

Non-catalytic steam reforming requires high temperatures. Methane cracks above 1000°C into radicals leading to the formation of ethylene, acetylene and coke [31]. The radicals may react with steam radicals, but temperatures above 1500°C are necessary for significant conversion [267]. One approach to improve rates is the use of plasma technology [90], the key issue being the power consumption [90] [237] [513]. As thermal plasma reforming is not sensitive to sulphur poisoning, it may represent a solution for reforming of logistic fuels (jet fuel, diesel) [202].

The thermal cracking of higher alkanes becomes significant above 650°C [374] [532] with the formation of alkenes, aromatics and coke. This is applied in steam crackers in ethylene plants, where steam is added as a diluent and for minimising coke formation.

The product distribution in steam cracking is determined by the kinetic severity function (KSF) [374] [533], which is the integral of the rate constant and residence time, θ:

$$\text{KSF} = \int k_5 d\theta \quad (4.7)$$

Using the cracking of n-pentane as a reference point, k_5 depends on temperature and hence KSF depends on the temperature profile and residence time profile of the reactor. In practice, this is difficult to determine. However, the KSF can be determined empirically if the feed contains n-pentane simply by:

$$\text{KSF} = \ln\left\{\frac{C_{5,\text{inlet}}}{C_{5,\text{outlet}}}\right\} \quad (4.8)$$

Typical yield structures are shown in Figure 4.10 for cracking of naphtha. Ethylene plants are designed for high severity up to 4 with milliseconds residence time [374].

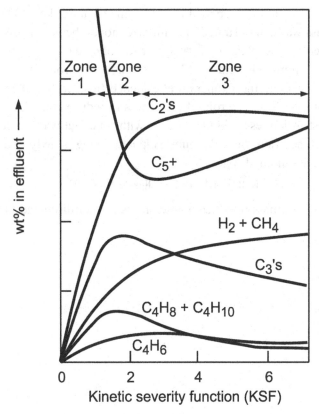

Figure 4.10 Naphtha cracking. Typical yields as a function of KSF. Zdonik et al. [533]. Reproduced with the permission of Elsevier.

In principle, the tubular reformer is a steam cracker filled with catalyst. Feed preheaters can also be considered steam crackers, each characterised by a KSF and corresponding olefin yields. As olefins are coke precursors in the steam reforming process, preheaters should be designed for a low KSF or the problem should be eliminated by installation of a prereformer (refer to Sections 1.2.3 and 5.3.4).

There have been attempts to improve the steam cracking process by installing a catalyst in the cracking tubes [374] [547]. This resulted in co-production of syngas and light alkenes from heavy gas oil and naphtha. The catalyst was a potassium-promoted zirconia support operating at

232   *Concepts in Syngas Manufacture*

conditions close to those applied in industrial steam crackers [374] [547]. The process was demonstrated in a full-size monotube pilot plant [547].

A reactor concept with a catalyst bed followed by a narrow tube resulted in improved yields [547]. Most likely, the role of the catalyst bed was to change the temperature/residence time profile (KSF) rather than to form olefins. The role of alkali was to reduce coke formation on the catalyst. Alkali escaped the catalyst with time, but was replaced by a continuous addition of potassium sulphate being slowly reduced to volatile potassium hydroxide:

$$K_2SO_4(\text{solid}) + 4H_2 \leftrightarrow 2KOH(\text{gas}) + H_2S + 2H_2O \qquad (4.9)$$

The coke formation in steam cracking is further discussed in Section 5.3.4.

5 Carbon and Sulphur

5.1 Secondary phenomena

Main constraints for optimum operation of the steam reformer are related to "secondary phenomena", the formation of carbon and poisoning by sulphur.

Carbon formation cannot be tolerated as it may result in breakdown of the catalyst, increased pressure drop, and uneven flow distribution leading to overheating of the tubes locally as "hot spots" or totally as "hot tubes". This will limit the plant capacity or the life of the catalyst tubes. Sulphur poisoning results in lower activity and hence higher tube-wall temperature (refer to Chapter 3). It may also provoke carbon formation as higher hydrocarbons may pass the deactivated catalyst and be exposed to pyrolysis which may lead to coke formation. In other situations sulphur may inhibit the formation of carbon.

This chapter will deal with an analysis of these phenomena and formulation of design for carbon-free operation.

5.2 Carbon formation

5.2.1 Routes to carbon

Carbon may be formed by three different mechanisms [389]:

- "whisker" carbon
- "gum" formation
- pyrolytic carbon

with the characteristics listed in Table 5.1:

Table 5.1 Routes to carbon [389].

Carbon type	Reaction (Table 5.2)	Phenomena	Critical parameters
Whisker carbon	R6-R9	Break-up of catalyst pellet	low H_2O/C ratio, high temperature, presence of olefins, aromatics
Gum	R50	Blocking of Ni surface	low H_2O/C ratio, absence of H_2, low temperature, presence of aromatics
Pyrolytic coke	R49	Encapsulation of catalyst pellet, deposits on tube wall	high temperature, long residence time, presence of olefins, sulphur poisoning

The main reactions are summarised in Table 5.2. Equilibrium constants for Reactions R6–R8 are listed in Appendix 2 based on data from Tables 1.4 and 1.5.

Table 5.2 Carbon-forming reactions.

Reaction		$-\Delta H^0_{298}\ \dfrac{kJ}{mol}$
R6	$CH_4 \leftrightarrow C + 2H_2$	-75
R7	$2CO \leftrightarrow C + CO_2$	172
R8	$CO + H_2 \leftrightarrow C + H_2O$	131
R9	$C_nH_m \rightarrow nC + 0.5mH_2$	-188 (n-C_7H_{16})
R49	$C_nH_m \rightarrow$ olefins $\rightarrow$ coke	
R50	$C_nH_m \rightarrow (CH_2)_n \rightarrow$ gum	

Whisker carbon is formed as characteristic fibres (nanotubes) [505] from CO, CH_4 and higher hydrocarbons and may result in breakdown of the catalyst pellet (Section 5.2.2). The low-temperature phenomenon "gum formation" involves blockage of the metal surface by a film of polymerised carbonaceous material (Section 5.3.3). Pyrolytic carbon is a result of thermal reactions (pyrolysis) as experienced in steam crackers and results in deposits on the tube wall or encapsulating the catalyst pellets (Sections 4.3.3 and 5.3.4).

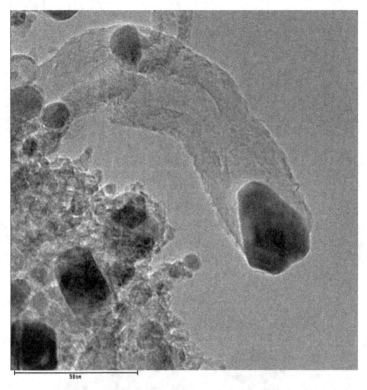

Figure 5.1 Carbon whisker. The nickel particles have a diameter of about 40 nm. Helveg [227]

Whisker carbon is formed by dissociation of hydrocarbons or carbon monoxide on the nickel surface [389]. The "whisker" typically grows as carbon fibre (nanotube) with a nickel crystal at the top as shown in Figure 5.1.

According to the classic model [33] [378] [383], adsorbed carbon atoms are dissolved in the metal particle. Carbon diffuses through the particle and nucleates into the fibre at the rear interface as illustrated in Figure 5.2 (as shown below this remains a simplification (refer to Figure 6.5)). Nickel carbide is not stable under steam reforming conditions.

236 Concepts in Syngas Manufacture

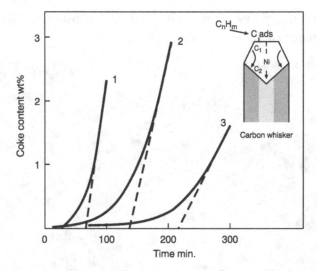

Figure 5.2 Rates of carbon formation. TGA studies [380]. Ni/MgO catalyst, 1 bar abs, 500°C, n-heptane. 1: $H_2O/C < 1.3$, 2: $H_2O/C=1.5$, 3: $H_2O/C=2.0$ [381]. Reproduced with the permission of Elsevier.

Figure 5.3 Topsøe TGA test unit 1970s. Thermogravimetic analysis (TGA) was used in many of the early studies on carbon formation, often in "home-made" units before standard equipment became available. (Ulla Ebert Petersen.)

The nickel crystal changes shape into a pear-like particle, leaving small fragments of nickel behind in the whisker. Whisker carbon is also observed on cobalt and iron catalysts [278] [498].

The whisker has high strength and destroys the catalyst particle when it hits the pore wall. Broken catalyst and carbon may have a serious impact on the operation of the reformer by maldistribution of feed and overheating of the tubes.

The rate of carbon formation depends strongly on the type of hydrocarbon as illustrated in Figure 5.4 with alkenes (and acetylene) being the most reactive [380] [381]. After an induction time, t_0, the carbon grows at a constant rate.

$$\frac{dC_w}{dt} = k_c(t - t_0) \qquad (5.1)$$

as shown by TGA measurements (Figure 5.2).

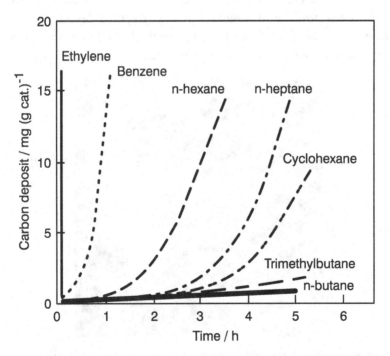

Figure 5.4 Rate of carbon formation from different hydrocarbons [384] [389]. TGA measurements. ($H_2O/C=2$ mol/atom, 1 bar abs, 500°C). Reproduced with the permission of Springer.

238   *Concepts in Syngas Manufacture*

The growth mechanism appears to be the same, irrespective of the type of hydrocarbon or whether it results from the endothermic dissociation of methane or the exothermic dissociation of carbon monoxide [378] [381] (refer to Table 5.2). However, the resulting morphology and degree of graphitisation depend on parameters such as type of hydrocarbon, metal, particle size and temperature. Hence, there might not be a unique growth mechanism for the formation of carbon fibres and nanotubes [146] [412] [415].

Olefins and acetylene show rapid dissociation and the diffusion of carbon through nickel has been suggested for the rate-determining step [33] [383] which is in line with the activation energy for carbon formation being close to that for the diffusion of carbon through nickel.

In situ high-resolution transmission electron microscopy (HRTEM) [226] demonstrated that the process of carbon formation is very dynamic involving drastic changes of the nickel crystal during the nucleation. It has also been possible to observe how the graphene layers grow from surface steps which are created during the nucleation and which moves as the graphene layers grow. This is illustrated in Figure 5.5.

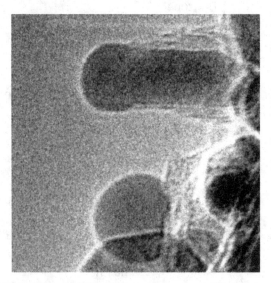

Figure 5.5 Nucleation of whisker carbon. Helveg *et al.* [226]. *In situ* HRTEM. CH_4/H_2=1/1, 2.1 mbar, 536°C. The nickel crystal (approximately 60 nm) is prolonged during the nucleation. Reproduced with the permission of Nature.

The role of step sites was confirmed by DFT calculations [48] [415]. These calculations also indicated that the "classic" mechanism (Figure 5.2) is a simplification [7] (refer to Chapter 6).

The addition of potassium results in a significant increase in the induction time for methane decomposition, as shown in Figure 5.6 [415]. This implies a retarding role of potassium on the dissociation of methane and the nucleation of carbon (refer to Chapter 6).

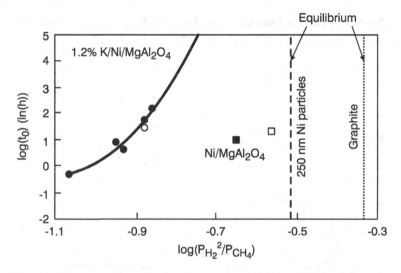

Figure 5.6 Decomposition of methane on supported nickel at 500°C and 1 bar: the influence of potassium on the induction time, t_o (refer to Equation 5.1) [415]. Filled symbols: carbon formation. Open symbols: no carbon formation. Potential for carbon at H_2/CH_4 ratios smaller than the equilibrium ratios shown (refer to Section 5.2.3). Reproduced with the permission of Elsevier.

The nickel particle size has an impact on the nucleation of carbon. The smaller the crystals, the more difficult the initiation of carbon formation. This result was demonstrated in TGA experiments summarised in Table 5.3 with two catalysts having the same activity but different dispersions of nickel [415] [425]. The catalysts were heated at a fixed rate, and it was shown that the onset temperature was approximately 100°C higher for the catalyst with small nickel crystals (approximately 7 nm) than for that with large crystals (approximately 100 nm).

240 Concepts in Syngas Manufacture

Table 5.3 Carbon formation and Ni particle size. [425]. TGA studies Ni/MgAl$_2$O$_4$. (65°C min^{-1}). H$_2$O/C$_4$H$_{10}$=2.8 mol mol^{-1}.

Ni (wt%)	d$_{Ni}$ (nm)	Ni area (m^2 g^{-1})	Reforming rate 500°C (mol g^{-1}h^{-1})	Temperature for carbon formation (°C)
15	102	0.9	0.29	450
0.92	7	1.0	0.26	>550

The rate of carbon formation was found to be far less on noble metals than on nickel as illustrated in Figure 5.7 [396].

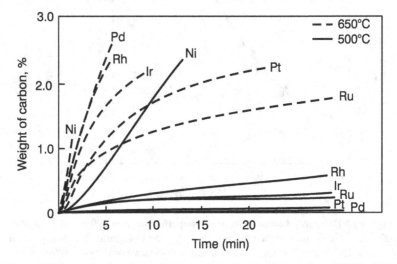

Figure 5.7 Rates of carbon formation on various metals. TGA studies. CH$_4$/H$_2$=95/5. [396]. Reproduced with the permission of Elsevier.

This result is explained by the fact that the noble metals do not dissolve carbon [304] (a more likely explanation is discussed in Section 6.2). The carbon formed on the noble metals was observed to be of a structure that was difficult to distinguish from the catalyst structure [304]. On a ruthenium catalyst, high-resolution electron microscopy revealed a structure which looked like a few atomic layers of carbon covering most of the surface.

The whisker growth mechanism is also blocked by sulphur poisoning of the nickel surface. When formed, the carbon has a typical "octopus" structure with several fibres growing from one nickel crystal [390]. As shown in Figure 5.8, a similar structure is formed on Ni/Cu catalysts with high copper contents [49].

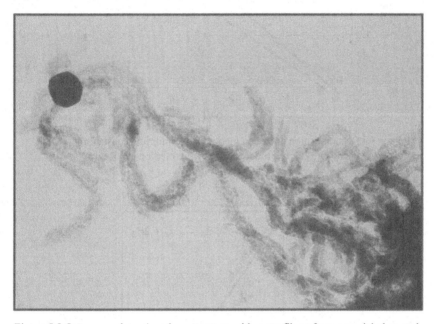

Figure 5.8 Octopus carbon. A carbon structure with more fibres from one nickel crystal. Bernardo et al. [49]. Reproduced with the permission of Elsevier.

5.2.2 Carbon from reversible reactions

The whisker carbon has a higher energy than graphite, which is reflected in lower equilibrium constants for the reversible decomposition reactions of carbon monoxide (Reaction R7 in Table 5.2, the Boudouard reaction) and of methane (Reaction R6 in Table 5.2) as shown for the two reactions in Figures 5.9 and 5.10, respectively [49] [149] [150] [378] [381].

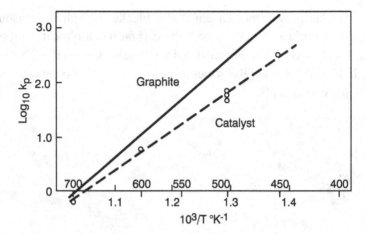

Figure 5.9 Decomposition of carbon monoxide. Equilibrium constant for Ni/MgO catalyst [378]. Reproduced with the permission of Elsevier.

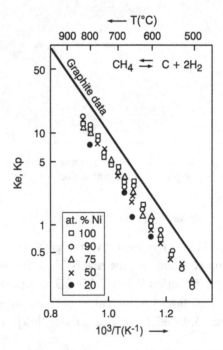

Figure 5.10 Decomposition of methane on supported nickel catalysts. Equilibrium constants for Ni/SiO_2 and $Ni,Cu/SiO_2$ catalysts from TGA measurements. The copper content of the catalyst does not affect the equilibrium constant. Bernardo et al. [49]. Reproduced with the permission of Elsevier.

The impact of nickel particle size on the equilibrium constant is also illustrated in Figure 5.11 showing data for various metal catalysts [396]. Ni-a and Ni-b curves represent data for catalysts with particle sizes of 300 nm and 10 nm, respectively. The noble metals have even smaller equilibrium constants. This is due to the metal particle effect or to other structural effects (μ^* in Equation 5.3). Palladium shows a different behaviour than the other metals, probably due to the formation of an interstitial solid solution and mobility of carbon in palladium [540].

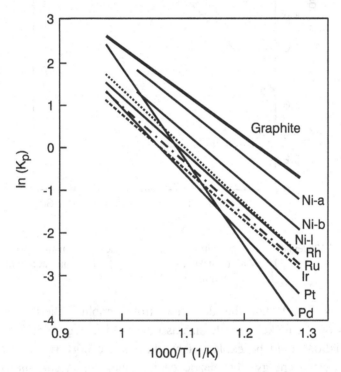

Figure 5.11 Decomposition of methane. Equilibrium constants for various metal catalysts [396].

In conclusion, equilibrium constants should be measured on the individual catalysts to be studied. Equilibrium constants for a catalyst with large nickel crystals are shown in Appendix 2.

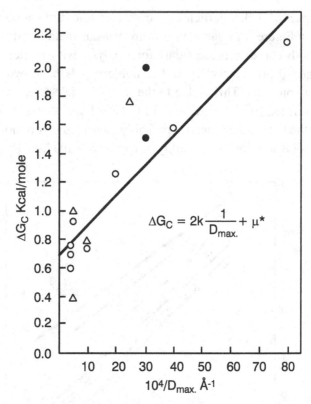

Figure 5.12 Deviation from graphite data and Ni particle size (Equation 5.10). CH_4 decomposition, 500°C. Various nickel catalysts [378], [381]. Reproduced with the permission of Elsevier.

It was found that the deviation from graphite thermodynamics depends on the nickel particle size as illustrated in Figure 5.12 and that the deviation could be explained by the extra energy required by the higher surface energy, the elastic energy, and defect structure of the carbon filaments. The correlation between free energy and surface tension is expressed by the Kelvin equation. In a simplified model [378] [381], the Kelvin equation becomes:

$$\mu - \mu_o = \frac{\sigma \cdot M_w}{r_W \cdot \rho} \quad (5.2)$$

in which μ and μ_o are the free energies of a whisker with radius r_W and no curvature, respectively. σ is the surface tension.

The deviation from graphite data may be expressed by

$$\Delta G_C = \mu - \mu_0 + \mu^* \quad (5.3)$$

in which μ^* is the contribution from structural defects compared to graphite.

From (5.2) and (5.3) the straight-line equation can be written as

$$\Delta G_C = k\frac{1}{r_W} + \mu^* \quad (5.4)$$

The exothermic reactions for decomposition of carbon monoxide, (Reactions R7 and R8 in Table 5.2) means that for a given gas composition and pressure, there is a temperature below which there is a thermodynamic potential for carbon formation. Likewise for the endothermic decomposition of methane, (Reaction R6 in Table 5.2), there is a temperature above which there is a thermodynamic potential for carbon formation. These carbon limits assume that there is no reaction during cooling or heating the gas. This is the situation when no catalyst is present such as in heat exchangers, boilers and convective reformers. Carbon formation may lead to fouling of the equipment or to metal dusting corrosion [211].

In principle, metal dusting could also be initiated by olefins, which are very reactive for formation of whisker carbon (Figure 5.4), but in most industrial plants (steam crackers, FCC units), olefins-containing streams contain sulphur, which will inhibit the nucleation of carbon.

The decomposition of carbon monoxide (Reactions R7 and R8, Table 5.2) may take place without catalyst on the surfaces of the equipment (e.g. heat exchangers). This may also lead to metal dusting corrosion [121] [211] [535]. Reaction R7 in Table 5.2 appears to be involved [211] [293]. It means that for a given gas composition (and pressure) there will be potential for metal dusting below the carbon limit temperature (see Example 5.1). At very low temperature, the rate will be too small.

Low alloy steels which are not protected by a dense oxide layer may be attached by metal dusting. Carbon which is formed via the whisker mechanism reacts with alloy components to carbides resulting in a disintegration of the alloy (Figure 5.13) into dust, which is blown away in the gas flow.

Example 5.1
A syngas leaving an ATR reactor may have the following composition (vol%): H_2=48.3, H_2O=17.9, CO=25.1, CO_2=4.9, CH_4=0.7 and inerts=2.9. The pressure is 28.7 bar abs.

When cooling this gas, the carbon limit temperature is calculated from the equilibrium quotient for Reaction R8 in Table 5.2:

$$Q_r = \frac{p_{H_2O}}{p_{CO} \cdot p_{H_2}} = \frac{0.179}{0.251 \cdot 0.483 \cdot 28.7} = 0.051$$

which corresponds to Kp at 868°C. This means that there is potential for nucleation of carbon (metal dusting) below this temperature.

The mechanism appears to be equivalent to that described for formation of whisker carbon on the reforming catalyst [415] [534], where chemisorbed sulphur blocks the nucleation of carbon with the formation of carbon on the free Fe-Ni surface of the construction material.

The sulphur present in the SPARG product gas (refer to Section 5.5) inhibits the nucleation of metal dusting corrosion in the outlet system, even when the operating conditions predict that carbon formation is possible. Alloying components [431] (Cu,Sn,Au.....) may also retard the corrosion (refer to Section 5.3.2).

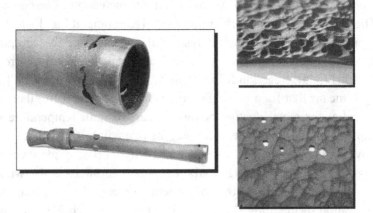

Figure 5.13 Example of equipment hit by metal dusting.

However, the limits for metal dusting corrosion are dictated by several parameters and in a complex way. The risk of corrosion is highly influenced by the selection of construction materials and passivation methods.

5.2.3 Principle of equilibrated gas

The formation of whisker carbon cannot be tolerated in a tubular reformer. The important question is whether or not carbon is formed, and not the rate at which it may be formed. In terms of the growth mechanism, it means to extend the induction period (t_0 in Equation 5.1) to infinity. This is achieved by keeping the steady-state activity of carbon smaller than one (refer to Section 5.2.4). The carbon formation depends on the kinetic balance between the surface reaction of the adsorbed hydrocarbon with oxygen species and the further dissociation of the hydrocarbon into adsorbed carbon atoms, which can nucleate to whisker carbon. However, this approach is complex and there is a need for simple guidelines using simple thermodynamic calculations.

The decomposition reactions of methane and carbon monoxide are reversible and the risk of carbon formation can be determined by thermodynamics. One may apply the *principle of equilibrated gas* [389] stating: carbon formation is to be expected on a nickel catalyst if the gas shows affinity for carbon after the establishment of the methane reforming and the shift equilibrium. This implies that the composition used to calculate the reaction quotient is calculated after establishing the reforming and shift equilibrium as shown in Section 1.2.1. The calculated equilibrium temperatures for the methane decomposition and the Boudouard reactions will be different unless graphite is considered and all reactions are in equilibrium. In the actual case the most severe reaction of the two must then be considered. For steam reforming of methane this is the methane decomposition reaction as used in the following.

By use of Reaction R6 in Table 5.2 the carbon formation check for the methane decomposition reaction can be written as:

$$CH_4 \leftrightarrow C + 2H_2$$

$$-\Delta G_c = RT \ln\left[\frac{K_{eq}}{Q_{r.eq}}\right] = RT \ln K_{eq} \left[\frac{p_{CH_4}}{p_{H_2}^2}\right]_{equilibrated\ gas} \qquad (5.5)$$

$$-\Delta G_c > 0 \Rightarrow Carbon$$

The carbon limits are a function of the atomic ratios O/C, H/C and inert/C and of total pressure. As the gas is at equilibrium, it is necessary to consider only one of the carbon-forming Reactions R6–R8 in Table 5.2, provided that the whisker structure (μ^*) is independent of the carbon-forming reaction.

The principle of equilibrated gas is justified by the low effectiveness factor of the reforming reaction, which implies that the gas inside most of the catalyst particle is nearly at thermodynamic equilibrium. The data in Table 5.4 supports this assumption. A series of TGA tests were carried out to determine the critical H_2O/CH_4 for onset of carbon formation [389]. The results shown in the table support the "principle of equilibrated gas". The calculated affinity for carbon formation ($-\Delta G_c^o$) from the equilibrated gas (approximately 4 kJ/mol) from graphite data is comparable with the deviation to be expected from the effect of the whisker structure.

Table 5.4 Carbon formation and equilibrated gas [389].
Thermogravimetric studies at 1 bar abs. CO_2/CH_4. The CH_4 flow was increased stepwise until on-set of carbon formation. Then the CH_4 flow was decreased for removal of carbon. $(H_2O/C)_{exit}$ was determined by interpolation.

Catalyst temperature (°C)	613	704	812	662
H_2O/CH_4 (critical)	1.2	1.1	0.88	1.75 [a]
$p_{H_2}^2 / p_{CH_4}$ actual gas (bar)	0.034	0.055	0.161	0.438
$p_{H_2}^2 / p_{CH_4}$ equilibrated gas (=Qr,eq) (bar)	2.21	7.22	10.99	3.06
K_p, graphite data	2.622	8.18	24.64	4.993
$-\Delta G_c^0 = RT \ln(K_p / Q_{r,eq})$ (kJ mol^{-1})	1.4	1.0	(7.3)	3.8

a) CO_2/CH_4

The equilibrium calculations result in no unambiguous upper or lower carbon limits. There may be upper as well as lower carbon limits for certain conditions. In general, however, as illustrated in Figure 5.14, there is a tendency for upper carbon limits from steam reforming and lower carbon limits for CO_2 reforming.

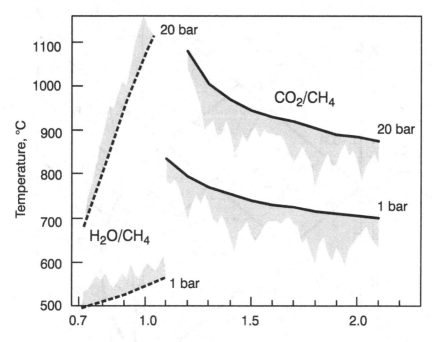

Figure 5.14 Carbon limits for steam and CO_2 reforming. Grey area indicates conditions for carbon formation.

One way of representing areas for potential for carbon formation is demonstrated in Figure 5.15 [152]. The diagram illustrates that CO_2 reforming is more critical than steam reforming.

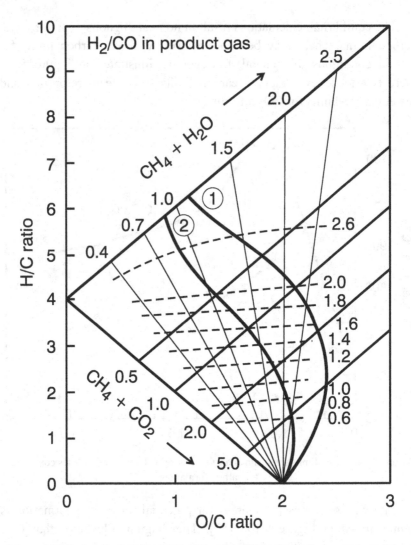

Figure 5.15 Carbon limit diagram. Principle of equilibrated gas. Carbon is formed for conditions to the left of the curves. Curves 1 and 2 represent graphite data and whisker carbon (Appendix 2), respectively. The dotted lines show feed gas compositions leading to product gas with indicated H_2/CO ratios [152]. Reproduced with the permission of Gulf Publishing.

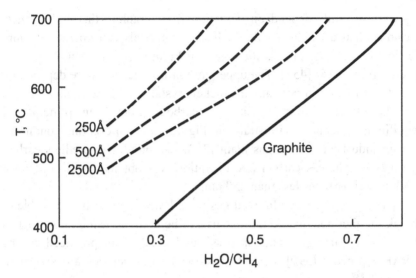

Figure 5.16 Carbon limits at various nickel particle sizes. Preformer conditions. Thermodynamic potential for carbon to the left of the curves [451]. Reproduced with the permission of ACS.

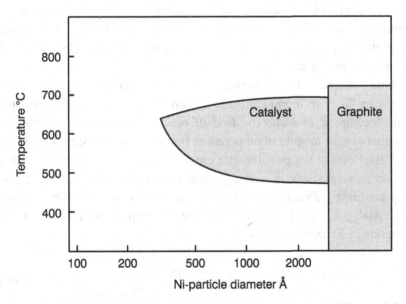

Figure 5.17 Principle of equilibrated gas. Carbon limit temperatures and Ni-crystal size [382]. Conditions: H_2O/CH_4=1.8, CO_2/CH_4=2.2, P=18 bar abs.

Carbon limits depend on deviation from graphite thermodynamics, meaning that a catalyst with small nickel crystals can operate at more critical conditions. This is illustrated in Figure 5.16 for conditions in a prereformer [415] [451]. The upper carbon limit temperature depends on the steam-to-carbon ratio and the nickel crystal size of the catalyst.

The effect of particle size on carbon limits from principle of equilibrated gas is also illustrated in Figure 5.17, representing conditions for an industrial oxo-syngas plant [382]. Graphite data predicts carbon formation, whereas carbon-free operation was obtained with a catalyst with nickel particles less than 250 nm.

The principle of equilibrated gas is no law of nature. It is possible to break the thermodynamic limit. This can be done by using noble metals [396] or by using a sulphur passivated catalyst as practiced in the SPARG process [390] (refer to Section 5.5). Examples are shown in Figure 2.18.

5.2.4 Principle of actual gas and steady-state equilibrium

The principle of equilibrated gas predicts conditions where carbon formation is expected (except for noble metals and SPARG). It does not guarantee that carbon is not formed if the principle predicts no potential in the equilibrated gas.

One example is the formation of carbon in high flux reformers [389] operating far from thermodynamic carbon limits. It means that methane may decompose to carbon instead of reacting with steam to form the required syngas in spite of no potential for carbon in the equilibrated gas. This is of course not possible in a closed system, but in an open system carbon may be stable in a steady state and the accumulation of carbon may continue [389]. This risk may be assessed by the so-called *criteria of actual gas*, which for the methane decomposition reaction as in Equation (5.5) can be written as:

$$-\Delta G_c = RT \ln\left[\frac{K_{eq}}{Q_{r,actual\,gas}}\right] = RT \ln K_{eq}\left[\frac{p_{CH_4}}{p_{H_2}^2}\right]_{actual\,gas} \quad (5.6)$$

$$-\Delta G_c > 0 \Rightarrow \text{Carbon}$$

This approach does not apply to the steam reforming of higher hydrocarbons, because in this case the decomposition into carbon is irreversible.

The analysis of the risk of carbon formation should take account of the strong radial temperature gradients in the reformer tube (refer to Chapter 3). As there are almost no radial concentration gradients, the situation is equivalent to heating up a gas with fixed composition to a temperature close to that at the tube wall. This means a higher potential for the endothermic decomposition of methane when approaching the tube wall compared to the values based on the average gas temperature. The potential for carbon formation can be analysed in detail by means of a two-dimensional reactor model as illustrated in Figure 5.18.

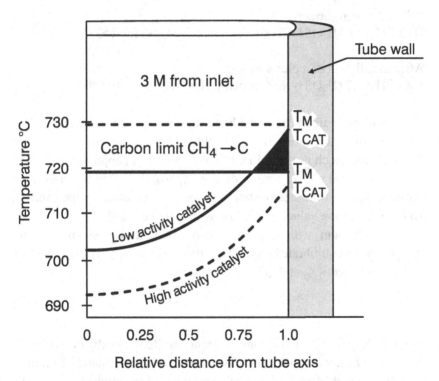

Figure 5.18 Radial gradients and carbon limits [389]. Steam reforming of methane (H_2O/CH_4=3.5, P=33 bar) [389]: 3 meters from tube inlet. T_M is the carbon limit temperature for methane decomposition. There is potential for carbon formation when the catalyst temperature $T_{CAT} > T_M$. Reproduced with the permission of Springer.

For the high activity catalyst, T_{CAT} remains below T_M, i.e. there is no potential for carbon formation. For the low activity catalyst, the mean catalyst temperature (not shown) is less than T_M, but the actual catalyst temperature, T_{CAT}, becomes higher than T_M close to the tube wall, indicating the risk of carbon formation.

The principle of actual gas predicts carbon at inlet conditions of a conventional steam reformer (500°C). However, when a prereformer is installed, there is no potential for whisker carbon at the catalyst inlet, as illustrated below.

Example 5.2
Inlet layer – tubular reformer (P=31 bar abs).

Without a prereformer:
H_2O/CH_4=1.8 (H_2O/H_2=81) which yields $-\Delta G_c$>0 for T>375°C

After installation of a prereformer:
H_2O/CH_4=1.7 (H_2O/H_2=6.9) which yields ΔG_c>0 for T>540°C

In practice, carbon from methane may not be formed below 600–650°C because of low reaction rate, although predicted by the principle of actual gas, which is a simplified (but conservative) approach.

As mentioned above, a more detailed approach should be based on a kinetic analysis of the steady-state activity, $a_{c,s}$, of carbon on the catalyst. Above a saturation value, $a_{c,s}$, carbon will nucleate [389].

At equilibrium with a given composition at the gas phase (not necessarily an equilibrated gas), the carbon activity $a_{c,eq}$ is expressed by the ratio between $K_{p,qr}$ and Q for the reaction:

$$a_{c,eq} = K_{eq} \frac{p_{CH_4}}{p_{H_2}^2} \tag{5.7}$$

in which K_{eq} is the equilibrium constant for Reaction R6 in Table 5.2 (based on whisker carbon). Although, in principle, $a_{c,eq}$ should be referred to carbon adsorbed on the step site on nickel, for simplicity it will be referred to graphite in the following.

The steady-state activity, $a_{c,s}$, can be expressed by balancing the rate of carbon formation without the presence of steam with the rate of the gasification of adsorbed carbon atoms [389]:

$$R_c = k_c p_{CH_4} p_{H_2}^{\alpha_{H_2}} - k_{-c} a_c p_{H_2}^{2+\alpha_{H_2}} = k_{-c} p_{H_2}^{2+\alpha_{H_2}} (a_{c,eq} - a_c) \qquad (5.8)$$

$$R_g = k_g a_c [\text{mari}] = k_g a_c \frac{K_w (p_{H_2O}/p_{H_2})}{1 + K_w (p_{H_2O}/p_{H_2})} = k_g a_c K_w' \left(\frac{p_{H_2O}}{p_{H_2}}\right)^\alpha \qquad (5.9)$$

in which [mari] is the surface concentration of the most abundant reaction intermediate, v.i.z. O-*.

TGA studies on methane decomposition [389] indicated a kinetic order for hydrogen $\alpha_{H2}=-3.5$. A simplified expression can be derived for $a_{c,s}$, i.e. when $R_g=R_c$. With $a_c=a_{c,s}$ and $\alpha_{H2}=-3.5$, Equations (5.8) and (5.9) yield:

$$a_{c,s} = \frac{a_{c,eq}}{1 + \frac{k_g}{k_{-c}} K_w' \left(\frac{p_{H_2O}}{p_{H_2}}\right)^\alpha p_{H_2}^{1.5}} \qquad (5.10)$$

Carbon may be formed when $a_{c,s}>1$. This means that carbon may be eliminated even when the actual gas shows affinity ($a_{c,eq}>1$) for decomposition of methane. An alternative approach was made by Froment et al. based on the parameters determining the rate of whisker growth being zero [469].

The different carbon limits discussed are summarised in Figure 5.19 [17]. Above a limit A', there is no potential for carbon in the actual gas. This is never the situation if higher hydrocarbons are present (refer to Section 5.3).

At a given temperature and for a given hydrocarbon feed, carbon will be formed when the steady-state activity for carbon $a_{c,s}>1$ (Equation 5.10) or below a critical steam-to-carbon ratio indicated by the carbon limit A in Figure 5.19. This critical steam-to-carbon ratio increases with temperature. The principle of actual gas is a simple tool for assessing the risk of carbon formation. It is conservative as it neglects the denominator in Equation (5.10). By promotion of the catalyst, it is possible to push this limit to the thermodynamic carbon limit B, reflecting the principle of equilibrated gas.

The thermodynamic carbon limit B is a function of the composition of the feed gas (atomic ratio O/C and H/C) and total pressure. Limit B can be shifted by using a catalyst with smaller nickel crystals, thus allowing operation under conditions not otherwise possible. By using noble metals and sulphur passivation, it is possible to push the carbon limit beyond limit B to carbon limit C.

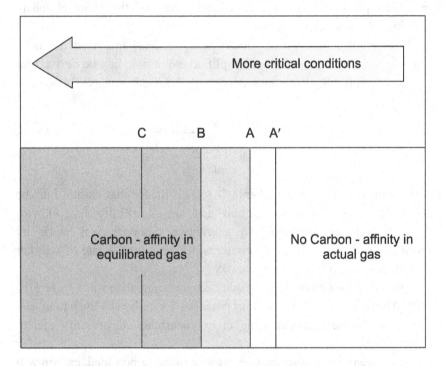

Figure 5.19 Summary of carbon limits in steam reforming [17].
A': no potential for carbon in actual gas. A: critical steam-to-carbon ratio (carbon formation on Ni). B: thermodynamic limit potential for carbon in equilibrated gas. C: carbon limit with "ensemble control" or noble metals. Reproduced with the permission of Elsevier.

For a given gas composition and pressure and no catalyst present, there is a temperature below which there is a tendency for CO decomposition and initiation of metal dusting.

5.3 Steam reforming of higher hydrocarbons

5.3.1 Whisker carbon in tubular reformer

The higher hydrocarbons may lead to carbon formation by all three mechanisms (Table 5.1). Thermodynamics will predict carbon formation as long as the higher hydrocarbons are present. Carbon may be stable in a steady state in spite of the principle of equilibrated gas (refer to Section 5.2.4).

The risk of carbon formation may be assessed by the critical steam-to-hydrocarbon ratio [384] [389]. This decreases with temperature and depends on the type of hydrocarbon and the type of catalyst.

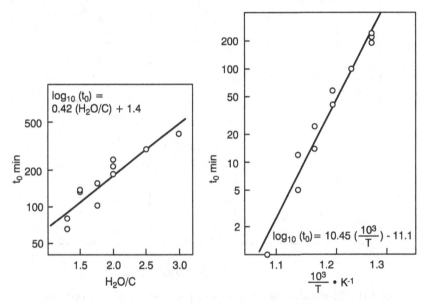

Figure 5.20 Carbon from n-heptane. TGA studies [381].
(a) Induction period t_0 and H_2O/C. (b) Induction period, t_0 and temperature.

The risk of carbon formation can be assessed by analysing the parameters determining the induction time for nucleation of carbon (Equation (5.1), Figure 5.4) by requiring this to be infinite. If so, the

TGA data in Figure 5.20 [381] for the impact of steam-to-hydrocarbon ratio and temperature leads to the empirical expression [389]:

$$(H_2O/C_nH_m)_{crit} = -\frac{A}{T} + B \qquad (5.11)$$

Carbon formation is to be expected if the actual steam-to-hydrocarbon ratio is lower than the critical ratio as illustrated in Figure 5.21 [384] [389].

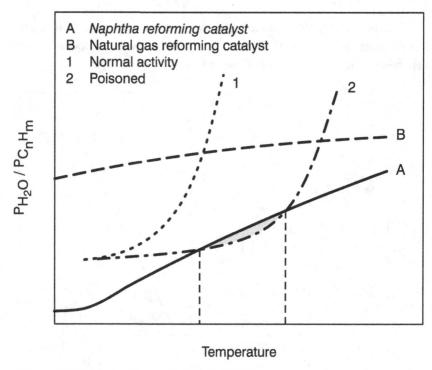

Figure 5.21 Critical and actual H_2O/C_nH_m [384]. Reproduced with the permission of Chem.Eng.Prog.

The critical ratio increases with temperature (Equation 5.11) and depends on the type of catalyst. The actual ratio increases as the hydrocarbons are being converted. As for the principle of actual gas, the comparison of actual and critical ratios should be carried out for any position in the reformer tube. One example is shown in Figure 5.22 [386].

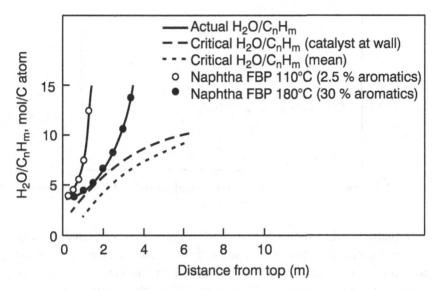

Figure 5.22 Actual critical steam-to-carbon ratios in NH_3 plant steam reformer [386].

The critical steam-to-carbon ratio depends on feedstock composition. As shown in Figure 5.4 (Section 5.2.1), the contents of aromatics and olefins may be harmful. The feedstock will rarely contain olefins, but even traces of ethylene may be critical. Ethylene may be formed by thermal cracking in preheaters, from oxidative coupling of methane if oxygen (air) is added to the feed or from dehydration of ethanol if added to the feed.

A kinetic analysis [389] [415] [425] of the simplified reaction sequence in Figure 5.23 leads to an expression for the steady-state activity, $a_{c,s}$ of carbon on the nickel surface being less than one:

$$a_{c,s} = \frac{k_A}{k_H} \frac{k_c}{k_g \cdot K_w} \cdot \frac{p_{C_nH_m}}{p_{H_2O}} \cdot p_{H_2}^{\alpha_{H2}} \qquad (5.12)$$

Reactants		Product
$C_nH_m + 2*$	$\xrightarrow{k_A}$	$C_nH_y*_2$
$C_nH_y*_2$	$\xrightarrow{k_H}$	$C* \longrightarrow$ whisker carbon
$C* + OH*$	$\xrightarrow{k_g}$	Gas
$H_2O + *$	$\xleftrightarrow{K_W}$	$OH* + \tfrac{1}{2}H_2$

Figure 5.23 Simplified reaction sequence [384], [415]. Reproduced with the permission of Elsevier.

Equation (5.12) shows that the risk of carbon is increased by high adsorption rate (k_A) as for alkenes and low hydrocracking rate as for aromatics. The other constants are related to the catalyst properties. Carbon formation is depressed by a strong adsorption of steam (high K_W) reflected by negative orders with respect to steam. k_c may be decreased by blocking the nucleation sites for carbon as discussed in Sections 5.3.2 and 5.5.

5.3.2 Catalyst promotion

The formation of whisker carbon may be eliminated by catalyst promotion [17] [425]. Steam adsorption is enhanced by the promotion with potassium [379] [381] as reflected by a negative reaction order with respect to water, $α_{H2O}$ (Section 3.5.2), or high value of K_w (Equation 5.12). It is well known that alkali promotes gasification of carbon and coke precursors [224] [374] and alkali-promoted nickel catalysts are efficient for carbon-free reforming of higher hydrocarbons [22] [370] [389]. A similar effect of alkali is observed for suppressing carbon formation in CO_2 reforming [241].

Potassium carbonate is the stable compound on the catalyst, but it will be decomposed by reaction with steam resulting in a partial pressure of the hydroxide over the catalyst and in the reformer effluent. In severe circumstances the loss of alkali may result in plugging of downstream equipment and also lead to stress corrosion [389]. To minimise the alkali loss, reformers are typically loaded with a combi-charge with a low alkali (or alkali-free) catalyst in the bottom. There are different methods of reducing the evaporation of alkali. This may be by binding a reservoir alkali such as $KAlSiO_4$ [22] or later as β-alumina [370] which hydrolyses slowly, or bound to a calcium aluminate support.

As discussed in Section 4.3.1 the presence of alkali results in a significant decrease in the activity reflected by a lower pre-exponential factor. This is not the case with active magnesia, which shows the promoting effect of enhanced steam adsorption (although smaller than that for alkali promotion), but without the loss of activity [380] [381]. As a result, a Ni/MgO catalyst is able to process liquid hydrocarbon feedstocks, even kerosene and diesel if properly desulphurised [405] [427]. The method of preparation of the Ni/MgO catalyst is critical to achieving the promoter effect [381] [415] as illustrated in Figure 5.24.

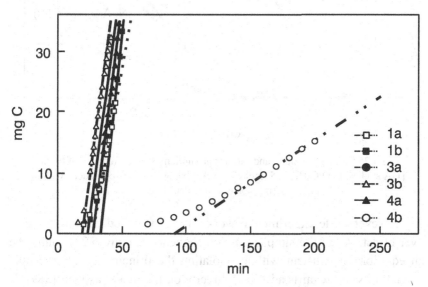

Figure 5.24 Carbon formation (H_2O/nC_7H_{16}=2.0 mol/C-atom). Ni/MgO catalysts. Various preparation methods [381], [542]. Method 4b appears advantageous.

The high activity of the alkali-free catalyst means that the higher hydrocarbons are converted before the process gas reaches the temperature level for thermal cracking, whereas the alkali-promoted catalyst may cope with coke precursors formed by cracking.

Similar promoter effects have been reported for La_2O_3 and Ce_2O_3 for CO_2 and steam reforming as shown in Figure 5.25 [403]. Just as for MgO, spill-over of OH species from the support may result in increased coverage of OH on the nickel surface [79] [178]. An indication of spill-over was found in CO_2 reforming of methane over Pt/ZrO_2 catalyst [55], as the rate correlated with the length of the metal support perimeter.

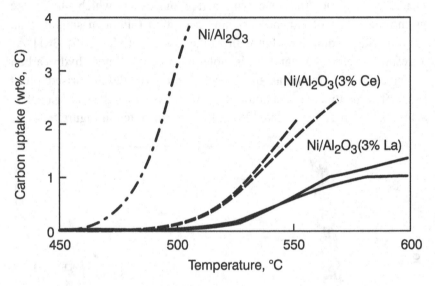

Figure 5.25 Carbon formation and catalyst promotion. Steam reforming of n-butane. TGA studies. $H_2O/C_nH_{10}=1.5$ (mol/C-atom), 1 bar abs [403]. Reproduced with the permission of Japan Petr. Inst.

However, little experimental work has been carried out on the spill-over effect. K_w in the simple reaction sequence (Figure 5.23) cannot be an equilibrium constant without violating the principle of microscopic reversibility as steam is adsorbed directly on the nickel as well [389]. If one reaction step is at equilibrium, this must be true for the others involved in the spill-over as well as illustrated in Table 5.5. This corresponds to micro-calorimetric measurements and isotopic exchange of studies [17] showing that the best supports are those with high *rate* for dissociated steam adsorption rather than those with high adsorption equilibrium constants.

Table 5.5 Sequence for spill-over of steam [17].

Reaction
H_2O + *support $\leftrightarrow$ H_2O * support
H_2O + *support + *support $\leftrightarrow$ OH * support + H * support
OH * support + *Ni $\leftrightarrow$ OH * Ni + support
H_2O + *Ni + *support $\leftrightarrow$ OH * Ni + H * support

Another approach for inhibition of carbon formation is to retard the full dissociation of the hydrocarbon into adsorbed carbon atoms. The presence of potassium on a nickel catalyst results in a significant increase in the induction period for nucleation of carbon whiskers (refer to Figure 5.6). It means that potassium works as promoter also without the presence of water.

The retarded dissociation of the hydrocarbon was the explanation in a series of studies of the impact of a number of oxides such as La, Ce, Ti, Mo, W [70] [80] [101]. For a Pt/TiO_2 catalyst it was concluded that TiO_x layers on the platinum surface were suppressing carbon formation, probably by ensemble control.

Ensemble control is also involved in carbon-free steam reforming on a sulphur passivated catalyst [390] (Section 5.5). Ensemble control was also reported for the addition of Bi [500] or B [526] to nickel and for bi-metallic catalysts such as Ni,Au [50], Pt,Re [367], Pt,Sn [474], and Ni,Sn [217] [247] [431] [456] [545]. Alloying nickel with copper [49] [16] can also decrease the rate of carbon formation, but it is not possible to achieve the same high surface coverage with copper as with sulphur and gold, because copper and nickel forms a bulk alloy with a fixed surface concentration of copper over a wide range as alloy composition.

It was shown earlier (Figure 5.11) that noble metals show higher resistance to carbon formation from methane. This is also true from higher hydrocarbons as illustrated in Figure 5.26 [403].

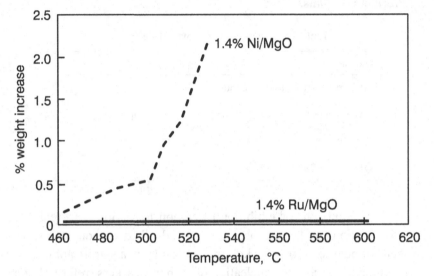

Figure 5.26 Carbon formation. TGA studies. $H_2O/C_4H_{10}=0.5$ on nickel and ruthenium catalysts [403]. Reproduced with the permission of Japan Petr. Inst.

5.3.3 "Gum formation" in prereformers

At low temperatures (<500°C and as applied in prereformers), the rate of adsorption of hydrocarbons (E_a=approximately 40 kJ/mol) becomes higher than that of hydrocracking (E_a=approximately 160 kJ mol^{-1}) [386] [405] as illustrated in Figure 5.27. Hydrocarbon fractions may then accumulate on the nickel surface and slowly be converted into a non-reactive layer ("gum") blocking the nickel surface [322].

The composition of the "gum layer" was examined by extracting the deactivated catalyst with trichlorine methane. After brief laboratory experiments [254] [453], a paraffinic structure of -CH$_2$- chains was identified, whereas extracts from industrially used catalysts showed a high content of poly-aromatics [31] [54]. This indicates that the "gum" deposits are slowly aged to less reactive deposits (Figure 5.28).

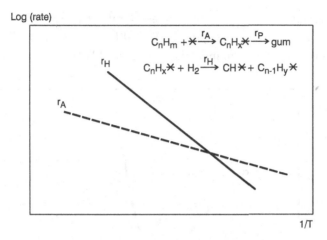

Figure 5.27 Adsorption and hydrocracking of hydrocarbons on nickel surface [529]. For $r_H > r_A$, gum cannot accumulate on the surface.

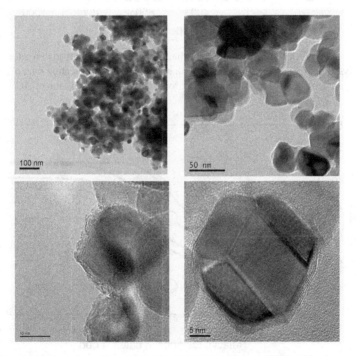

Figure 5.28 Gum formation, Ni-catalyst. HRTEM. Helveg [227]. This picture is consistent with observations [314] of the formation of different types of adsorbed carbon in the methanation reaction (α and β carbon) as also revealed in TPD studies in hydrogen.

The reactivity of the "gum" layer is reflected by TPD studies in hydrogen as illustrated in Figure 5.29 [108] [110]. Similar TPD results were reported earlier from methanation studies [314].

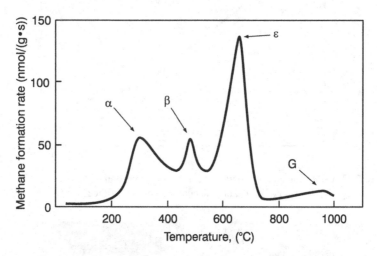

Figure 5.29 TPR (Temperature Programmed Reaction) of an adiabatic prereforming catalyst after six months of operation with naphtha feedstock. Heating rate: 6°C per minute. Christensen et al. [108]. Reproduced with the permission of ACS.

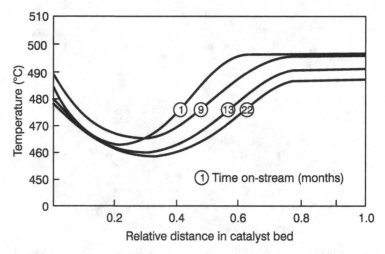

Figure 5.30 Prereforming of naphtha in hydrogen plant. Temperature profiles H_2O/C, p=26 bar. Christensen [110]. Reproduced with the permission of Elsevier.

As a result of the encapsulating gum deposits, the temperature profile in the adiabatic prereformer will move in the flow direction as illustrated in Figure 5.30 [110] for a naphtha-based plant. The movement of the temperature profile will be overlapped by the deactivation caused by sulphur poisoning.

The deactivation rate can be assessed from the profile as a resistance number, R (kg feed processed (g cat)$^{-1}$) [392]. For prereforming, the resistance number R will increase with temperature and hydrogen partial pressure as hydrocracking of surface species takes over. It decreases with the final boiling-point temperature of the feedstock and in particular with the content of aromatics in the feed [322] as shown in Figure 5.31 [453].

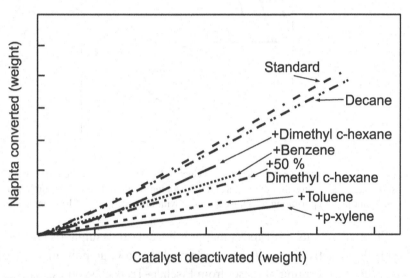

Figure 5.31 Prereforming of various hydrocarbons. Sehested [453]. Amount of naphtha processed versus the amount of catalyst deactivated for experiments with 25% (one experiment with 50%) of the naphtha substituted by another molecule. Naphtha IBP/FBP 80°C/179°C, P=31 bar abs, T_{in}=465–470°C, H_2O/C=2.5, H_2O/H_2=10. Reproduced with the permission of the author.

The movement of the profile may be followed by plotting the length of the reaction zone Z_{90} (corresponding to 90% of the temperature increase as shown in Figure 5.32) as a function of time or the amount of feedstock. The resistance number, R, proportional to the inverse slope, can be expressed as weight of feed per weight of deactivated catalyst.

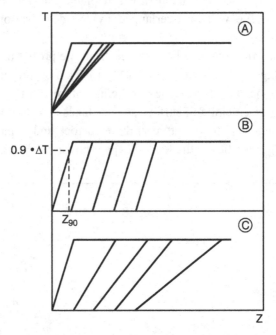

Figure 5.32 Typical temperature profile movements. Profiles at given time intervals and fixed load. A: sintering at high pressures. B: Poisoning from inlet layer. C: Combined sintering and poisoning [392]. Reproduced with the permission of Elsevier.

When processing feedstocks that are difficult to desulphurise such as diesel, the deactivation will be overlapped by sulphur poisoning [350] [415]. Steam reforming of diesel from Fischer–Tropsch synthesis is easy, as it is sulphur-free and contains no aromatics.

The reactivity of the "gum layer" with hydrogen is reflected by lower deactivation rates (high R) when operating with hydrogen-rich feed gas (for instance by recycling product gas [140]) and as illustrated in Figure 5.33.

R may not be constant with time (as in Figure 5.32 (B)) and it may be useful to make the plot in a double logarithmic diagram [392] to identify the nature of development. The typical situation results in nearly constant values of R after an initial period with faster deactivation. Typical values of R are within the range of 5–20 kg/g. Smaller values would reflect the need for catalyst replacement more than once a year for typical space time yields. A similar deactivation trend is observed also for methanation for SNG (β-deactivation) [205], and for methane to gasoline synthesis [492] [531].

The risk of carbon (or gum) formation from higher hydrocarbons can be analysed separately from the potential for carbon predicted by the "principle of equilibrated gas".

Note
A pilot test was carried out with naphtha as feed. The critical H_2O/C_nH_m was assessed to be around 1.3. As shown in Figure 5.33, it was possible to operate at a $H_2O/C_nH_m=1.0$ in the feed gas by using a hot (ejector) recycle, which brought H_2O/C_nH_m up to 1.6. The principle of equilibrated gas predicts carbon-free operation above O/C=0.9.

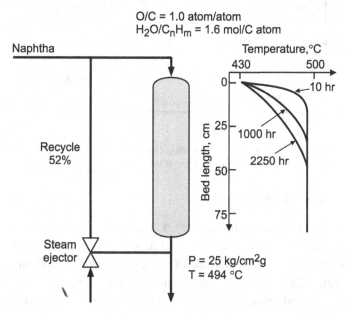

Figure 5.33 Prereforming, bench-scale test (duration 2300 hours) [386]. The low temperatures imply a low H_2O/C_nH_m ratio. Recycle operation results in an overall O/C ratio of 1.0 close to the carbon limits: O/C=0.9 (as calculated from principle of equilibrated gas).

In conclusion, there is a temperature window [405] for carbon-free operation on higher hydrocarbons as illustrated in Figure 5.34. For a given steam-to-carbon ratio, there is a temperature above which whisker carbon (or pyrolytic carbon) (refer to Section 5.3.4) is formed and a temperature below which the catalyst is deactivated by gum formation.

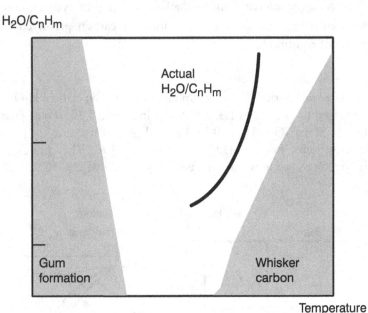

Figure 5.34 Temperature "window" for carbon-free operation [405]. Reproduced with the permission of Elsevier.

5.3.4 Carbon from pyrolysis

At high temperatures (above 600°C), the thermal cracking of higher hydrocarbons becomes significant and as a result pyrolytic carbon may be deposited on the outer surface of the catalyst pellets and on the tube wall [389]. This is typically the situation when the catalyst has become sulphur poisoned. Non-converted hydrocarbons pass unconverted over the deactivated catalyst to the hotter part of the reformer tube where they are cracked.

The formation of pyrolytic coke is complex [200] [499]. In general, it is initiated by gas-phase reaction forming unsaturated molecules and radicals which may react further by polymerisation and dehydrogenation reactions. Gas-phase soot as observed in partial oxidation (Section 1.3.1) may be formed. If the partial pressure of the macro-molecules increases, this may lead to condensation of droplets which are further dehydrogenated into solid coke [499]. In tubular reformers, the tar-like macro-molecules are deposited on the tube surface where they may be converted to dense shales or as deposits encapsulating the catalyst particles and eventually filling out the void between the particles as shown in Figure 5.35.

Note
The sample in Figure 5.35 was taken from the bottom of the reactor above the exit grid which forces the gas to leave close to the tube wall. The conic shape of the coke reflects the hydrodynamics with the coke being deposited in the area with shadowing effect from the grid – i.e. high residence time and high KSF (refer to Section 4.3.3).

Figure 5.35 Pyrolytic coke formed in catalyst bed. ZrO_2/K catalyst for "catalytic steam cracking" (refer to Section 4.3.3).

The tube surface as well as the surface acidity of the catalyst may enhance the initial deposits of coke [389] [499].

The formation of coke by pyrolysis increases strongly with temperatures as indicated in Figure 5.36 [405] which shows results from TGA measurements on cracking of ethylene. With no catalyst, the activation energy was estimated to 458 kJ/mol. In the presence of an alkali-promoted support, (ZrO_2, 0.5% K) (refer to Section 4.3.3), the coking was retarded, probably by the promotion of alkali of the reverse carbon-forming reaction R8 in Table 5.2 as shown in Reaction (5.13) below:

$$C + H_2O \leftrightarrow CO + H_2 \quad (-\Delta H^o_{298} = -132 kJ/mol) \tag{5.13}$$

A similar mechanism may work when using an alkali-promoted nickel catalyst [22].

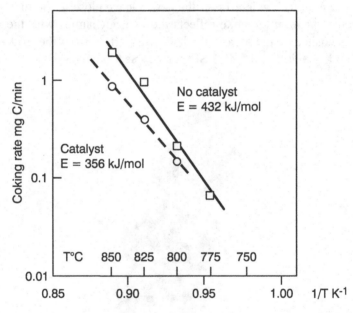

Figure 5.36 Coking from ethylene. TGA measurements [405]. Flows (mol/h): C_2H_4=0.2, H_2O=0.16, N_2=0.1, H_2S=10^{-4}). Tube diameter: 18 mm, basket diameter: 10 mm, 0.85 g catalyst (ZrO_2, 0.5% K). Reproduced with the permission of Elsevier.

The risk of coke formation is to be analysed in the same way as for a steam cracker [481]. In fact, a steam/naphtha reformer with a completely deactivated nickel catalyst will work as a steam cracker producing olefins as discussed in Section 4.3.3.

It is generally agreed that the gas film at the tube wall is overheated and acts as a source of radicals and coke precursors. In steam crackers, the tube skin temperature is the most important parameter determining the rate of coke formation. Moreover, the coking reactions are related to the so-called kinetic severity function (KSF) (refer to Section 4.3.3).

KSF describes the residence time – temperature history of the reactants in a way that is consistent with kinetics. This means that for a given temperature profile, the risk of carbon formation is increased with higher residence time (refer to Figure 5.35). The catalyst filling has an influence on these parameters, first by changing the film volume and secondly by influencing the residence time distribution via the void fraction.

5.3.5 Regeneration of coked catalyst

Rapid formation of whisker carbon may result in "spalling" of the catalyst pellets forming dust or complete breakdown of the pellet. However, in case of slow build-up of carbon, the catalyst pellet may stay intact – even with high amounts of carbon as illustrated in Figure 5.37.

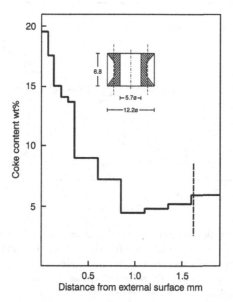

Figure 5.37 Radial carbon profile in ring-shaped reformer catalyst [381].

The catalyst pore system may, however, be blocked if the catalyst does not contain macropores [381].

The newly formed carbon whiskers are very reactive when there is a thermodynamic potential for gasification (reverse Reaction R6 in Table 5.2). This was demonstrated in the TGA tests close to equilibrium. *In situ* electron microscopy [33] has shown how nickel crystals may "eat" channels through carbon by a reverse whisker growth mechanism. Similar observations have been made for catalytic filters for car exhaust where cerium oxide crystals react with soot deposits [460]. The "gum" layer is also reactive in hydrogen [205].

Newly formed carbon may be removed by steaming (600–700°C) [381] according to Equation (5.13) which shows potential for gasification at a $H_2O/H_2=10$, which is sufficient to keep the nickel catalyst in a reduced state (refer to Chapter 4). In practice fresh whisker carbon can be removed by simply increasing the steam-to-carbon ratio.

With ageing, the whisker structure collapses and the reactivity diminishes. The same is true for the "gum" layer and eventually the coke deposits can only be removed by means of oxygen (air).

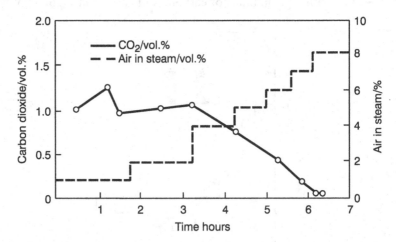

Figure 5.38 Regeneration for carbon formation in an industrial reformer [388] [389]. Analyses of dry exit gas. Approximately 0.4 tonne of steam per tonne of catalyst per hour. T_{cat}=450–600°C. P ~ 6 bar. The progress of the regeneration can easily be followed by analysis of the exit gas from the reformer. The additions of air is increased as the production of CO_2 decreases. In this way, overheating can be controlled. Reproduced with the permission of Brill and Springer Verlag.

It has been shown [440] [518] for coke deposits on cracking catalysts that the reactivity after oxidation depends mainly on the surface area of the coke and that the rate quickly becomes diffusion limited with a risk of overheating the catalyst pellet. In practice, the burn-off of coke can easily be performed by adding a few percent of air to the steam flow at temperatures above approximately 450°C as illustrated in Figure 5.38 [388] [389].

If whisker carbon has resulted in breakage of the catalyst pellets, a high pressure drop will remain after the removal of carbon.

Pyrolytic coke is dense with a very small surface area. Hence, it is not possible to remove after ageing by steaming [499] and it may still be difficult to burn off in air. Steam crackers are regenerated by frequent "on-line decoking" with steam [374] instead of off-line decoking with steam and air.

Gum formation on the prereforming catalyst can in principle be removed by high-temperature treatment in hydrogen [205], but in practice it is not possible to establish the required conditions.

5.4 Sulphur poisoning of reforming reactions

5.4.1 Chemisorption of hydrogen sulphide

Group VIII metals are susceptible to sulphur poisoning [39] and sulphur must be removed from the feed stream. Natural gas may be cleaned over active carbon, but normally all hydrocarbon feedstocks are treated by hydro-desulphurisation over CoMo (or Ni/Mo) catalyst followed by sulphur absorption on a zinc-oxide absorption mass (refer to Section 1.5.1).

Poisoning effects are often correlated with the poison concentration in the feed stream, which, of course, is the important parameter in practical operation. However, in a more detailed analysis this approach can hardly be justified for other than isothermal tests in gradientless reactors. The adsorption equilibrium depends on temperature and composition of the gas phase which varies through the reactor as well as within the single catalyst pellet. Therefore it appears more rational to correlate the deactivation with the amount of poison present on the catalyst rather than with the poison concentration in the feed stream. However, the correlation between sulphur in the feed and in the catalyst may be complex as illustrated below.

Under reforming conditions, all sulphur compounds will be converted to hydrogen sulphide, which is chemisorbed on the metal surface (Me):

$$H_2S + Me \leftrightarrow Me-S + H_2 \qquad (5.14)$$

This takes place at H_2S/H_2 ratios far below those required for formation of bulk sulphides [376].

Nickel is the most sensitive metal as illustrated in Figure 5.39 [523].

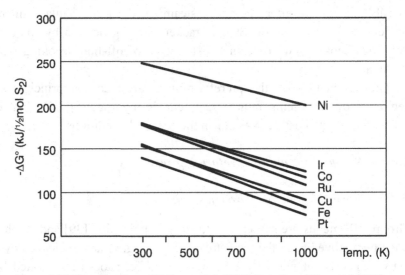

Figure 5.39 Chemisorption of H_2S on metals [425]. Data from Wise *et al.* [523]. Reproduced with the permission of Wiley.

Hydrogen sulphide chemisorbs dissociatively on nickel. Stable saturation uptakes of sulphur are observed at ratios of H_2S/H_2 from 10 10^{-6} up to close to 1000 10^{-6} above which bulk sulphide is formed.

The thermodynamics of nickel sulphide phases [372] indicates that the formation of a bulk phase sulphide (Ni_3S_2) at temperatures around 500–700°C requires a H_2S/H_2 concentration ratio in the order of 10^{-3}. This ratio is about 100–1000 times above what would normally cause poisoning at those temperatures.

The saturation layer has been described in Section 4.2.1. Below a certain H_2S/H_2 concentration ratio the saturation layers become unstable and the equilibrium coverage is dependent on the H_2S/H_2 concentration ratio and temperature [376]. This is normally described in terms of an adsorption isotherm and the isoteric heat of chemisorption.

5.4.2 Chemisorption equilibrium

Attempts to correlate data with a Langmuir-like isotherm have not been successful [376]. This failure is not surprising in view of the mechanism of the adsorption violating the assumptions for the Langmuir isotherm. However, the experimental data shown in Figure 5.40 may be represented by linear isobars [15] expressed by:

$$\theta_s = 1.45 - 9.53 \cdot 10^{-5} \cdot T + 4.17 \cdot 10^{-5} \cdot T \ln\left(\frac{p_{H_2S}}{p_{H_2}}\right) \quad (5.15)$$

This reflects a Temkin isotherm as shown below, but it is also apparent from Figure 5.40 that Equation 5.15 is not valid for θ_s close to one and at low coverage it represents a big extrapolation. The deviations at low temperatures may be explained by adsorption on the support [377].

Note
A Langmuir isotherm satisfies the ideal adsorption conditions and can be written as:

$$\frac{\theta_s}{1-\theta_s} = B \, e^{-\frac{\Delta H_{ads}}{RT}} \left(\frac{p_{H_2S}}{p_{H_2}}\right) \quad (5.16)$$

If it is assumed that the adsorption heat is independent of the sulphur coverage θ_s, a Temkin isotherm results:

$$-\Delta H_{ads} = -\Delta H_{o,ads}(1 - a\theta_s) \quad (5.17)$$

If these two equations are combined, the left-hand side still reflects the Langmuir assumption that the rate of adsorption and desorption is proportional to the number of free and occupied sites, respectively. For dissociative adsorption as in Equation (5.14) it is likely that the desorption rate of H_2S is proportional to the non-occupied sites available for hydrogen adsorption. If so, it is more appropriate to represent the isotherm as a Temkin isotherm, where the heat of adsorption is independently calculated from Equation (5.14) as:

$$\Delta G_{ads} = \Delta H_{ads} - T\Delta S_{ads} = -RT \cdot \ln\left(\frac{p_{H_2}}{p_{H_2S}}\right) \quad (5.18)$$

Insertion of (5.17) in Equation (5.18) and rearrangement give:

$$\ln\left(\frac{p_{H_2S}}{p_{H_2}}\right) = \Delta H_{o,ads}(1 - a\theta_s) - \frac{\Delta S_{ads}}{R} \quad (5.19)$$

Which after solving for θ_s can be written as

$$\theta_s = \frac{1}{a} - \frac{T\Delta S}{a \cdot \Delta H_{o,ads}} - \frac{RT}{a \cdot \Delta H_{o,ads}} \ln\left(\frac{p_{H_2S}}{p_{H_2}}\right) \quad (5.20)$$

Equation (5.15) corresponds to the constants: $a=0.69$, $\Delta H_{o,ads}=-289$ kJ/mol H_2S, and $\Delta S_{ads}=-0.019$ kJ/mol H_2S/K in Equation (5.20).

Equation (5.15) implies an entropy of adsorption being independent of θ_s which is in accordance with a "bulk"-like behaviour of the chemisorbed layer (i.e. a two-dimensional sulphide).

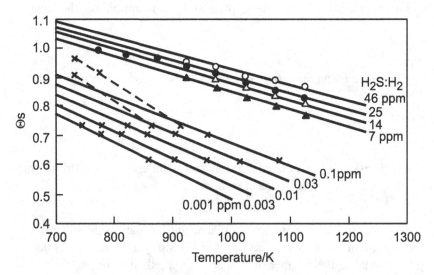

Figure 5.40 Isobars for chemisorption of H_2S on Ni catalysts [15]. The linear curves represent Equation (5.15). Reproduced with the permission of Elsevier.

The initial heat of chemisorption of 289 kJ/mol H_2S corresponds to 189 kJ/mol H_2S at $\theta_s=0.5$ (or 104 kJ/½ mol S_2). These values are in the range of what is reported elsewhere in literature [39] [523] and well above the heat of formation of bulk sulphur (Ni_3S_2), 83 kJ/mol H_2S. This demonstrates that the chemisorbed sulphur is strongly bound to the nickel surface and that the "two-dimensional sulphide layer" can be stable at conditions where bulk sulphide does not exist.

In principle, it should be expected that Reaction (5.14) is influenced by the competing chemisorption of water:

$$*+H_2O \leftrightarrow *-O+H_2 \qquad (5.21)$$

However, it was shown that the sulphur uptake was not affected by the presence of steam up to $H_2O/H_2=10$ [377]. These observations were confirmed by STM experiments [430], which illustrate how H_2S displaces adsorbed oxygen as illustrated in Figure 5.41.

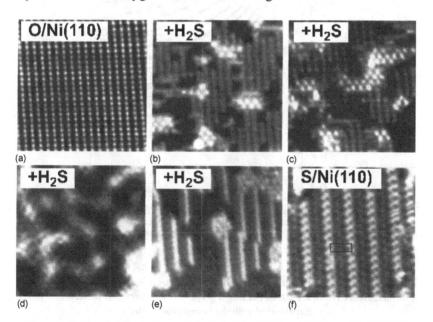

Figure 5.41 Competition between chemisorption of H_2S and H_2O on nickel. STEM studies. Ruan et al. [430]. Reproduced with the permission of the authors and Phys.Soc.

This result may appear surprising as the heat of chemisorption of oxygen is high (approximately 430 kJ/mol O_2). However, the value is less than the heat of chemisorption for sulphur, which is about 480 kJ/mol S_2. This means that Equation (5.15) can be applied to estimate sulphur coverages at equilibrium in the presence of steam as illustrated in Figure 5.42 for a typical ammonia plant reformer [387] [389].

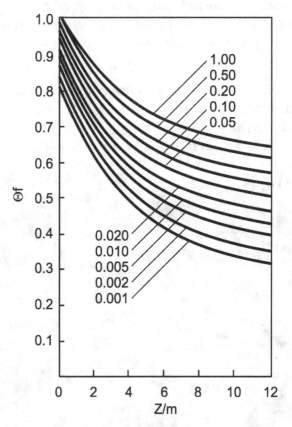

Figure 5.42 Sulphur coverage of nickel surface of catalyst in an ammonia plant tubular reformer, H_2O/CH_4=3.3, P_{exit}=34 bar abs, T_{in}/T_{exit}=505°C/800°C, no prereformer [387]. Reproduced with the permission of Brill.

It was shown by Auger spectroscopy [459] that adsorbed carbon does not overlap into sulphur domains, meaning that carbon is limited to regions that are not already occupied by sulphur [381].

5.4.3 Dynamics of sulphur poisoning

$\theta_S=0.5$ at 500°C corresponds to the concentration ratio $H_2S/H_2=1.6 \; 10^{-12}$. It is evident that a nickel-based prereformer will remove any trace of sulphur. In practice, it would take a very long time to arrive at the equilibrium coverage.

The dynamics of sulphur uptake in a prereformer is like a fixed-bed absorption as seen in a zinc-oxide bed (refer to Chapter 1). However, in a tubular reformer the pore diffusion restrictions in the sulphur adsorption in a single pellet has a complex influence on the transient sulphur profiles in the reactor and a mathematical model [112] [387] [389] is required to evaluate more exactly the time for full saturation and the breakthrough curves of sulphur.

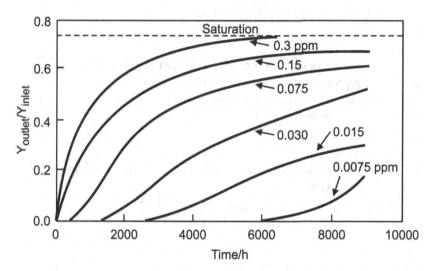

Figure 5.43 Breakthrough curves for H_2S in tubular reformer feed. Model for transient profiles in poisoning of tubular reformer. Naphtha-based ammonia plant. $H_2O/C=3.5$, $P_{exit}=34$ bar abs. $T_{exit}=800°C$. Curves for different S contents in naphtha, molar fractions at outlet/inlet of reformer [387]. Reproduced with the permission of Brill.

Figure 5.43 shows calculated breakthrough curves in a steam reformer of a naphtha-based ammonia plant with no prereformer [387] [389]. The relative concentration of hydrogen sulphide at the tube exit is plotted as a function of time. It is seen that the breakthrough occurs immediately for sulphur contents in the naphtha higher than 0.1 wt ppm, indicating the serous effect of a sulphur peak in the reformer feed on downstream catalysts.

Note
The rate of sulphur uptake in a single catalyst pellet can be described by means of a modified version of the equation for diffusion-controlled elution from a catalyst pellet derived by Goring and de Rosset [210] [381]:

$$\frac{s_t}{s_\infty} = 1 - \frac{6}{\pi^2} \sum_{n=1}^{\infty} \frac{1}{n^2} \exp\left(-\frac{n^2 \cdot D_{eff} \cdot \pi^2 \cdot t}{\varepsilon \cdot (B+1) \cdot R_{eq}^2}\right)$$

$$= 1 - \frac{6}{\pi^2} \sum_{n=1}^{\infty} \frac{1}{2} \exp(-n^2 \beta t)$$

(5.22)

Where
s_t sulphur uptake after a period of t (wt ppm)
s_∞ sulphur uptake at equilibrium (wt ppm)
D_{eff} effective diffusion coefficient (cm^2/s)
ε_p catalyst porosity
ρ_p catalyst particle density
t   run time (s)
B   adsorption coefficient in a linear isotherm mol S/mol H$_2$S in pellet void

$$B = \frac{s_\infty \cdot 10^{-6} \cdot \rho_p}{32} \cdot \frac{P_{H_2S}}{RT} \cdot \varepsilon_p$$

5.4.4   Regeneration for sulphur

The chemisorption process is reversible and in principle sulphur should be removed in a hydrogen stream. However, the driving forces are very small and decreases with decreasing θ_s. And the elution process is subject to diffusion restrictions [381] [389] as reflected by the Goring–de Rosset equation, Equation (5.22) [210]. Even without diffusion restrictions, the sulphur leaving the reactor in a hydrogen stream will decrease exponentially with time as illustrated below [377] [381].

Note
Assuming a linear adsorption isotherm

$$s = B \cdot \frac{P_{H_2S}}{P_{H_2}} = B \cdot \frac{F_{H_2S}}{F_{H_2}} \quad (5.23)$$

in which
s   sulphur content on the catalyst
s_1 sulphur content before regeneration
F   component flow

If s_m is the number of moles of sulphur on a catalyst weight, the sulphur flow leaving the catalyst can be calculated from the differential equation as:

$$-\frac{ds_m}{dt} = \frac{1}{B} F_{H_2} \cdot s = k' \cdot s \quad (5.24)$$

This equation can be integrated from the initial amount s_1 as:

$$\frac{s}{s_1} = e^{-k't} \quad (5.25)$$

Sulphur may be removed by oxidation in steam under controlled conditions [377]. As illustrated in Figure 5.44, no improvement is observed when increasing the H_2O/H_2 ratio, as also expected from equilibrium studies (refer to Section 5.4.2).

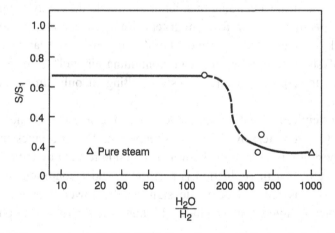

Figure 5.44 Regeneration in H_2O/H_2 [377]. Reproduced with the permission of Elsevier.

However, at H_2O/H_2 above 150–250, a significant change in the regeneration degree of the catalyst is obtained as shown in Figure 5.44. Increased sulphur removal is achieved at a H_2O/H_2 ratio which may be close to the equilibrium constant for the oxidation of the catalyst.

The exit gas shows the presence of sulphur dioxide as well as hydrogen sulphide [377], which may indicate the following reaction pattern:

$$Ni-S+H_2O \leftrightarrow NiO+H_2S$$
$$H_2S+2H_2O \leftrightarrow SO_2 +3H_2 \quad -\Delta H^o_{298} = -207 \text{ kJ/mol}$$
(5.26)

This sequence does not include the chemisorption equilibrium for H_2S/Ni, Equation (5.14), linked to an exponential decrease in sulphur removal, Equation (5.24), and this may be the reason for the improved rate of regeneration.

Equilibrium constants for Reaction (5.26) are shown in Appendix 2. The equilibrium constant at 700°C is $3.3 \cdot 10^{-8}$, which means that even a small amount of hydrogen will inhibit the conversion of hydrogen sulphide. Therefore, sulphur removal following this reaction pattern requires a total oxidation of the catalyst. If some part of the nickel surface is still exposed to the gas, hydrogen formed by Equation (5.14) will cause hydrogen sulphide to be retained at the surface by the chemisorption reaction [377] [389].

A strong temperature dependence of the sulphur removal by steaming [377] may indicate a bonding of sulphur dioxide to the catalyst support at low temperatures. Therefore, high-temperature steaming is normally required to obtain sufficient removal of the sulphur on the catalyst. When the catalysts were exposed to steam containing air (vol%), formation of sulphates took place on all catalysts resulting in only limited sulphur removal.

In principle, it should be possible to remove sulphur retained by the support (or promoter) by means of a reduction. This presumes first that the support will not retain the hydrogen sulphide formed by the reduction and secondly that the reduction is performed under conditions at which nickel oxide is reduced at a rate significantly lower than the rate of formation of hydrogen sulphide. If not so, hydrogen sulphide is chemisorbed at the nickel surface [377] [389]. This might be solved by a two-step regeneration [389].

5.4.5 Impact of sulphur on reforming reactions

Sulphur has a strong impact on the reforming activity of nickel catalysts. The deactivation is strong at small coverages of sulphur and the catalyst is completely deactivated at full coverage ($\theta_s=1$) [389]. The non-linear behaviour is demonstrated in Figure 5.45 [390]. The intrinsic rate of the poisoned catalyst is around two sizes of orders less than that of the non-poisoned catalyst. The rates are compared by referring to free nickel surface and using a simple Maxted model for poisoning [390]

$$R_{sp} = R_{sp}^o (1-\theta_s)^n \tag{5.27}$$

As shown in Figure 5.45, n=3 leads to a reasonable agreement for $r^o{}_{sp}$ with the data for sulphur-free catalyst extrapolated to higher temperature.

Note
The activation energy in the SPARG reaction in the presence of sulphur in Figure 5.45 is estimated to 227 kJ/mol and the value of the sulphur-free test is 110 kJ/mol. This can be explained by the heat of chemisorption of hydrogen sulphide by comparing the rate expressions for the sulphur-poisoned and the sulphur-free catalysts.

The intrinsic rate of the poisoned catalyst could be fitted [390] to the kinetic expression:

$$R_{sp} = k_s e^{\frac{-32100}{T}} \cdot p_{CH_4}^{0.8} \cdot p_{H_2}^{0.3} \cdot p_{H_2S}^{-0.9} \tag{5.28}$$

Rate expressions for sulphur-free catalyst are discussed in Section 3.5.2. Here the expression by Bodrov [59] is used, and the activation energy from Table 3.5 [379] [389]:

$$R_{sp}^o = k_0 e^{\frac{-13000}{T}} \cdot p_{CH_4} \cdot p_{H_2}^{\alpha_{H2}} \tag{5.29}$$

It was shown [390] that the Equation (5.15) for sulphur coverage in the range of interest can be simplified to:

$$1-\theta_S = 0.293 e^{-\frac{4300}{T}} \left(\frac{p_{H_2S}}{p_{H_2}}\right)^{-0.3} \tag{5.30}$$

By combining (5.29) and (5.30):

$$R_{sp} = R_{sp}^{o}(1-\theta_s)^3 = k\, e^{\frac{-25900}{T}} \cdot p_{CH_4} \cdot p_{H_2}^{\alpha_{H2}+0.9} \cdot p_{H_2S}^{-0.9} \qquad (5.31)$$

which corresponds to an activation energy of 215 kJ/mol not too far from the experimental value of 229 kJ/mol, as discussed in [390]. Equation (5.31) is also in reasonable agreement with reaction orders in Equation (5.28).

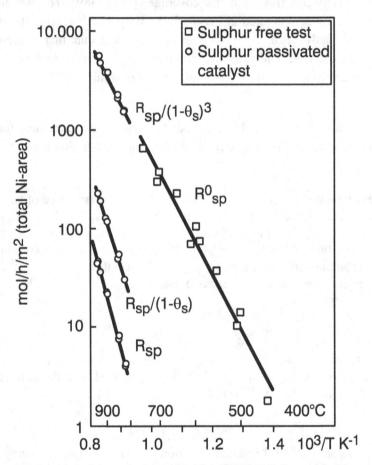

Figure 5.45 Specific activity and sulphur poisoning [390]. Ni/Al$_2$O$_3$ catalyst. The rate R_{sp} is referred to total nickel area, whereas $R^o{}_{sp}$ is referred to free nickel area. Experiment with sulphur: H$_2$O/CH$_4$=1.0, H$_2$O/$_2$=5, H$_2$S/H$_2$=28 10^{-6}, 1 bar abs. Experiment without sulphur. H$_2$O/CH$_4$=0.94, H$_2$O/H$_2$=2.5, 1 bar abs. Reproduced with the permission of Elsevier.

CPO over noble metals is less sensitive to sulphur poisoning than steam reforming, as illustrated by the reactions in Table 5.6 [424]. In the presence of oxygen, sulphur is oxidised to SO_2, but nickel will also be oxidised, which is not the case for rhodium, a typical catalyst for CPO. It means that rhodium remains active as long as oxygen is present. After depletion of oxygen, SO_2 will be reduced to H_2S, which will be chemisorbed on the catalyst and downstream catalysts and anodes. The task is then reduced to removing hydrogen sulphide from the product gas. It means, however, that the endothermic steam reforming reaction is poisoned when the oxygen has been consumed. This is reflected by a temperature increase in the CPO reactor when operating in the presence of sulphur [496].

Table 5.6 Sulphur reactions [424].

Reaction	
R51	$Ni + H_2S \leftrightarrow Ni-S + H_2$
R52	$Ni-S + 1.5O_2 \leftrightarrow NiO + SO_2$
R53	$SO_2 + 3H_2 \leftrightarrow H_2S + 2H_2O$
R54	$Rh + H_2S \leftrightarrow Rh-S + H_2$
R55	$Rh-S + O_2 \rightarrow Rh + SO_2$

The sulphur resistance of the CPO process is in particular an advantage when using heavy fuels (diesel) as feedstock. These will contain sulphur compounds, which are difficult to remove at low pressure. This is crucial for the operation of low-temperature fuel cells, but may not be the case for high-temperature fuel cells, as demonstrated by results on a SOFC stack [421].

Sulphur poisoned the internal reforming as expected, whereas the electrochemical reaction is less sensitive [66] [421]. It was possible to operate an SOFC stack in the presence of 50 ppm H_2S with little loss in power output, whereas methane in the feed gas passed unconverted [421]. Hydrogen sulphide may be converted into sulphur dioxide by the electrochemical reaction. An analysis of literature data [66] shows that the loss in power output, W_{loss}, could be correlated by:

$$W_{loss}=k\,(\theta_s-\theta_{min}) \tag{5.32}$$

using Equation (5.15) to calculate θ_s. θ_{min} is a constant.

5.5 Sulphur passivated reforming

TGA studies [390] showed that the rate of the carbon formation was more sensitive to sulphur than the reforming rate, as illustrated in Figure 5.46. In the simple Maxted model for poisoning (refer to Equation 5.33), this reflects that the nucleation of carbon requires a larger ensemble (6–7 atoms) than does the reforming reaction (3 atoms) [21].

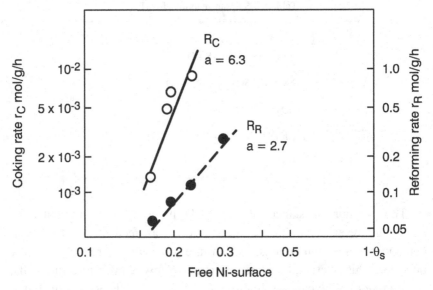

Figure 5.46 TGA test on sulphur passivated catalyst [390]. Reproduced with the permission of Elsevier.

Equilibrium studies for CH_4 decomposition [390] in the presence of sulphur showed the formation of carbon at close to the equilibrium point for sulphur-free gas. However, at high coverage (θ_s about 0.7) the whisker carbon structure was replaced by "octopus" carbon (refer to

Figure 5.8). Therefore, the sulphur passivation is not able to inhibit carbon formation from an equilibrated gas with affinity for carbon formation, i.e. the principle of equilibrated gas works if the gas is equilibrated.

Therefore the retarding effect of sulphur is a dynamic phenomenon. This means that carbon may be formed under certain conditions in spite of sulphur passivation – although at markedly reduced rates. However, the thermodynamic potential (approximately 30 kJ/mol) required to initiate carbon [389] formation was found to depend on the sulphur coverage as illustrated in Figure 5.47. This energy is much higher than that (approximately 2 kJ/mol) (refer to Table 5.4) required to initiate carbon formation on sulphur-free catalysts.

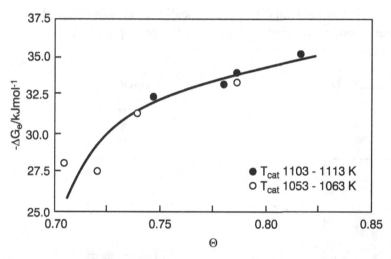

Figure 5.47 Sulphur coverage and supersaturation ($-\Delta G_c$) for start of carbon formation [389]. TGA studies. $-\Delta G_c$ calculated from equilibrated gas at $(H_2O/CH_4)_{exit}$ [389]. Reproduced with the permission of Springer.

Example 5.3
Laboratory experiments [390] were carried out at conditions for carbon formation:
H_2O/CH_4=0.7, H_2O/H_2 (inlet)=30
T_{in}/T_{exit}=515°C/905°C, P_{exit}=3 bar abs

> The principle of equilibrated gas shows potential for carbon > 540°C (whisker carbon data). The potential for carbon increases throughout the catalyst bed and carbon is formed when the gas approaches equilibrium. Carbon was formed from the equilibrated gas at the centre of the catalyst pellet above 750°C and 850°C with 7 and 18 vol ppm H_2S in the feed gas, respectively. In the outer shell of the pellet the steady-state situation prevails, resulting in $-\Delta G_e$ less than the approximately 30 kJ/mol necessary for nucleation of carbon.

In principle, there are two situations for sulphur passivated reforming as illustrated in Figure 5.48 (in accordance with the trend in Figure 5.15):

(1) For gases with low H/C, there is a lower carbon limit above which there is no thermodynamic potential for carbon formation.

(2) For gases with high H/C (typically steam reforming), there is an upper temperature limit above which the principle of equilibrated gas predicts carbon.

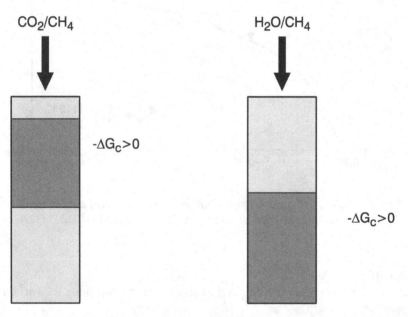

Figure 5.48 Carbon limits and sulphur passivated reformer. The principle of equilibrated gas predicts an upper carbon limit for gases with high H/C and a lower carbon limit for gases with low H/C.

The first case is represented for instance by the Midrex process for reducing gas (refer to Section 2.4.3). Table 5.7 shows results [390] from laboratory tests simulating the Midrex process [490]. The actual feed gas shows potential for carbon ($-\Delta G_c$ at inlet>0 for T<840°C). The principle of equilibrated gas predicts carbon ($-\Delta G_c$<0 for T<890°C). The results indicate that carbon formation is eliminated above θ_s=approximately 0.8.

The feed gas in the heating up zone is far from equilibrium because of nearly complete sulphur coverage. When the reaction starts, the interior of the pellets is filled with equilibrated gas, but the gas has no longer potential for carbon.

Table 5.7 Sulphur passivated reforming. Optional sulphur content [390]. T_{in}/T_{exit}=520°C/945°C, P_{exit}, 3 bar abs. Ni/Al$_2$O$_3$. Feed (vol%): CH$_4$=CO=CO$_2$=20.8, H$_2$=30.7, H$_2$O=6.9, $-\Delta G_c$>0 for T<840°C.

Exp. No	1	2	3	4
S in feed, vol ppm H$_2$S	28	14	5	1
Duration hours	47	42	95	5
CH$_4$ (dry exit), vol%	0.71	0.70	0.36	-
θ_s (equilibrated gas), 850°C	0.86	0.88	0.78	0.70
Carbon formation	No	No	No	yes

In the case of steam reforming (the second situation for sulphur passivated reforming), it is not possible to operate under conditions where the principle of equilibrated gas shows potential for carbon because the centre of the pellet will have equilibrated gas (refer to Example 5.3). However, with the sulphur passivation it is possible to operate at conditions under which the principle of actual gas predicts carbon because of the high ($-\Delta G_c$) overpotential required for nucleation of carbon (Figure 5.47).

The sulphur passivated reforming was demonstrated in a series of tests in a full-size monotube reformer in Figure 3.6 [152]. Some of the tests are referred to in Figure 5.49 and Table 2.6 (Section 2.4.2). A typical temperature profile is shown in Figure 5.49 with a fast heat up of the gas to the reaction temperature for the sulphur passivated catalyst.

The scale-up from fundamental studies to pilot testing led to the introduction of the SPARG process to industry [503].

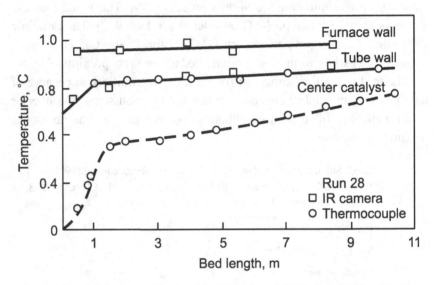

Figure 5.49 Sulphur passivated reforming. Dibbern et al. [152]. Axial temperature from full-size monotube pilot plant $H_2O/NG=0.96$ mol/carbon. $CO_2/NG=0.64$ mol/C-atom, P=7 bar abs. Reproduced with the permission of Gulf Publishing.

Sulphur also inhibits the dissociation of carbon monoxide (the Boudouard reaction). TGA studies [385] on nickel catalysts showed that methanation as well as carbon from the Boudouard reaction were strongly inhibited with increasing sulphur coverage. Other studies [186] found that sulphur eliminates the dissociation of carbon monoxide at a H_2S/Ni coverage above 0.33.

Sulphur was also reported [314] to promote the conversion of the reactive α-carbon formed by dissociation of carbon monoxide into the less active β-carbon as also seen for gum formation [108] (refer to Section 5.3.3).

The sulphur passivation of the CO dissociation is applied for inhibition of metal dusting corrosion (refer to Section 5.2.2) on high alloy steel in the presence of syngas with high contents of carbon monoxide. Sulphur passivation is also used for retarding initiation of coke formation on tube walls in steam crackers [407] (refer to Section 5.3.4).

5.6 Other poisons

Chlorine typically originates from the process feed water (or from air in coastal areas) or as organic chlorine compounds. Chlorine does not adsorb on nickel catalysts, on the contrary any chlorine content in the fresh catalyst will leave the catalyst during operation [389] [391]. A nickel catalyst may even be used for steam reforming of chlorine carbons [129]. However, chlorine is withheld if the catalyst contains alkali. If so, chlorine is reported to deactivate the reforming catalyst [85]. A catalyst with pre-impregnation of chlorine showed less of a decrease in the initial activity than was the case for sulphur [389].

Chlorine leaving the catalyst or passing through the reformer may, however, poison downstream copper containing catalysts (low-temperature shift, methanol synthesis). Chlorine causes a drastic sintering of the copper crystals [391] and chromatographs through the catalyst bed, leaving the deactivated catalyst behind.

Arsenic may come from the feed, impurities and some zinc-oxide masses for desulphurisation, or from the CO_2-removal system (Vetro-coke). In contrast, calcium and zinc transferred from the zinc-oxide vessel has no deactivating effect on the reforming catalyst.

Arsenic may poison nickel catalysts by forming a less active Ni-As alloy [389], although the poisoning effect is much less than that observed for sulphur. The reformer tube may also pick up arsenic, which may then be transferred to subsequent catalyst charges [86]. A similar effect may be experienced with phosphor. A test reactor made from a steel alloy containing phosphor led to steady deactivation of the reforming catalyst [29].

In addition the catalyst may be deactivated or fouled by impurities of iron, silica, etc. in the feedgas.

6 Catalysis of Steam Reforming

6.1 Historical perspective

The development of the steam reforming process has not been based on initial understanding of the catalysis. The progress was driven mainly by an "inductive approach" with feedback from industrial operation and pilot tests [417].

The industrial breakthrough of the steam reforming process was created mainly by reaction engineering and mechanical engineering with a pragmatic and empirical approach. This was followed by a period in the 1970s and 1980s with a strong input from reaction kinetics and reactor modelling accompanied by systematic studies of promoter effects. This work was strengthened by the input from surface science during the 1990s.

The initial vague ideas about the mechanism were formulated by industrial groups around 1970 [22] [381], whereas only few academic scientists were involved in studies of the reaction. This included the groups around Temkin [58], Balandin [35], Borowiecki [36], Trimm [304] and Ross [373]. The activation of methane was studied in the 1960s by Kemball [270] by deuterium exchange studies also applied by Frennet [198].

Dowden (ICI) [162] [163] had ideas based on "the collective properties" of the solids. These ideas were strongly opposed by Sachtler who argued in terms of the properties of the "individual surface atom" [433] or groups (ensembles) of atoms [354].

The ICI group [446] thought that olefins were intermediates, but the studies were carried out at high temperatures and thus influenced by thermal cracking. Other work by Andrew [22] dealt with the function of alkali in terms of regasification. The Topsøe Group [366] [381] [384] argued that alkali promotes the adsorption of water by spill-over of water from the support to the metal surface. At the same time, early research dealt with understanding the mechanism for carbon formation [33] [378] [380] [381] [383] using mainly TGA and electron microscopy.

Research on nickel catalysts was mainly related to hydrogenolysis [311] [462] and later methanation [37] [314] became of interest in SNG from coal in the late 1970s. Sinfelt's work on bimetallic catalysts [462] was of interest to steam reforming, in particular as a direct correlation was found [379] [407] between the activity for steam reforming of ethane and that of hydrogenolysis of ethane as illustrated in Figure 6.1.

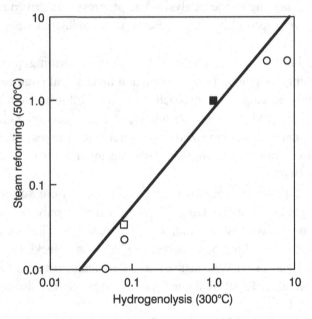

Figure 6.1 Ethane reactions on Ni catalysts. Relative TOFs [379] [407]. Steam reforming: H_2O/C_2H_6=8, H_2O/H_2=10, 500°C, 1 bar. Hydrogenolysis: H_2/C_2H_6=4, N_2=75 vol%, 300°C, 1 bar. Reproduced with the permission of Elsevier.

The strong depressing effect of alkali on the catalyst activity was similar for the two reactions (refer to Table 6.1). It means that the retarding effect of alkali on the activity is independent of the presence of water. The impact on the activity is not related to K_w as discussed in Sections 3.5.2 and 4.3.1.

The reforming activity also correlates with the activity for methanation as shown in Table 6.1 comparing relative TOFs for a number of reactions

[379] [425]. The impact of alkali is much stronger for methanation than for reforming and hydrogenolysis, probably caused by enhanced adsorption of CO. In contrast, alkali has no impact on the decomposition of ammonia [379].

Table 6.1 Relative specific activities at atmospheric pressure [379].

Catalyst	Alkali (wt%)	Reforming 500°C		Hydrogenolysis of C_2H_6, 300°C	Methanation of CO, 250°C	Decomposition of NH_3, 500°C
		C_2H_6	CH_4			
A: Ni/MgO	0.07Na	1.0	1.0	1.0	1.0	1.0
B: Ni/MgO		2.4	2.0	4.2	9.1	0.9
C: Ni/MgO	0.5K	0.03	0.09	0.08	0.002	0.9
D: Ni/MgAl$_2$O$_4$		3.0	1.4			3.2
E: Ni/Al$_2$O$_3$		0.7			2.7	0.4
F: Ni/Al$_2$O$_3$		2.5		8.7	20.6	0.5
G: Ni/Al$_2$O$_3$	5.8K	0.09			0.001	0.5

Sinfelt's early work on Ni, Cu alloys [462] showed that the TOF for hydrogenolysis referred to the nickel surface area dropping with increasing copper content, whereas dehydrogenation was unaffected. This may be compared with the difference between reforming and hydrogenolysis and ammonia decomposition shown in Table 6.1

Boudart [74] referred to hydrogenolysis as a demanding reaction with the need for large ensembles as suggested by Ponec [354] and Martin [311]. This work supported ideas on ensemble control in the SPARG process (refer to Section 5.5). The importance of inhomogeneities was expressed in the role of kink and step sites as suggested by Somerjai et al. [56] and in terms of B_5 sites introduced by van Hardefeld et al. [507]. The activity of nickel catalysts for steam reforming [379] [381] correlated with the presence of B_5 sites as illustrated in Figure 6.2.

There is a strong correlation with the N_2 capacity, whereas the correlation with total nickel area is less clear. It was also shown that promotion of the catalyst with alkali removed the B_5 sites causing a drastic decrease in activity [379] [381].

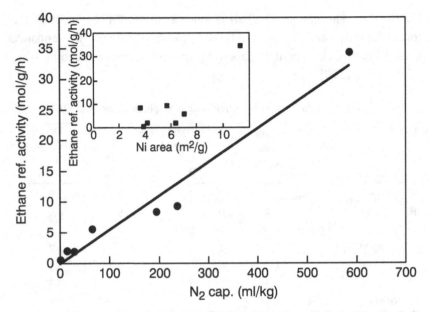

Figure 6.2 Nickel activity and B_5 sites [415]. Ni/Al$_2$O$_3$ catalysts (including K-promoted catalysts). B_5 sites measured by nitrogen adsorption [415]. Reproduced with the permission of Elsevier.

The development of ultra-high vacuum methods in the 1970s followed by *in situ* (*in operandi*) techniques and the progress in theoretical methods have led to a more detailed understanding of the loose ideas resulting from the early studies.

6.2   The role of step sites

The early indications of the importance of surface heterogenity for steam reforming and related reactions described in Section 6.1 were supported by measurements on single metal crystals. It was found that methane is activated more easily on open nickel surfaces such as Ni(110) than on close-packed surfaces [45] [47] [290]. There was also evidence that methane is dissociated in a direct mechanism [98] [415] rather than via a CH$_4$ intermediate on the nickel surface.

Egebjerg et al. [179] investigated the effect of steps on the dissociation CH_4 on Ni(111) and Ru (0001). The different activation energies reported for the dissociation could be ascribed to the varying presence of step sites in the experiments [415].

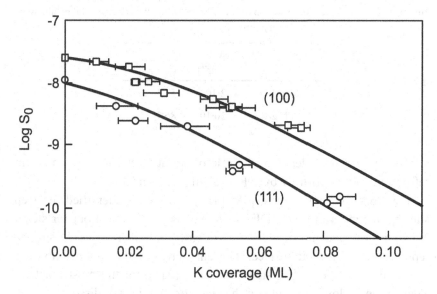

Figure 6.3 The initial sticking coefficient, S_0, for CH_4 chemisorption on Ni(100) and Ni(111) at 475 K vs potassium coverage. CH_4 chemisorption at 500 K. Bengaard et al. [47]. Reproduced by the permission of the authors and Elsevier.

Bengaard et al. [47] studied the effect of potassium promotion (in the absence of oxygen) on the dissociative sticking of methane to Ni(111) and observed that 0.1 ML (monolayer) of potassium reduces the sticking coefficient by more than two orders of magnitude as shown in Figure 6.3.

Other investigations of methane activation by Sehested [415] showed a strong impact of alkali on the methane/deuterium exchange reaction as illustrated in Table 6.2.

A micro-kinetic analysis [415] showed that the activation energy increases by 34 kJ/mol on the alkali-promoted catalyst. At the same time, alkali causes the pre-exponential factor to increase by a factor of 15. This could be explained by potassium blocking the highly active step site, leaving the reaction to take place on the less active open planes; however, with many more sites available. This supports the TGA data in Figure 5.6, demonstrating that alkali retards the dissociation of methane.

Table 6.2 CH_4/D_2 exchange studies (Sehested *et al.* [415]). Parameters obtained for the first two steps in the dehydrogenation of methane by fitting the experimental results with a micro-kinetic model.

Catalyst	A (bar^{-1})	E (kJ mol^{-1})
Ni/MgAl$_2$O$_4$	$2 \cdot 10^{-7}$	53.1
K/Ni/MgAl$_2$O$_4$	$30 \cdot 10^{-7}$	86.7

The empirical evidence of the role of surface heterogeneity in steam reforming was confirmed by DFT calculations and further experiments by the Nørskov/Topsøe team [48] [415]. This work identified the step sites as playing a key role. DFT calculations were carried out for steam reforming of methane on the Ni(111) surface and the Ni(211) surface representing a stepped surface. The energetics of the full reaction are shown in Figure 6.4. The step sites of Ni(211) are the most reactive. They have a lower activation barrier for the initial dissociation of methane and all the intermediates are more strongly bound than on the close-packed surface of Ni(111). Methane may be activated on any single surface atom [264] [508], but stabilises on high coordination sites at steps having a lower barrier for further dissociation of methane. The activation of water does not depend strongly on the coordination number of the surface metal atom [506].

For both reaction paths, the highest barrier is the surface reaction for formation of CO. However, the methane activation may remain the rate determining step, since the sticking coefficient for the methane dissociation is much lower ($\sim 10^7$) than the pre-exponential factor of a surface reaction [104] [264].

The data in Figure 6.4 shows that carbon is bound stronger on the step sites than on the flat surface. It means that the step sites may also be the sites for nucleation of carbon [48] [415].

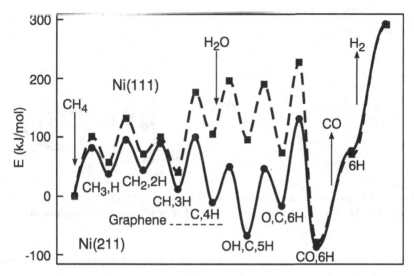

Figure 6.4 Energy diagram for steam reforming of methane. DFT calculations. Bengaard et al. [48]. Reproduced with the permission of the authors and Elsevier.

The energy of a graphene layer is lower than that of the adsorbed carbon atoms, thus creating a driving force for the nucleation of carbon. This is true only for large graphene islands. Calculations [415] show that islands below 2.5 nm should not be stable, which is consistent with the observation that carbon is not formed on catalysts with small nickel crystals (Section 5.2.3). In situ HRTEM by Helveg et al. [226] confirmed the theoretical conclusions. It was possible to observe how graphene layers grow from the surface steps, which are created during the nucleation and which moves as the graphene layers grow.

Abild-Petersen et al. [7] used DFT calculations to analyse the various diffusion processes which may be involved in the formation of whisker carbon. It was concluded that diffusion through the bulk phase is very unlikely, whereas surface diffusion and perhaps diffusion through the sub-surface layer of nickel are the most favoured routes. This is in accordance with the observations by Helveg et al. [226]), as illustrated in Figure 6.5 and in contrast to the mechanisms based on bulk diffusion of carbon, whether based on assumptions of temperature gradients [33] or concentration gradients [383].

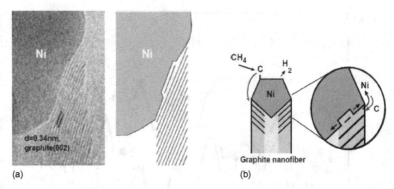

Figure 6.5 Growth mechanism for whisker carbon. a: HRTEM observation. Helveg et al. [226]. b: Mechanism. Abild-Pedersen et al. [7]. Reproduced with the permission of Amer. Phys. Soc.

Now the difficult carbon formation on most noble metals remains to be explained. So far this was explained based on the lack of bulk diffusion due to low solubility of carbon in these metals. It is more likely that the weaker bonding of carbon to the noble metals [57] results in a low coverage of carbon being insufficient for the nucleation of graphene.

The step sites are not only the most stable sites for carbon, but also for a number of ad-atoms as illustrated by the results from DFT calculations [48] [415] shown in Table 6.3.

Table 6.3 Calculated adsorption energies for ad-atoms (kJ/mol) [48].

Surface	K*	S**	N_2	(Au)
Ni(111)	-131	-162	-8	(0)
Ni(211)	-206	-210	-49	-36

* in presence of oxygen, ** $H_2S(g)+Ni^*=Ni-S+H_2(g)$

The preferential adsorption of nitrogen explains the basis for determination of B_5 sites [379] [507]. The preferential binding of potassium, sulphur and gold explains the promoter effects discussed in Section 5.3.2 involving blockage of the step sites for nucleation of carbon.

This also explains the strong deactivation effects of potassium on the activity for steam reforming and hydrogenolysis (refer to Table 6.1) which are independent of the presence of steam.

Gold is also preferentially located on step sites and it was demonstrated on a model catalyst that addition of gold to nickel could eliminate carbon formation [50]. DFT calculations showed a weaker carbon bonding in the vicinity of gold and a strong tendency for gold to decorate the steps. Similar effects may be expected from other bi-metallic catalysts as discussed in Chapter 4. The role of gold or silver on blocking step sites has been demonstrated, also for other reactions [135] [509].

The blockage of sites for carbon nucleation was the idea of the SPARG process (refer to Section 5.5), with chemisorbed sulphur "passivating the ensembles" for nucleation of carbon.

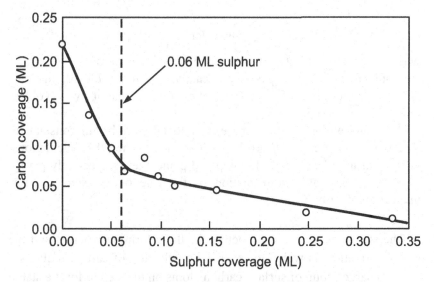

Figure 6.6 Sulphur poisoning of decomposition of methane on Ni (14 13 13) surface. 227°C. Abild-Pedersen et al. [5]. Reproduced with the permission of the authors and Elsevier.

The results from the DFT calculations in Table 6.3 are confirmed by STM observations by Besenbacher et al. [51] [430] (Figure 5.41) and from the data in Figure 6.6 [5], showing the strong deactivating effect of sulphur on methane decomposition on a stepped nickel surface. The most active sites are blocked first.

In the SPARG process, there is a competition between sulphur and carbon for the sites required for nucleation of carbon and the growth of graphene islands above the critical size, as reflected by the reaction

$$CH_4 + S* \leftrightarrow C* + H_2S + H_2 \tag{6.1}$$

This is reflected by the supersaturation required for on-set of the whisker growth depending on sulphur coverage [390] as illustrated in Figure 5.47.

Example 6.1
It is interesting to make a rough comparison of the experimental value of the equilibrium constant for Reaction (6.1) that derived from the DFT calculations [415].

$$K_{eq} = \frac{a_{C*} \cdot a_{H_2S} \cdot a_{H_2}}{a_{S*} \cdot a_{CH_4}} \tag{6.2}$$

where a is the activity of the compound in equation (6.1). From experiments on sulphur passivated catalyst ([390] Table 6), the gas composition for onset of carbon formation at 850°C bar abs was (vol%): CH_4=43.0, H_2O=9.5, H_2=47.5.

If so, and assuming $\theta_s = \theta_c$ the experimental equilibrium constant in Reaction (6.1) becomes K_p (6.1)=7 10^{-7}, whereas the DFT calculations [48] result in 10^{-7} and 10^{-11} for step and plane sites, respectively [415]. This is a reasonable confirmation of the role of the step site for nucleation of carbon.

The step sites are the most active for the reforming reaction and the carbon formation as well. However, the nucleation of carbon requires a critically large group of surface carbon atoms on the step to form a stable carbon island above approximately 2.5 nm as outlined above. This may explain the different dependency of the sulphur coverage of the two reactions shown in Figure 5.46.

One effect of the role of step sites is that the number of highly coordinate sites should increase with decreasing particle size with a resulting higher activity. This was observed on noble metals by Wei and

Iglesia [517] and confirmed by the Nørskov/Topsøe team [264] as illustrated in Figure 6.7. The TOF varies linearly with dispersion which indicates the role of steps and might suggest a more dominant effect with increasing dispersion. In a simpler analysis [264], the linear dependency should reflect steps, whereas corners should result in a square relation. The data are in contrast to observations on industrial nickel catalysts [389] with dispersions less than 5%, as described in Section 4.3.1. Even the strongly sintered catalysts described in Section 4.2.3 showed no significant change in TOF.

This discrepancy may be due to the presence of step sites also on large nickel crystals created by faceting. This is a subject for further studies.

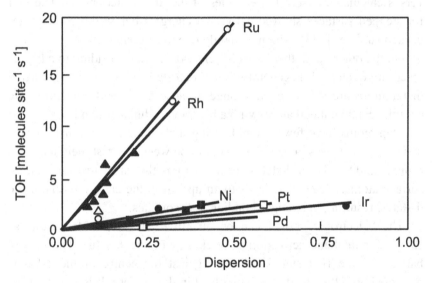

Figure 6.7 Steam reforming rate and dispersion. Jones et al. [264].
H_2O/CH_4=3.9. H_2O/H_2=10, 500°C. Reproduced with the permission of Elsevier.

6.3 Geometric or electronic effects

Discussions of catalytic phenomena have often been based on whether the effects should be related to geometric or electronic effects. The discussion has illustrated that both phenomena are involved in the catalysis of steam reforming. This may be further analysed by DFT calculations.

Purely geometrical effects have been identified for dissociations (and associations) for di-atomic molecules (N_2, CO, NO, O_2). The reason is that the transition state complexes for these molecules are quite extended and at the step it is possible to involve more metal atoms in the stabilisation of them [334].

In general terms, it appears that transition states for dissociative adsorption are very similar. For a given substrate geometry, it is essentially independent of the molecule and transition metal in question [218]. This was shown by Nørskov et al. [57] by linear plots (Brønsted Polyani plots) of calculated activation energies for diatomic molecules versus calculated adsorption energies of the dissociated atoms. The plot for stepped surfaces showed activation energies below those for close-packed surfaces [333], which is mainly due to a geometric effect.

On the other hand, there are also electronic effects leading to a higher reactivity of steps. It is generally found that steps with under-coordinated metal atoms are able to bond stronger to adsorbates (and transition state complexes) than metal atoms with a higher coordination number.

Step atoms have fewer neighbouring metal atoms than other surface atoms. This causes a smaller overlap between nearest neighbour d-orbitals and results in d-states for step atoms that are narrower than on surface atoms. [422]. This leads to an upshift in the energy levels of the d-states, resulting in stronger bonding to adsorbates.

The "electronic" effect can be quantified by DFT calculations as shown for methane activation on nickel in Figure 6.8 [6]. The energy barrier for the first step is plotted against the centre of the d-bands projected onto the metal atoms involved in the process. It is seen that the d-band centre not only rationalises the effect of steps, but also the effect of alloying with Au, the effect of pre-adsorbed C-atoms and S-atoms, and the effect of strain. The straight line in Figure 6.8 reflects that the difference in activation energy can be related to electronic effects only.

Hence, the role of step sites is supported by experimental data as well as DFT calculations. The higher reactivity appears to be related to effects that are given for steps and low coordinated surface geometries (or to the electronic properties of the "individual surface atom" as formulated by Sachtler more than 30 years ago (refer to Section 6.1)).

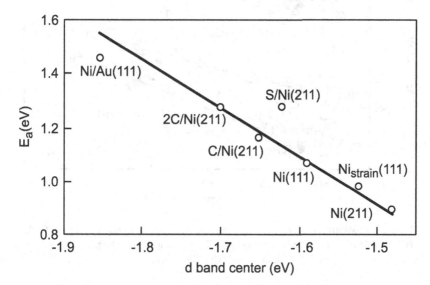

Figure 6.8 The electronic factor. The activation energy for CH_3-H bond breaking over various surfaces is shown as a function of the local d-band centre for the metal atoms involved in the stabilisation of the dissociating molecule. Abild-Petersen et al. [6]. Reproduced with the permission of the authors and Elsevier.

6.4   Metal activity. Micro-kinetics

The Nørskov group was able to push the DFT analysis further by the so-called *scaling model* introduced by Abild-Pedersen et al. [8]. The aim was to provide a simple tool to estimate the bonding energy assumed to be the key quantity describing the surface reaction.

It was found that the adsorption energy of a molecule, ΔH_x, was linearly correlated with the adsorption energy of the atom A and that the slope was related to the "valency" of the absorbate as illustrated for CH_x in Figure 6.9.

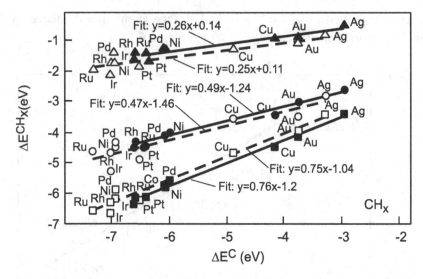

Figure 6.9 Adsorption energies of CH_x intermediates and adsorption energy of C. $\Delta E (CH_x) = \gamma \Delta E(C) + \xi$. Abild-Pedersen et al. [8]. Reproduced with the permission of the authors and Amer.Phys.Soc.

It was further shown that the constant ξ (Figure 6.9) can be obtained from any transition metal as it is related to the sp states which are essentially the same for the transition metals. This means that it is possible to estimate adsorption energy for all other transition metals.

By combining this simple scaling model with the Brøndsted–Evans–Polyani type correlations, it is possible to estimate the full potential energy diagram (refer to Section 6.2) for a surface reaction on any transition metal on the basis of a calculation for a single metal.

This approach was applied by Jones et al. [264] for the steam reforming reaction. Since the data by Bengaard et al. [48] in Figure 6.4 indicated that the methane activation as well as the surface reaction leading to carbon monoxide may be rate-determining, the analysis was expanded by applying the scaling principles for activation of methane and carbon monoxide and at the same time using values for the free energy and not only the total energy for the reaction scheme.

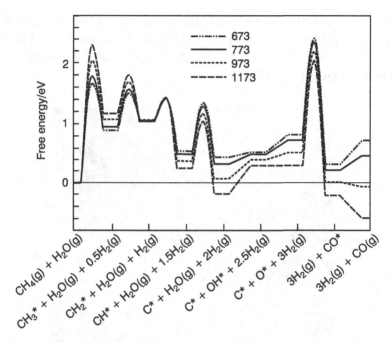

Figure 6.10 Reaction scheme illustrating the temperature dependence of the steam reforming reaction over Ni determined from DFT total energies (1 bar abs). Jones et al. [264]. Reproduced with the permission of Elsevier.

The results in Figure 6.4 are based on total energies. For steam reforming, it appears more relevant to apply the free energy as the entropy term becomes dominating at the high reaction temperatures (refer to Figure 1.7).

The energy diagram for a variety of metals was established using the scaling methods [264] and shows how silver and copper are inactive, whereas tungsten binds oxygen and carbon too strongly. The Group VIII metals appear the most suitable as expected. Figure 6.10 shows how the energy varies for nickel, indicating that a two-step mechanism as postulated in Section 3.5.2 is likely. It was shown for nickel that the barrier for CO formation decreases with increasing temperature, which supports the kinetic data by Iglesia et al. [517] of high temperature (550–800°C). At these temperatures, however, industrial reforming will show gas compositions close to equilibrium (refer to Section 3.4.3) and therefore the kinetics appears more important around 400–500°C. At this temperature level, it appears reasonable to refer to a two-step mechanism and two-step kinetics to describe the activity in terms of the binding energy of carbon and oxygen as well. Results for 500°C are shown in the 2D plot in Figure 6.11 [264].

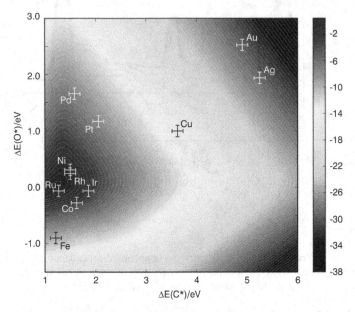

Figure 6.11 Activity for steam reforming ($\log_{10}$, TOF) as a function of C and O adsorption energies, 500°C, 1 bar abs. Jones *et al.* [264]. Reproduced with the permission of Elsevier.

The maximum is close to binding energies estimated for nickel, rhodium and ruthenium. The calculations (Figure 6.11) by Jones *et al.* [264] result in the following sequence of activity of the Group VIII metals:

Ru ≥ Rh > Ni ≥ Ir > Pt ~ Pd

This is in agreement with earlier results [272] [356] [379] (refer to Section 4.3.1), although measurements were not carried out at controlled dispersions, but in contrast to recent work by Iglesia *et al.* [517], who observed platinum to be the most active metal. Jones *et al.* [264] discuss various reasons for this discrepancy.

6.5 The parallel approach

The understanding of the catalysis of steam reforming was achieved in parallel with the industrial developments described in Chapters 3, 4 and 5. The knowledge obtained from this research has been essential to strengthen the know-how gained from the process development and from the feedback from the industrial practice. It has provided a rational basis to cope with the secondary problems. The analysis has also demonstrated that the progress in catalysis to a large extent is related to the development of new characterisation techniques and new theoretical methods. This is an obvious field for collaboration between scientists in industry and in academia.

Figure 6.12 Members of the Nørskov/Topsøe team on steam reforming (Jens Sehested, Jens K. Nørskov, J.R. Rostrup-Nielsen, Bjerne S. Clausen, Stig Helveg, Frank Abild-Pedersen).

You never finish a book. You quit it.
Lanny D. Schmidt, 2010

Appendix 1
Enthalpy of formation

Enthalpy coefficients in H (kJ/mol) $H_i = \sum_{k=1}^{5} E_k \cdot T^{k-1}$

Data sources [137] [375].

	E_1	E_2	E_3	E_4	E_5
O_2	$-7.962663\ 10^0$	$2.427982\ 10^{-2}$	$9.023752\ 10^{-6}$	$-3.098198\ 10^{-9}$	$4.552447\ 10^{-13}$
H_2	$-8.838190\ 10^0$	$3.016572\ 10^{-2}$	$-2.325840\ 10^{-6}$	$2.027647\ 10^{-9}$	$-3.426771\ 10^{-13}$
H_2O	$-2.513828\ 10^2$	$3.106223\ 10^{-2}$	$2.641890\ 10^{-6}$	$2.466168\ 10^{-9}$	$-6.372514\ 10^{-13}$
N_2	$-8.270338\ 10^0$	$2.702524\ 10^{-2}$	$2.168403\ 10^{-6}$	$8.503154\ 10^{-10}$	$-3.202777\ 10^{-13}$
CO	$-1.186737\ 10^2$	$2.621832\ 10^{-2}$	$3.759823\ 10^{-6}$	$2.129007\ 10^{-11}$	$-1.733685\ 10^{-13}$
CO_2	$-4.029275\ 10^2$	$2.448608\ 10^{-2}$	$2.651733\ 10^{-5}$	$-9.591309\ 10^{-9}$	$1.423390\ 10^{-12}$
CH_4	$-8.211111\ 10^1$	$1.405654\ 10^{-2}$	$3.554687\ 10^{-5}$	$-2.975929\ 10^{-9}$	$-1.066366\ 10^{-12}$
C_2H_6	$-9.266902\ 10^1$	$-4.343278\ 10^{-4}$	$1.007521\ 10^{-4}$	$-3.255931\ 10^{-8}$	$4.787174\ 10^{-12}$
C_4H_{10}	$-1.407352\ 10^2$	$-2.401684\ 10^{-3}$	$1.952472\ 10^{-4}$	$-6.701212\ 10^{-8}$	$1.002040\ 10^{-11}$
C_7H_{16}	$-2.131201\ 10^2$	$-4.456496\ 10^{-3}$	$3.353877\ 10^{-4}$	$-1.191684\ 10^{-7}$	$1.837205\ 10^{-11}$
CH_3OH	$-2.109931\ 10^2$	$2.014756\ 10^{-2}$	$4.225563\ 10^{-5}$	$4.233056\ 10^{-10}$	$-4.017644\ 10^{-12}$
$(CH_3)_2O$	$-1.966958\ 10^2$	$1.646284\ 10^{-2}$	$9.291303\ 10^{-5}$	$-2.199279\ 10^{-8}$	$1.358183\ 10^{-12}$

Table values with enthalpies of formation at selected temperatures, kJ/mol
O_2, H_2, H_2O, N_2, CO, CO_2, CH_4, C_2H_6, C_4H_{10}, C_7H_{16} Methanol and DME.

Temp. °C	O_2	H_2	H_2O	N_2	CO	CO_2
0	-0.72	-0.73	-242.65	-0.71	-111.23	-394.45
25	0.00	0.00	-241.83	0.00	-110.52	-393.51
50	0.73	0.73	-240.99	0.71	-109.81	-392.55
75	1.46	1.46	-240.15	1.43	-109.09	-391.57
100	2.20	2.19	-239.31	2.15	-108.37	-390.57
125	2.95	2.92	-238.46	2.88	-107.64	-389.54
150	3.71	3.65	-237.60	3.61	-106.91	-388.50
175	4.47	4.38	-236.74	4.34	-106.17	-387.43
200	5.24	5.11	-235.86	5.08	-105.43	-386.35
225	6.02	5.84	-234.99	5.82	-104.69	-385.25
250	6.80	6.57	-234.10	6.56	-103.94	-384.13
275	7.59	7.30	-233.21	7.31	-103.18	-382.99
300	8.38	8.03	-232.32	8.06	-102.43	-381.83
325	9.18	8.76	-231.41	8.81	-101.66	-380.66
350	9.99	9.50	-230.50	9.57	-100.90	-379.48
375	10.80	10.23	-229.58	10.33	-100.13	-378.28
400	11.62	10.96	-228.65	11.10	-99.35	-377.06
425	12.44	11.70	-227.72	11.87	-98.57	-375.83
450	13.27	12.43	-226.78	12.64	-97.79	-374.59
475	14.10	13.17	-225.83	13.42	-97.00	-373.34
500	14.93	13.91	-224.88	14.20	-96.21	-372.07
525	15.77	14.65	-223.91	14.98	-95.41	-370.79
550	16.62	15.39	-222.94	15.77	-94.61	-369.50
575	17.47	16.13	-221.96	16.56	-93.81	-368.20
600	18.32	16.88	-220.98	17.36	-93.00	-366.89
625	19.17	17.63	-219.98	18.16	-92.19	-365.57
650	20.03	18.37	-218.98	18.96	-91.38	-364.24
675	20.90	19.12	-217.97	19.77	-90.56	-362.90
700	21.76	19.88	-216.95	20.58	-89.73	-361.55
725	22.63	20.63	-215.93	21.39	-88.91	-360.19
750	23.51	21.39	-214.89	22.21	-88.08	-358.83
775	24.38	22.15	-213.85	23.03	-87.25	-357.46
800	25.26	22.91	-212.80	23.86	-86.41	-356.08
825	26.14	23.67	-211.75	24.68	-85.57	-354.69
850	27.02	24.44	-210.68	25.51	-84.73	-353.30
875	27.91	25.20	-209.61	26.35	-83.88	-351.90
900	28.80	25.97	-208.53	27.18	-83.04	-350.50
925	29.69	26.75	-207.44	28.03	-82,18	-349.09
950	30.58	27.52	-206.35	28.87	-81.33	-347.67
975	31.48	28.30	-205.25	29.72	-80.47	-346.25
1000	32.38	29.08	-204.14	30.56	-79.61	-344.82

Appendix 1: Enthalpy of Formation 315

Temp. °C	CH_4	C_2H_6	C_4H_{10}	C_7H_{16}	CH_3OH	$(CH_3)_2O$
0	-75.69	-85.91	-128.13	-191.64	-202.35	-185.71
25	-74.85	-84.67	-125.79	-187.65	-201.25	-184.10
50	-73.97	-83.33	-123.27	-183.36	-200.10	-182.40
75	-73.05	-81.91	-120.59	-178.78	-198.90	-180.61
100	-72.09	-80.40	-117.73	-173.92	-197.65	-178.73
125	-71.09	-78.81	-114.72	-168.79	-196.35	-176.77
150	-70.06	-77.13	-111.55	-163.39	-195.00	-174.72
175	-68.98	-75.37	-108.23	-157.74	-193.60	-172.58
200	-67.87	-73.53	-104.76	-151.85	-192.16	-170.37
225	-66.72	-71.61	-101.15	-145.71	-190.67	-168.07
250	-65.53	-69.63	-97.40	-139.35	-189.13	-165.70
275	-64.31	-67.56	-93.52	-132.76	-187.55	-163.25
300	-63.05	-65.43	-89.51	-125.95	-185.92	-160.73
325	-61.76	-63.24	-85.37	-118.94	-184.25	-158.14
350	-60.43	-60.97	-81.12	-111.73	-182.53	-155.47
375	-59.07	-58.65	-76.75	-104.32	-180.78	-152.74
400	-57.67	-56.26	-72.26	-96.72	-178.98	-149.94
425	-56.24	-53.81	-67.67	-88.95	-177.14	-147.08
450	-54.77	-51.30	-62.97	-80.99	-175.26	-144.15
475	-53.28	-48.73	-58.17	-72.88	-173.35	-141.16
500	-51.75	-46.12	-53.27	-64.59	-171.40	-138.11
525	-50.19	-43.44	-48.28	-56.16	-169.41	-135.00
550	-48.60	-40.72	-43.19	-47.57	-167.39	-131.83
575	-46.99	-37.95	-38.02	-38.84	-165.33	-128.61
600	-45.34	-35.13	-32.76	-29.96	-163.24	-125.34
625	-43.66	-32.26	-27.42	-20.96	-161.12	-122.01
650	-41.96	-29.35	-22.00	-11.82	-158.97	-118.63
675	-40.23	-26.39	-16.51	-2.56	-156.79	-115.21
700	-38.47	-23.39	-10.94	6.81	-154.58	-111.73
725	-36.68	-20.35	-5.30	16.31	-152.35	-108.22
750	-34.87	-17.27	0.41	25.91	-142.19	-104.65
775	-33.04	-14.15	6.18	35.62	-147.81	-101.05
800	-31.18	-10.99	12.01	45.43	-145.51	-97.40
825	-29.30	-7.80	17.91	55.35	-143.19	-93.72
850	-27.40	-4.57	23.87	65.35	-140.85	-90.00
875	-25.47	-1.31	29.88	75.45	-138.50	-86.24
900	-23.52	1.98	35.95	85.63	-136.13	-82.44
925	-21.56	5.31	42.07	95.90	-133.74	-78.62
950	-19.57	8.67	48.24	106.25	-131.35	-74.76
975	-17.56	12.06	54.46	116.68	-128.94	-70.87
1000	-15.54	15.47	60.72	127.19	-126.53	-66.95

Appendix 2
Chemical equilibrium constants

Table A2.1. Reforming reactions
Coefficients in equilibrium function:

$$\ln(K_{eq,j}) = C_{1,j} \cdot \ln(T) + \frac{C_{2,j}}{T} + C_{3,j} + C_{4,j} \cdot T + C_{5,j} \cdot T^2 + C_{6,j} \cdot T^3$$

Data sources [137] [375].

	$CH_4+H_2O=CO+3H_2$	$CO+3H_2=CH_4+H_2O$	$CO+H_2O=CO_2+H_2$
C_1	8.611124 10^0	-8.611124 10^0	-3.161652 10^{-1}
C_2	-2.264801 10^4	2.264801 10^4	5.016474 10^3
C_3	-2.898252 10^1	2.898252 10^1	-4.104367 10^0
C_3 P (bar)	-2.895619 10^1	2.895619 10^1	-4.104367 10^0
C_4	-4.980062 10^{-3}	4.980062 10^{-3}	2.139623 10^{-3}
C_5	3.977411 10^{-7}	-3.977411 10^{-7}	-6.044372 10^{-7}
C_6	2.013436 10^{-11}	-2.013436 10^{-11}	7.582519 10^{-11}

	$CO_2+H_2=CO+H_2O$	$CH_4+CO_2=2CO+2H_2$	$C_2H_6+2H_2O=2CO+5H_2$
C1	3.161652 10-1	8.927289 100	1.702761 101
C2	-5.016474 103	-2.766448 104	-3.775312 104
C3	4.104367 100	-2.487816 101	-5.358552 101
C3 P (bar)	4.104367 100	-2.485183 101	-5.353287 101
C4	-2.139623 10-3	-7.119686 10-3	-1.324748 10-2
C5	6.044372 10-7	1.002178 10-6	2.273621 10-6
C6	-7.582519 10-11	-5.569083 10-11	-2.234181 10-10

Note:
The constant C_3 (bar) is calculated by inserting the reference pressure so that the pressure unit of measurement must be in bar abs.

Equilibrium constants in reforming reactions at selected temperatures.
Pressure unit is bar.

Temp. °C	$CH_4+H_2O=$ $CO+3H_2$	$CO+3H_2=$ CH_4+H_2O	$CO+H_2O=$ CO_2+H_2	$CO_2+H_2=$ $CO+H_2O$	$CH_4+CO_2=$ $2CO+2H_2$	$C_2H_6+2H_2O$ $=2CO+5H_2$
100	$2.6892\ 10^{-18}$	$3.7186\ 10^{17}$	$3.5870\ 10^{3}$	$2.7878\ 10^{-4}$	$7.4971\ 10^{-22}$	$3.8926\ 10^{-26}$
150	$8.1934\ 10^{-15}$	$1.2205\ 10^{14}$	$7.6626\ 10^{2}$	$1.3050\ 10^{-3}$	$1.0693\ 10^{-17}$	$2.8934\ 10^{-20}$
200	$4.8681\ 10^{-12}$	$2.0542\ 10^{11}$	$2.2943\ 10^{2}$	$4.3586\ 10^{-3}$	$2.1218\ 10^{-14}$	$1.3665\ 10^{-15}$
225	$7.4724\ 10^{-11}$	$1.3383\ 10^{10}$	$1.3802\ 10^{2}$	$7.2453\ 10^{-3}$	$5.4140\ 10^{-13}$	$1.3610\ 10^{-13}$
250	$8.9260\ 10^{-10}$	$1.1203\ 10^{9}$	$8.7377\ 10^{1}$	$1.1445\ 10^{-2}$	$1.0215\ 10^{-11}$	$8.8782\ 10^{-12}$
275	$8.5795\ 10^{-9}$	$1.1656\ 10^{8}$	$5.7805\ 10^{1}$	$1.7300\ 10^{-2}$	$1.4842\ 10^{-10}$	$4.0128\ 10^{-10}$
300	$6.8220\ 10^{-8}$	$1.4658\ 10^{7}$	$3.9731\ 10^{1}$	$2.5169\ 10^{-2}$	$1.7171\ 10^{-9}$	$1.3168\ 10^{-8}$
325	$4.5925\ 10^{-7}$	$2.1775\ 10^{6}$	$2.8235\ 10^{1}$	$3.5417\ 10^{-2}$	$1.6265\ 10^{-8}$	$3.2617\ 10^{-7}$
350	$2.6689\ 10^{-6}$	$3.7468\ 10^{5}$	$2.0662\ 10^{1}$	$4.8398\ 10^{-2}$	$1.2917\ 10^{-7}$	$6.3030\ 10^{-6}$
375	$1.3613\ 10^{-5}$	$7.3462\ 10^{4}$	$1.5517\ 10^{1}$	$6.4445\ 10^{-2}$	$8.7726\ 10^{-7}$	$9.7702\ 10^{-5}$
400	$6.1800\ 10^{-5}$	$1.6181\ 10^{4}$	$1.1924\ 10^{1}$	$8.3866\ 10^{-2}$	$5.1829\ 10^{-6}$	$1.2441\ 10^{-3}$
425	$2.5279\ 10^{-4}$	$3.9559\ 10^{3}$	$9.3521\ 10^{0}$	$1.0693\ 10^{-1}$	$2.7030\ 10^{-5}$	$1.3284\ 10^{-2}$
450	$9.4142\ 10^{-4}$	$1.0622\ 10^{3}$	$7.4704\ 10^{0}$	$1.3386\ 10^{-1}$	$1.2602\ 10^{-4}$	$1.2105\ 10^{-1}$
475	$3.2213\ 10^{-3}$	$3.1044\ 10^{2}$	$6.0662\ 10^{0}$	$1.6485\ 10^{-1}$	$5.3101\ 10^{-4}$	$9.5608\ 10^{-1}$
500	$1.0208\ 10^{-2}$	$9.7962\ 10^{1}$	$4.9994\ 10^{0}$	$2.0002\ 10^{-1}$	$2.0418\ 10^{-3}$	$6.6330\ 10^{0}$
525	$3.0168\ 10^{-2}$	$3.3147\ 10^{1}$	$4.1757\ 10^{0}$	$2.3948\ 10^{-1}$	$7.2247\ 10^{-3}$	$4.0903\ 10^{1}$
550	$8.3663\ 10^{-2}$	$1.1953\ 10^{1}$	$3.5303\ 10^{0}$	$2.8326\ 10^{-1}$	$2.3698\ 10^{-2}$	$2.2654\ 10^{2}$
575	$2.1890\ 10^{-1}$	$4.5684\ 10^{0}$	$3.0177\ 10^{0}$	$3.3138\ 10^{-1}$	$7.2537\ 10^{-2}$	$1.1372\ 10^{3}$
600	$5.4296\ 10^{-1}$	$1.8418\ 10^{0}$	$2.6056\ 10^{0}$	$3.8379\ 10^{-1}$	$2.0838\ 10^{-1}$	$5.2172\ 10^{3}$
625	$1.2823\ 10^{0}$	$7.7985\ 10^{-1}$	$2.2705\ 10^{0}$	$4.4043\ 10^{-1}$	$5.6477\ 10^{-1}$	$2.2033\ 10^{4}$
650	$2.8945\ 10^{0}$	$3.4548\ 10^{-1}$	$1.9952\ 10^{0}$	$5.0120\ 10^{-1}$	$1.4507\ 10^{0}$	$8.6210\ 10^{4}$
675	$6.2665\ 10^{0}$	$1.5958\ 10^{-1}$	$1.7669\ 10^{0}$	$5.6597\ 10^{-1}$	$3.5467\ 10^{0}$	$3.1438\ 10^{5}$
700	$1.3052\ 10^{1}$	$7.6615\ 10^{-2}$	$1.5758\ 10^{0}$	$6.3459\ 10^{-1}$	$8.2829\ 10^{0}$	$1.0741\ 10^{6}$
725	$2.6229\ 10^{1}$	$3.8126\ 10^{-2}$	$1.4146\ 10^{0}$	$7.0690\ 10^{-1}$	$1.8541\ 10^{1}$	$3.4543\ 10^{6}$
750	$5.0980\ 10^{1}$	$1.9615\ 10^{-2}$	$1.2776\ 10^{0}$	$7.8271\ 10^{-1}$	$3.9902\ 10^{1}$	$1.0503\ 10^{7}$
775	$9.6062\ 10^{1}$	$1.0410\ 10^{-2}$	$1.1603\ 10^{0}$	$8.6182\ 10^{-1}$	$8.2787\ 10^{1}$	$3.0306\ 10^{7}$
800	$1.7585\ 10^{2}$	$5.6867\ 10^{-3}$	$1.0593\ 10^{0}$	$9.4403\ 10^{-1}$	$1.6601\ 10^{2}$	$8.3291\ 10^{7}$
825	$3.1333\ 10^{2}$	$3.1915\ 10^{-3}$	$9.7169\ 10^{-1}$	$1.0291\ 10^{0}$	$3.2245\ 10^{2}$	$2.1873\ 10^{8}$
850	$5.4436\ 10^{2}$	$1.8370\ 10^{-3}$	$8.9532\ 10^{-1}$	$1.1169\ 10^{0}$	$6.0800\ 10^{2}$	$5.5046\ 10^{8}$
875	$9.2360\ 10^{2}$	$1.0827\ 10^{-3}$	$8.2838\ 10^{-1}$	$1.2072\ 10^{0}$	$1.1149\ 10^{3}$	$1.3312\ 10^{9}$
900	$1.5326\ 10^{3}$	$6.5248\ 10^{-4}$	$7.6941\ 10^{-1}$	$1.2997\ 10^{0}$	$1.9919\ 10^{3}$	$3.1011\ 10^{9}$
925	$2.4907\ 10^{3}$	$4.0149\ 10^{-4}$	$7.1723\ 10^{-1}$	$1.3943\ 10^{0}$	$3.4727\ 10^{3}$	$6.9749\ 10^{9}$
950	$3.9691\ 10^{3}$	$2.5195\ 10^{-4}$	$6.7084\ 10^{-1}$	$1.4907\ 10^{0}$	$5.9166\ 10^{3}$	$1.5178\ 10^{10}$
975	$6.2091\ 10^{3}$	$1.6105\ 10^{-4}$	$6.2943\ 10^{-1}$	$1.5887\ 10^{0}$	$9.8645\ 10^{3}$	$3.2018\ 10^{10}$
1000	$9.5454\ 10^{3}$	$1.0476\ 10^{-4}$	$5.9232\ 10^{-1}$	$1.6883\ 10^{0}$	$1.6115\ 10^{4}$	$6.5590\ 10^{10}$

Appendix 2: Chemical Equilibrium Constants 319

Table A2.2. Other syngas reactions
Coefficients in equilibrium function:

$$\ln(K_{eq,j}) = C_{1,j} \cdot \ln(T) + \frac{C_{2,j}}{T} + C_{3,j} + C_{4,j} \cdot T + C_{5,j} \cdot T^2 + C_{6,j} \cdot T^3$$

Data sources [137] [375].

	$CO+2H_2=CH_3OH$	$2CH_3OH=$ $(CH_3)_2O+H_2O$	$N_2+3H_2=2NH_3$
C_1	-7.986366 10^0	8.695637 10^{-1}	1.372121 10^1
C_2	8.977513 10^3	3.138192 10^3	1.548699 10^4
C_3	2.381310 10^1	-9.077292 10^0	-1.178399 10^2
C_3 P (bar)	2.378677 10^1	-9.077292 10^0	-1.177873 10^2
C_4	5.189461 10^{-3}	1.328249 10^{-3}	3.052080 10^{-3}
C_5	-2.196947 10^{-7}	-1.225171 10^{-6}	2.960918 10^{-7}
C_6	-1.266439 10^{-10}	3.510445 10^{-10}	-8.885442 10^{-11}

	$COS+H_2O=H_2S+CO_2$	$H_2S+2H_2O=SO_2+3H_2$
C_1	-1.198941 10^0	3.450676 10^0
C_2	3.199162 10^3	-2.388664 10^4
C_3	7.283090 10^0	-1.638488 10^1
C_3 P (bar)	7.283090 10^0	-1.637172 10^1
C_4	1.207978 10^{-3}	3.945482 10^{-4}
C_5	-1.912639 10^{-7}	-5.425752 10^{-7}
C_6	1.258838 10^{-11}	9.033646 10^{-11}

Note:
The constant C_3 (bar) is calculated by inserting the reference pressure so that the pressure unit of measurement must be in bar abs.

Equilibrium constants in syngas reactions at selected temperatures.
Pressure unit is bar.

Temp. °C	$CO+3H_2=$ CH_3OH	$CH_3OH=$ $(CH_3)_2O+H_2O$	$N_2+H_2=$ $2NH_3$	$COS+H_2O=$ H_2S+CO_2	$H_2S+2H_2O=$ SO_2+3H_2
100	$1.1584\ 10^1$	$1.2465\ 10^2$	$4.6793\ 10^2$	$9.7124\ 10^3$	$9.9322\ 10^{-27}$
150	$3.1673\ 10^{-1}$	$5.2833\ 10^1$	$2.2917\ 10^1$	$3.1985\ 10^3$	$2.9551\ 10^{-23}$
200	$1.7636\ 10^{-2}$	$2.7184\ 10^1$	$2.6115\ 10^0$	$1.3258\ 10^3$	$1.6898\ 10^{-20}$
225	$5.0970\ 10^{-3}$	$2.0578\ 10^1$	$1.1115\ 10^0$	$9.1098\ 10^2$	$2.5385\ 10^{-19}$
250	$1.6455\ 10^{-3}$	$1.6031\ 10^1$	$5.3474\ 10^{-1}$	$6.4838\ 10^2$	$2.9659\ 10^{-18}$
275	$5.8490\ 10^{-4}$	$1.2801\ 10^1$	$2.8556\ 10^{-1}$	$4.7578\ 10^2$	$2.7881\ 10^{-17}$
300	$2.2621\ 10^{-4}$	$1.0440\ 10^1$	$1.6676\ 10^{-1}$	$3.5854\ 10^2$	$2.1689\ 10^{-16}$
325	$9.4244\ 10^{-5}$	$8.6740\ 10^0$	$1.0518\ 10^{-1}$	$2.7658\ 10^2$	$1.4292\ 10^{-15}$
350	$4.1940\ 10^{-5}$	$7.3234\ 10^0$	$7.0927\ 10^{-2}$	$2.1779\ 10^2$	$8.1366\ 10^{-15}$
375	$1.9793\ 10^{-5}$	$6.2710\ 10^0$	$5.0693\ 10^{-2}$	$1.7466\ 10^2$	$4.0690\ 10^{-14}$
400	$9.8453\ 10^{-6}$	$5.4371\ 10^0$	$3.8124\ 10^{-2}$	$1.4238\ 10^2$	$1.8130\ 10^{-13}$
425	$5.1343\ 10^{-6}$	$4.7665\ 10^0$	$2.9985\ 10^{-2}$	$1.1776\ 10^2$	$7.2860\ 10^{-13}$
450	$2.7942\ 10^{-6}$	$4.2197\ 10^0$	$2.4533\ 10^{-2}$	$9.8685\ 10^1$	$2.6687\ 10^{-12}$
475	$1.5806\ 10^{-6}$	$3.7686\ 10^0$	$2.0786\ 10^{-2}$	$8.3683\ 10^1$	$8.9909\ 10^{-12}$
500	$9.2603\ 10^{-7}$	$3.3923\ 10^0$	$1.8168\ 10^{-2}$	$7.1724\ 10^1$	$2.8083\ 10^{-11}$
525	$5.6020\ 10^{-7}$	$3.0752\ 10^0$	$1.6325\ 10^{-2}$	$6.2075\ 10^1$	$8.1894\ 10^{-11}$
550	$3.4896\ 10^{-7}$	$2.8056\ 10^0$	$1.5037\ 10^{-2}$	$5.4201\ 10^1$	$2.2433\ 10^{-10}$
575	$2.2329\ 10^{-7}$	$2.5745\ 10^0$	$1.4161\ 10^{-2}$	$4.7710\ 10^1$	$5.8035\ 10^{-10}$
600	$1.4644\ 10^{-7}$	$2.3749\ 10^0$	$1.3604\ 10^{-2}$	$4.2308\ 10^1$	$1.4248\ 10^{-9}$
625	$9.8254\ 10^{-8}$	$2.2014\ 10^0$	$1.3306\ 10^{-2}$	$3.7774\ 10^1$	$3.3336\ 10^{-9}$
650	$6.7320\ 10^{-8}$	$2.0495\ 10^0$	$1.3226\ 10^{-2}$	$3.3937\ 10^1$	$7.4619\ 10^{-9}$
675	$4.7030\ 10^{-8}$	$1.9158\ 10^0$	$1.3341\ 10^{-2}$	$3.0667\ 10^1$	$1.6033\ 10^{-8}$
700	$3.3452\ 10^{-8}$	$1.7975\ 10^0$	$1.3636\ 10^{-2}$	$2.7860\ 10^1$	$3.3173\ 10^{-8}$
725	$2.4195\ 10^{-8}$	$1.6923\ 10^0$	$1.4106\ 10^{-2}$	$2.5435\ 10^1$	$6.6271\ 10^{-8}$
750	$1.7774\ 10^{-8}$	$1.5983\ 10^0$	$1.4752\ 10^{-2}$	$2.3328\ 10^1$	$1.2815\ 10^{-7}$
775	$1.3248\ 10^{-8}$	$1.5141\ 10^0$	$1.5581\ 10^{-2}$	$2.1487\ 10^1$	$2.4042\ 10^{-7}$
800	$1.0009\ 10^{-8}$	$1.4382\ 10^0$	$1.6606\ 10^{-2}$	$1.9869\ 10^1$	$4.3849\ 10^{-7}$
825	$7.6577\ 10^{-9}$	$1.3697\ 10^0$	$1.7845\ 10^{-2}$	$1.8442\ 10^1$	$7.7894\ 10^{-7}$
850	$5.9286\ 10^{-9}$	$1.3076\ 10^0$	$1.9321\ 10^{-2}$	$1.7177\ 10^1$	$1.3500\ 10^{-6}$
875	$4.6410\ 10^{-9}$	$1.2512\ 10^0$	$2.1064\ 10^{-2}$	$1.6050\ 10^1$	$2.2863\ 10^{-6}$
900	$3.6710\ 10^{-9}$	$1.1998\ 10^0$	$2.3108\ 10^{-2}$	$1.5042\ 10^1$	$3.7891\ 10^{-6}$
925	$2.9322\ 10^{-9}$	$1.1529\ 10^0$	$2.5497\ 10^{-2}$	$1.4138\ 10^1$	$6.1531\ 10^{-6}$
950	$2.3636\ 10^{-9}$	$1.1100\ 10^0$	$2.8280\ 10^{-2}$	$1.3324\ 10^1$	$9.8026\ 10^{-6}$
975	$1.9218\ 10^{-9}$	$1.0706\ 10^0$	$3.1519\ 10^{-2}$	$1.2588\ 10^1$	$1.5338\ 10^{-5}$
1000	$1.5753\ 10^{-9}$	$1.0345\ 10^0$	$3.5285\ 10^{-2}$	$1.1920\ 10^1$	$2.3593\ 10^{-5}$

Appendix 2: Chemical Equilibrium Constants   321

Table A2.3. Carbon-forming reactions.
Coefficients in equilibrium function:

$$\ln(K_{eq,j}) = C_{1,j} \cdot \ln(T) + \frac{C_{2,j}}{T} + C_{3,j} + C_{4,j} \cdot T + C_{5,j} \cdot T^2 + C_{6,j} \cdot T^3$$

Data sources [137] [375] for graphite carbon and [425] for whisker carbon.

	$CH_4=$ $C+2H_2$	$CH_4=$ $C(\text{whisker}) +2H_2$	$2CO=$ $C+CO_2$	$CO+H_2=$ $C+H_2O$
C_1	5.291666 10^0	0	-3.635623 10^0	-3.319458 10^0
C_2	-7.610846 10^3	-1.0779 10^4	2.005364 10^4	1.503716 10^4
C_3	-2.449759 10^1	1.268 10^1	3.805679 10^{-1}	4.484935 10^0
C_3 P (bar)	-2.448443 10^1	1.269 10^1	3.674049 10^{-1}	4.471772 10^0
C_4	-2.023153 10^{-3}	0	5.096533 10^{-3}	2.956910 10^{-3}
C_5	-1.593520 10^{-7}	0	-1.161530 10^{-6}	-5.570931 10^{-7}
C_6	7.797205 10^{-11}	0	1.336629 10^{-10}	5.783769 10^{-11}

Note:
The constant C_3 (bar) is calculated by inserting the reference pressure so that the pressure unit of measurement must be in bar abs.

Equilibrium constants in carbon (graphite) reactions at selected temperatures.
Pressure unit is bar.

Temperature, °C	$CH_4=C+2H_2$	$2CO=C+CO_2$	$CO+H_2=C+H_2O$
100	$6.0590\ 10^{-7}$	$8.0817\ 10^{14}$	$2.2530\ 10^{11}$
150	$1.1808\ 10^{-5}$	$1.1043\ 10^{12}$	$1.4411\ 10^{9}$
200	$1.2832\ 10^{-4}$	$6.0475\ 10^{09}$	$2.6358\ 10^{7}$
225	$3.5822\ 10^{-4}$	$6.6166\ 10^{8}$	$4.7938\ 10^{6}$
250	$9.1347\ 10^{-4}$	$8.9421\ 10^{7}$	$1.0234\ 10^{6}$
275	$2.1530\ 10^{-3}$	$1.4506\ 10^{7}$	$2.5094\ 10^{5}$
300	$4.7364\ 10^{-3}$	$2.7584\ 10^{6}$	$6.9427\ 10^{4}$
325	$9.8056\ 10^{-3}$	$6.0286\ 10^{5}$	$2.1351\ 10^{4}$
350	$1.9237\ 10^{-2}$	$1.4893\ 10^{5}$	$7.2078\ 10^{3}$
375	$3.5974\ 10^{-2}$	$4.1007\ 10^{4}$	$2.6427\ 10^{3}$
400	$6.4449\ 10^{-2}$	$1.2435\ 10^{4}$	$1.0429\ 10^{3}$
425	$1.1110\ 10^{-1}$	$4.1102\ 10^{3}$	$4.3949\ 10^{2}$
450	$1.8497\ 10^{-1}$	$1.4678\ 10^{3}$	$1.9648\ 10^{2}$
475	$2.9840\ 10^{-1}$	$5.6194\ 10^{2}$	$9.2634\ 10^{1}$
500	$4.6780\ 10^{-1}$	$2.2911\ 10^{2}$	$4.5827\ 10^{1}$
525	$7.1448\ 10^{-1}$	$9.8895\ 10^{1}$	$2.3683\ 10^{1}$
550	$1.0655\ 10^{0}$	$4.4960\ 10^{1}$	$1.2735\ 10^{1}$
575	$1.5545\ 10^{0}$	$2.1430\ 10^{1}$	$7.1013\ 10^{0}$
600	$2.2226\ 10^{0}$	$1.0666\ 10^{1}$	$4.0935\ 10^{0}$
625	$3.1194\ 10^{0}$	$5.5233\ 10^{0}$	$2.4326\ 10^{0}$
650	$4.3034\ 10^{0}$	$2.9664\ 10^{0}$	$1.4867\ 10^{0}$
675	$5.8432\ 10^{0}$	$1.6475\ 10^{0}$	$9.3244\ 10^{-1}$
700	$7.8175\ 10^{0}$	$9.4382\ 10^{-1}$	$5.9894\ 10^{-1}$
725	$1.0316\ 10^{1}$	$5.5641\ 10^{-1}$	$3.9332\ 10^{-1}$
750	$1.3441\ 10^{1}$	$3.3684\ 10^{-1}$	$2.6365\ 10^{-1}$
775	$1.7303\ 10^{1}$	$2.0901\ 10^{-1}$	$1.8013\ 10^{-1}$
800	$2.2028\ 10^{1}$	$1.3270\ 10^{-1}$	$1.2527\ 10^{-1}$
825	$2.7752\ 10^{1}$	$8.6065\ 10^{-2}$	$8.8572\ 10^{-2}$
850	$3.4622\ 10^{1}$	$5.6944\ 10^{-2}$	$6.3601\ 10^{-2}$
875	$4.2796\ 10^{1}$	$3.8384\ 10^{-2}$	$4.6336\ 10^{-2}$
900	$5.2443\ 10^{1}$	$2.6328\ 10^{-2}$	$3.4218\ 10^{-2}$
925	$6.3744\ 10^{1}$	$1.8356\ 10^{-2}$	$2.5593\ 10^{-2}$
950	$7.6887\ 10^{1}$	$1.2995\ 10^{-2}$	$1.9372\ 10^{-2}$
975	$9.2069\ 10^{1}$	$9.3333\ 10^{-3}$	$1.4828\ 10^{-2}$
1000	$1.0950\ 10^{2}$	$6.7946\ 10^{-3}$	$1.1471\ 10^{-2}$

Notation and Abbreviations

Formulas

A	surface area (m^2)
a	activity
$a_{c,s}$	steady-state carbon activity
C_A	concentration of component A ($kmol/m^3$)
C_p	gas heat capacity (kJ/kg/K)
C_t	centre-to-centre reformer tube distance
d	diameter, m
d_p	equivalent sphere particle diameter (m)
d_{eq}	equivalent diffusion particle diameter (m)
d_{Ni}	mean nickel particle diameter (nm)
D_i	dispersion of nickel (%)
D_K	Knudsen diffusion coefficient (m^2/s)
D_{eff}	effective diffusion coefficient (m^2/s)
$D_{r,eff}$	effective radial diffusivity (m^2/s)
ΔEx	exergi (kW)
E	voltage
f	fugacity (bar)
f_m	friction factor in pressure drop equation
F	total flow (mol/s)
$\mathcal{F}$	Faraday's constant
G	gas mass velocity ($kg/s/m^2$ bed)
ΔG	Gibbs free energy of formation (kJ/mol)
ΔG_c	Gibbs energy deviation from graphite (kJ/mol)
h	local channel heat transfer coefficient ($kJ/m^2/K$)
h_{surf}	catalyst particle surface film heat transfer coefficient (kJ/m^2 particle/s/K)
H	enthalpy of formation (kJ/mol)
$-\Delta H^o_{298}$	heat of reaction at 25°C (kJ/mol)
ΔH	enthalpy difference in a steady-state flow (kJ/s)

i	component index
j	reaction index
K_i	phase equilibrium ratio
K_a	adsorption constant in rate expression
K_{eq}	chemical equilibrium constant
k_{int}	rate constant for intrinsic kinetics per kg cat.
$k_{int,vol}$	rate constant for intrinsic kinetics per pellet volume
$k_{surf,i}$	catalyst particle surface film mass transfer coefficient (kmol i/m² particle/s)
K_w	water adsorption constant
M_w	molecular weight (kg/kmol)
N_i	mass flux of component i (kmol i/m²/s)
Nu	Nusselt heat transfer number, $Nu = \alpha_{wall} d_p/\lambda_g$
p_i	partial pressure of component i (atm.abs)
P	pressure (atm.abs)
$\Delta \mathcal{P}$	profit function
Pr	Prandtl number, $Pr = C_p \mu/\lambda$
Pe	Peclet number for turbulent radial mixing, Pe_H for heat transfer, Pe_M for mass transfer
q	heat flux (kW/m²)
Q_r	reaction quotient
Q	heat duty (kJ/s)
R	carbon resistance number
Re	Reynolds number, $Re = G d_p/\mu$
r	radial distance from centre of tube (m) or radial distance in catalyst pellet
r_p	pellet radius (m)
R_A	rate of production of A (kmol/kg cat/s)
R	gas constant, $=0.082056$ atm.m³/kmol/K $= 8.3144$ J/mol/K
R_j	effective rate of production of j originating in reaction j only and with stoichiometric coefficient equal to +1 (kmol/kg cat/s)
$R_{int,j}$	intrinsic rate of production of j originating in reaction j only and with stoichiometric coefficient equal to +1 (kmol/kg cat/s)
$R_{jnt,surf}$	rate of production of j originating in reaction j only and with stoichiometric coefficient equal to +1 (kmol/m² cat/s)
R_p	pore radius

R_{sp}	methane reforming rate of sulphur poisoned catalyst (mol/h/m² total Ni area)
R^0_{sp}	methane reforming rate of sulphur poisoned catalyst (mol/h/m² free Ni area)
R_t	tube radius (m)
S	surface area (m²)
S_p	catalyst particle surface area (m²)
s	catalyst sulphur content (μgS /gNi)
s_0	catalyst sulphur capacity (μgS /gNi)
ΔS	entropy (kJ/mol/K)
SQ	sum of squares of deviations in data reconciliation
t	time (s)
t_0	induction time (s)
T	temperature (K)
T_s	surface temperature (25°C)
ΔT_j	temperature approach to equilibrium for reaction j
U	overall heat transfer coefficient (kJ/m²/K)
u	gas velocity (m/s)
v_i	molar volume (m³/kmol)
V_p	catalyst particle volume (m³)
W	work (kW)
x	dimensionless radius
z	axial distance (m)
y_i	gas mole fraction of component i

Greek letters

α_{wall}	heat transfer coefficient at tube wall (kJ/m²/K)
β	equilibrium quotient (Q_r/K_{eq})
ε	void fraction of packed bed, dimensionless
ε_t	tube-wall emissivity, dimensionless
ξ	conversion of a component originating in reaction j only and with stoichiometric coefficient equal to +1 (mol/s/mol feed)
η	catalyst effectiveness factor
η	efficiency
φ	fugacity coefficient

ϕ	Thiele modulus $\phi = d_{eq}\sqrt{k_{int,vol}/D_{eff}}$
ψ	catalyst surface shape factor
γ	phase equilibrium activity coefficient
λ	thermal conductivity (kJ/m/s/K)
λ_p	particle thermal conductivity (kJ/m/s/K)
$\lambda_{r,eff}$	effective radial thermal conductivity (kJ/m/s/°C)
λ	excess air ratio in combustion
$\nu_{i,j}$	stoichiometric coefficient of component i in reaction j
ν^{-1}	inverted stoichiometric coefficient
ρ	gas density (kg/m³)
ρ_{bulk}	catalyst bulk density (kg/m³ bed)
ρ_p	catalyst particle density (kg/m³ cat)
μ	viscosity (kg/m/s)
μ	chemical potential (Chapter 5)
μ_L	pyrometer wave length
Δ	difference
σ	Stefan–Bolzman constant (Section 3.3.6)
σ	standard deviation (Section 3.2.4)
σ	surface tension (Chapter 5)
τ	tortuosity factor
θ	residence time (Section 4.3.3)
θ	sulphur coverage (Chapter 5)

Subscripts

ads	adsorption
app	temperature approach to equilibrium
av	average
bulk	bulk conditions
cat	property in catalyst bed
eff	effective, pseudo-homogeneous cat bed property
eq	chemical equilibrium
f	furnace
g	gas
i	component No.
ideal	ideal

inner	inner wall of catalyst bed
int	intrinsic kinetics
j	reaction No.
key	key component
lost	lost thermodynamic work in a process
max	maximum
meas	measured
outer	outer wall of catalyst bed
p	catalyst particle
rad	radiation
ref	steam reforming of methane
rec	reconciled
s	furnace surface
shf	water-gas-shift reaction
stag	stagnant gas
surf	surface
t	tube
turb	gas with turbulent flow
wall	property of confining catalyst wall
vol	volume based

Superscripts

º	reference state at 25°C and 1.01325 bar abs
res	residual thermodymamic property
sat	saturation vapour pressure

Abbreviations

APR	aqueous phase reforming
ASU	air separation unit
ATR	autothermal reforming
CCS	carbon capture and storage
CETS	chemical energy transmission systems
CFD	computational fluid dynamics
CPO	catalytic partial oxidation

CSTR	continuous stirred tank reactor
DFT	density function theory
DME	dimethylether
EOS	equation of state
GTL	gas to liquids
HDS	hydrodesulphurisation
HRTEM	high-resolution transmission electron microscopy
HTER	Topsøe convective reformer
HTS	high-temperature shift
HHV	higher heating value
IGCC	integrated gasification combined cycle
ITM	ion transport membrane
KSF	kinetic severity function
LEED	low energy electron diffraction
LHV	lower heating value
LNG	liquefied natural gas
LPG	liquefied petroleum gas
LTS	low-temperature shift
MCFC	molten carbonate fuel cell
MEA	methylethylamine
ML	monolayer
MTBE	methyl t-butyl ether
MTG	methanol to gasoline
MTS	mid temperature shift
PAFC	phosphoric acid fuel cell
PEMFC	polymer membrane fuel cell
POX	partial oxidation
PROX	preferential oxidation
PSA	pressure swing adsorption
RKNR	Topsøe alkali-free catalyst for steam reforming of naphtha
S/C	steam-to-carbon ratio
SMR	steam methane reforming
SNG	synthetic (substitute) natural gas
SOFC	solid oxide fuel cell
SPARG	sulphur passivated reforming
STEM	scanning electron microscopy

SRK	Soave–Redlich–Kwong
TBR	Topsøe bayonet reformer
TGA	thermogravimetric analysis
TEM	transmission electron microscopy
TIGAS	Topsøe integrated gasoline synthesis
TOF	turnover frequency
TPR	temperature programmed reaction
WGS	water gas shift
WSA	Topsøe wet gas sulphuric acid process
XRD	X-ray diffraction

References

Text in italics refer to sections in which the reference is used.

1 Aasberg-Petersen, K.; Stub Nielsen, C.; Lægsgaard Jørgensen, S.; Catal. Today 46 (1998) 193 *(Section 2.2.4)*
2 Aasberg-Petersen, K.; Bak Hansen, J.-H.; Christensen, T.S.; Dybkjær, I.; Seier Christensen, P.; Stub Nielsen, C.; Winter Madsen, S.E.L.; Rostrup-Nielsen, J.R.; Appl. Catal., A. 22 (2001) 379 *(Sections 1.3.3, 1.3.4)*
3 Aasberg-Petersen, K.; Stub Nielsen, C.; Dybkjær, I.; Stud. Surf. Sci. Catal. 167 (2007) 243 *(Sections 1.3.2, 2.6.1)*
4 Aasberg-Petersen, K.; Stub Nielsen, C.; Dybkjær, I.; Perregaard, J.; (unpublished results) (2008) *(Section 2.6.2)*
5 Abild-Pedersen, F.; Lytken, O.; Engbæk, J.; Nielsen, G.; Chorkendorff, I.; Nørskov, J.K.; Surf. Sci. 590 (2005) 127 *(Section 6.2)*
6 Abild-Pedersen, F.; Greeley, J.; Nørskov, J.K.; Catal. Lett. 105 (2005) 9 *(Section 6.3)*
7 Abild-Pedersen, F.; Nørskov, J.K.; Rostrup-Nielsen, J.R.; Sehested, J.; Helveg, S.; Phys. Rev. Lett. B73 (11) (2006) 115419 *(Sections 5.2.1, 6.2)*
8 Abild-Pedersen, F.; Greeley, J.; Studt, F.; Rossmeisl, J.; Munter, T.R.; Moses, P.G.; Skúlason, E.; Bligaard, T.; Nørskov, J.K.; Phys. Rev. Lett. 99 (2007) 016105 *(Section 6.4)*
9 Adhikari, S.; Fernando, S.; Haryanto, A.; Catal. Today 129 (2007) 355 *(Section 1.4.1)*
10 Adris, A.M.; Pruden, B.B.; Lim, C.J.; Grace, J.R.; Can. J. Chem. Eng. 74 (1996) 177 *(Section 1.2.6)*
11 Agnew, J.B.; Rev. Pure Appl. Chem. 19 (1969) 205 *(Section 3.3.3)*
12 Aguiar, P.; Adjiman, C.S.; Brandon, N.P.; J. Power Sources, 138 (1–2) (2004) 120 *(Section 2.3.2)*
13 Akers, W.W.; Camp, D.P.; AIChE J. 1 (1955) 471 *(Section 3.5.2)*
14 Alberton, A.L.; Schwaab, M.; Fontes, C.E.; Bittencourt, R.C.; Pinto, J.C.; Ind. Eng. Chem. Res. 48 (2009) 9369 *(Section 3.4)*
15 Alstrup, I.; Rostrup-Nielsen, J.R.; Røen, S.; Appl. Catal. 1 (1981) 303 *(Section 5.4.2)*
16 Alstrup, I.; Tavares, M.T.; J. Catal. 135 (1992) 147 *(Section 5.3.2)*
17 Alstrup, I.; Clausen, B.S.; Olsen, C.; Smits, R.H.H.; Rostrup-Nielsen, J.R.; Stud. Surf. Sci. Catal. 119 (1998) 5 *(Sections 5.2.4, 5.3.2)*

18   Alzarmora, L.E.; Ross, J.R.H.; Kruissink, E.C.; van Reijnen, L.L.; Chem. Soc. Faraday Trans. I. 77 (1981) 665 *(Section 3.1)*
19   Andersen, K.H.; Nielsen, N.K.; Ammonia Plant Saf. 46, (2005) 352 *(Section 3.2.5)*
20   Andersen, K.H.; Carstensen, J.H.; Hydrocarb. Eng. 14 (11), 2009, 78 *(Section 3.3.5)*
21   Andersen, N.T.; Topsøe, F.; Alstrup, I.; Rostrup-Nielsen, J.R.; J. Catal. 104 (1987) 454 *(Section 5.5)*
22   Andrew, S.P.S.; Ind. Eng. Chem. Prod. Res. Develop 8 (1969) 321 *(Sections 5.3.2, 5.3.4, 6.1)*
23   Aparicio, L.M.; J. Catal. 165 (1997) 262 *(Section 3.5.2)*
24   Aris, R.; "Elementary Chemical Reactor Analysis", Prentice-Hall, Englewood Cliffs, NJ (1969) *(Sections 3.3, 3.4)*
25   Armstrong, P.A.; Foster, E.P.; Gunnardson, H.H.; Henry Oil Symp. 2005, p. 317 *(Section 1.3.4)*
26   Arnstad, B.; Venvik, H.; Klette, H.; Walmsley, J.C.; Tucho, W.M.; Holmestad, R.; Holmen, A.; Bredesen, R.; Catal. Today 118 (2006) 63 *(Section 1.4.1)*
27   Ashcroft, A.T.; Cheetham, A.K.; Foord, J.S.; Green, M.L.H.; Grey, C.P.; Murrell, A.J.; Vernon, P.D.F.; Nature 344 (1990) 319 *(Section 1.3.3)*
28   Avetisov, A.K.; Rostrup-Nielsen, J.R.; Kuchaev, V.L.; Bak Hansen, J.-H; Zyskin, A.G.; Shapatina, E.N.; J. Mol. Catal. A: Chem. 315 (2009) 155 *(Sections 3.5.2, 3.5.4)*
29   Avetisov, A.K.; (private communication) (2000) *(Section 5.6)*
30   Azar, C.; Lindgren, K.; Larson, E.; Möllersten, K.; Climate Change 74 (2006) 47 *(Section 1.4.2)*
31   Back, M.H.; Back, R.A.; in "Pyrolysis, Theory and Practice", (Albright, L.F.; Grynes, B.L.; Corcooan, W.H.; eds), Academic Press, New York, 1983, chapt. 1, 1 *(Sections 4.3.3, 5.3.3)*
32   Badmaev, S.D.; Volkova, G. G.; Belyaev, V.D.; Sobyanin, V.A.; React. Kinet. Catal. Lett. 90 (2007) 205 *(Section 1.4.1)*
33   Baker, R.T.K.; Barber, M.A.; Feates, F.S.; Harris, P.S.; Waite, R.J.; J. Catal. 26 (1972) 51 *(Sections 5.2.1, 5.3.5, 6.1, 6.2)*
34   Bakkerud, P.; Gøl, J.N.; Aasberg-Petersen, K.; Dybkjær, I.; Stud. Surf. Sci. Catal. 147 (2004) 13 *(Section 2.6.5)*
35   Balashova, S.A.; Slovokhotova, T.A.; Balandin, A.A.; Kinet. Katal. 7 (1966) 303 *(Section 6.1)*
36   Barcicki, J.; Denis, A.; Grzegorzyk, W.; Nazimek, D.; Borowiecki, T.; React. Kinet. Catal. Lett. 5 (1976) 471 *(Section 6.1)*
37   Bartholomew, C.H.; Farrauto, R.J.; J. Catal. 45 (1976) 41. 605 *(Section 6.1)*
38   Bartholomew, C.H.; Pannell, R.B.; J. Catal. 65 (1980) 390 *(Section 4.2.1)*
39   Bartholomew, C.H.; Agrawal, P.K.; Katzer, J.R.; Advan. Catal. 31 (1983) 135 *(Sections 5.4.1, 5.4.2)*

40   Basile, A.; Gallucci, F.; Paturzo, L.; Catal. Today. 104 (2005) 244 *(Section 1.4.1)*
41   Basini, L.; Aasberg-Petersen, K.; Guarinoni, A.; Østberg, M.; Catal. Today 64 (2001) 9 *(Section 1.3.3)*
42   Bauer, R.; "Effective Radiale Wärmeleitfähigkeit Gasdurchsströmter Schüttungen", Dissertation, Karlsruhe (1976) *(Section 3.3.2)*
43   Bauer, R.; Schlünder, E.; Int. Chem. Eng. 18 (1978) 181 *(Section 3.3.2)*
44   Becker, M.; Harms, H.; Müller, W.D.; Proc. Intersoc. Energy Conv. Eng. Conf. (1986) 920 *(Section 2.7)*
45   Beebe Jr., T.P.; Goodman, D.W.; Kay, B.D.; Yates Jr., J.T.; J. Chem. Phys., 87 (1987) 2305 *(Section 6.2)*
46   Beeck, O.; Advan. Catal. 2 (1950) 151 *(Section 4.2.1)*
47   Bengaard, H.S.; Alstrup, I.; Chorkendorff, I.; Ulmann, S.; Rostrup-Nielsen, J.R.; Nørskov, J.K.; J. Catal. 187 (1999) 238 *(Section 6.2)*
48   Bengaard, H.S.; Nørskov, J.K.; Sehested, J.; Clausen, B.S.; Nielsen, L.P.; Molenbroek, A.M.; Rostrup-Nielsen, J.R.; J. Catal. 209 (2002) 365 *(Sections 5.2.1, 6.2, 6.4)*
49   Bernardo, C.A.; Alstrup, I.; Rostrup-Nielsen, J.R.; J. Catal. 96 (1985) 517 *(Sections 4.3.1, 5.2.1, 5.2.2, 5.3.2)*
50   Besenbacher, F.; Chorkendorff, I.; Clausen, B.S.; Hammer, B.; Molenbroek, A.M.; Nørskov, J.K.; Steensgaard, I.; Science 279 (1998) 1913 *(Sections 5.3.2, 6.2)*
51   Besenbacher, F.; Vang, R.T.; (unpublished results) (1992) *(Sections 4.2.1, 6.2)*
52   Beutier, D.; Renon, H.; Ind. Eng. Chem. Proc. Dev. 17, (1978) 220 *(Section 2.1.1)*
53   Bharadwaj, S.; Schmidt, L.D.; J. Catal. 146 (1994) 11 *(Section 1.3.3)*
54   Bhatta, K.S.M.; Dixon, G.M.; Trans. Faraday Soc. 63 (1967) 2217 *(Section 5.3.3)*
55   Bitter, J.H.; Seshan, K.; Lercher, J.A.; J. Catal. 171 (1997) 279 *(Section 5.3.2)*
56   Blakely, D.W.; Somerjai, G.A.; J. Catal. 42 (1976) 181 *(Section 6.1)*
57   Bligaard, T.; Nørskov, J.K.; Dahl, S.; Mathiesen, J.; Christensen, C.H.; Sehested, J.; J. Catal. 224 (2004) 206 *(Sections 6.2, 6.3)*
58   Bodrov, I.M.; Apel'baum, L.O.; Temkin, M.I.; Kinet. Catal., 5 (1964) 614 *(Sections 3.5.2, 6.1)*
59   Bodrov, I.M.; Apel'baum, L.O.; Kinet. Katal. 8 (1967) 379 *(Sections 3.5.2, 3.5.4, 5.4.5)*
60   Bøgild Hansen, J.; Joensen, F.; Stud. Surf. Sci. Catal. 61 (1991) 457 *(Section 2.6.4)*
61   Bøgild Hansen, J.; in "Handbook of Fuel Cells" (Vielstick, W.; Lamm, A.; Gasteiger, H.A.; eds)., vol. 3 (1) 141, Wiley, 2003 *(Sections 1.4.1, 2.3.1)*
62   Bøgild Hansen, J.; Proc. Eur. Fuel Cell Forum EFCF 081, Luzerne 2004 *(Sections 2.3.1, 2.3.2, 2.3.3)*

63 Bøgild Hansen, J.; Pålsson, J.; Proc. 1st European Fuel Cell Technology and Applications Conf. (2005) p. 49 *(Section 1.4.1)*
64 Bøgild Hansen, J.; Rostrup-Nielsen, J.R.; Højlund Nielsen, P.E.; Abstr. 2006 Fuel Cell Seminar (poster) Honolulu, HI (2006) *(Section 2.3.3)*
65 Bøgild Hansen, J.; Højlund Nielsen, P.E.; in "Handbook of Heterogeneous Catalysis", (Ertl, G.; Knözinger, H.; Schuth, F.; Weitkamp, J.; eds), vol. 6, Chap. 13.13, Weinheim (2008), 2920 *(Section 2.6.2)*
66 Bøgild Hansen, J.; Electrochem. Solid-State Lett. 11 (10) (2008) B178 *(Section 5.4.5)*
67 Bohlbro, H.; Acta. Chem. Scan. 17 (1963) 1001 *(Section 1.5.2)*
68 Bohlbro, H.; "An Investigation on the Kinetics of Carbon Monoxide with Water Vapour over Iron Oxide Based Catalyst, Gjellerup, Copenhagen, 1966, chapt. 7 *(Section 1.5.2)*
69 Bohlbro, H.; Jorgensen, M.H.; Chem. Eng. World 5 (1970) 46 *(Section 1.5.2)*
70 Borowiecki, T.; Golebiowski, A.; Stasinska, B.; Appl. Catal., A. 153 (1997) 141 *(Section 5.3.2)*
71 Bose, A.C.; (ed.) "Progress in Transport Membranes for Gas Separation Application", Springer, New York, 2009 *(Section 1.3.4)*
72 Bossen, B.S.; "Simulation and Optimisation of Ammonia Plants". PhD Thesis. Technical University of Denmark. (1995) *(Section 3.2.4)*
73 Boudart, M.; AIChE J. 18 (1972) 465 *(Section 3.5.2)*
74 Boudart, M.; Proc. 5th ICC, London, 1976, 1 (1977) 1 *(Section 6.1)*
75 Boudart, M.; Djéga-Mariadassou, G.; "Kinetics of Heterogeneous Catalytic Reactions", Princeton Univ. Press, Princeton (1984) *(Section 3.5.2)*
76 Boukis, N.; Galla, U.; Henningsen, T.; Dinjus, E.; Larsen, T.; Andersen, K.J.; Chem. Ing. Tech. 76 (2004) 1287 *(Section 1.4.3)*
77 BP Gas Making; BP, London (1972) *(Section 2.4.1)*
78 BP Statistical Review of World Energy June 2009 (www.bp.com/statisticalreview2009) *(Sections 1.1.1, 1.4.2)*
79 Bradford, M.C.J.; Vannice, M.A.; Appl. Catal., A. 142 (1996) 97 *(Section 5.3.2)*
80 Bradford, M.C.J.; Vannice, M.A.; J. Catal. 173 (1998) 157 *(Section 5.3.2)*
81 Bradford, M.C.J.; Vannice, M.A.; Catal. Rev.-Sci. Eng. 41 (1999) 1 *(Section 3.5.4)*
82 Braunschweig, B.; Gani, R.; "Software Architecture and Tools for Computer Aided Process Engineering", Elsevier (2002) *(Sections 2.1, 3.2.4)*
83 Breen, J.P.; Ross, J.R.H.; Catal. Today. 51 (1999) 521 *(Section 1.4.1)*
84 Bresler, S.A.; Ireland, J.O.; Chem. Eng. 1 (1972) 94 *(Section 2.6.6)*
85 Bridger, G.W.; Chincen, G.C.; in "Catalyst Handbook" Wolfe Scientific Books, London (1964) 64 *(Sections 4.1.1, 5.6)*

References

86 Bridger, G.W.; Wyrwas, W.; Chem. Eng. Prog. 48 (1967) 101 *(Section 5.6)*
87 Brigder, G.W.; Chem. Eng. Prog. 53 (1972) 39 *(Sections 2.4.1, 2.5, 3.1)*
88 Brightling, J.; Farnell, P.; Foster, C.; Beyer, F.; Ammonia Plant Saf. 46 (2005) 190 *(Section 3.1)*
89 Broman, E.; Carstensen, J.H.; Hydrocarb. Eng., 14 (2) (2009) 31 *(Section 3.3.5)*
90 Bromberg, L.; Cohn, D.R.; Rabinovich, A.; O'Brien, C.; Hochgreb, S.; Energy Fuels 12 (1998) 11 *(Section 4.3.3)*
91 Brown, A.; J. Petr. Technol. 61 (4) (2009) 62 *(Section 2.6.5)*
92 Bussani, G.; Chiari, M.; Grottoli, M.G.; Pierucci, S.; Faravelli, T.; Ricci, G.; Gioventu, G.; Comput. Chem. Eng., 19 (1995) S299 *(Section 3.2.4)*
93 Byrne Jr., P.J.; Gohr, R.J.; Haslam, R.T.; Ind. Eng. Chem. 24 (1932) 1129 *(Section 3.1)*
94 Cao, C.; Palo, D.R.; Tonkovich, A.Y.; Wang, Y.; Catal. Today. 125 (2007) 29 *(Section 1.4.1)*
95 Carolan, M.F.; Chen, C.M.; Miller, C.F.; Minford, E.; Steppan, J.J.; Waldron, W.E.; ECS Trans. 1(7) (2006) 335 *(Section 1.3.4)*
96 Carstensen, J.H.; Hansen, J.B.; Pedersen, P.S.; Ammonia Plant Saf. 30 (1990) 139 *(Section 1.5.2)*
97 Carstensen, J.H.; Hammershøj, B.S.; Ammonia Plant Saf. 3 (1999) 171 *(Section 1.5.2)*
98 Ceyer, S.T.; Yang, Q.Y.; Lee, M.B.; Beckerle, J.D.; Johnson, A.D.; Stud. Surf. Sci. Catal. 36 (1987) 51 *(Section 6.2)*
99 Chen, C.M.; Bennett, D.L.; Carolan, M.F.; Foster, E.P.; Schinski, W.L.; Taylor, D.M.; Stud. Surf. Sci. Catal. 147 (2004) 55 *(Section 1.3.4)*
100 Chen, Z.; Elnashaie, S.S.E.H.; Chem. Eng. Sci. 60 (2005) 4287 *(Section 1.3.3)*
101 Cheng, Z.; Wu, Q.; Li, J.; Zhu, Q.; Catal. Today 30 (1996) 147 *(Section 5.3.2)*
102 Chew, K.; in "Fundamentals of Gas to Liquids", Petroleum Economist (2003) p. 11 *(Section 1.1.1)*
103 Chin, Y.-H.; Dagle, R.; Hu, J.; Dohnalkova, A.C.; Wang, Y.; Catal. Today 77 (2002) 79 *(Section 1.4.1)*
104 Chorkendorff, I.; Niemantsverdriet, J.W.; "Concepts of Modern Catalysis and Kinetics", Wiley-VCH, Weinheim (2003) *(Section 6.2)*
105 Choudhary, V.R.; Rajput, A.M.; Prabhakar, B.; J. Catal. 139 (1993) 326 *(Section 1.3.3)*
106 Christensen, C.H.; Rass-Hansen, J.; Marsden, C.C.; Taarning, E.; Egeblad, K.; ChemSusChem. 1 (2008) 283 *(Section 1.1.1)*
107 Christensen, T.S.; Primdahl, I.I.; Hydrocarb. Process. 77 (3) (1994) 39 *(Section 1.3.2)*

108 Christensen, T.S.; Rostrup-Nielsen, J.R.; ACS Symp. Ser. 634 (1995) 186 *(Sections 5.3.3, 5.5)*
109 Christensen, T.S.; Dybkjær, I.; Hansen, L.; Primdahl, I.I.; Ammonia Plant Saf. 35 (1995) 205 *(Section 1.3.2)*
110 Christensen, T.S. ; Appl. Catal. A., 138 (1996) 285 *(Section 5.3.3)*
111 Christensen, T.S. ; Christensen, P.S. ; Dybkjær, I.; Bak Hansen, J.-H. ; Primdahl, I.I. ; (1998) Stud. Surf. Sci. Catal. 119 (1998) 883 *(Sections 1.3.2, 2.6.2)*
112 Christiansen, L.J.; Andersen, S.L.; Chem. Eng. Sci. 35 (1980) 314 *(Section 5.4.3)*
113 Christiansen, L.J.; Jarvan, J.E.; "Chemical Reactor Design and Technology" (de Lasa, H.I.; ed.), NATO ASI Series E, No. 110, Martin Nijhoff Publishers (1986), 35 *(Sections 3.4, 3.4.1)*
114 Christiansen, L.J.; Proc. CEC-Italian Fuel Cell Workshop, Taormina, Sicily (1987) 161 *(Section 3.3.5)*
115 Christiansen, L.J.; Comput. Chem. Eng., Giardini Naxos, Sicily (1987) 217 *(Section 2.1.3)*
116 Christiansen, L.J.; Comput. Chem. Eng. 16 (1992) S55 *(Section 2.1)*
117 Christiansen, L.J.; in "Ammonia" (Nielsen, A.; Dybkjær, I.; eds.), Springer-Verlag, Berlin (1995) 1 *(Section 2.1.2)*
118 Christiansen, L.J.; Brüniche-Olsen, N.; Carstensen, J.H.; Schrøder, M.; Comput. Chem. Eng. 21, (1997) S1179 *(Section 3.2.4)*
119 Christiansen, L.J.; (unpublished results) *(Sections 3.3.2, 3.3.3, 3.3.4)*
120 Christiansen, L.J.; Christensen, P.S.; Stub Nielsen, C.; Bak Hansen, J.-H.; (unpublished results) *(Sections 3.3.3, 3.3.6, 3.3.7)*
121 Chun, C.M.; Hershkowitz, F.; Ramanarayanan, T.A.; Ceramic Trans. 202 (2009) 129 *(Section 5.2.2)*
122 Clarigde, J.B.; York, A.P.E.; Brungs, A.J.; Marquez-Alvarez, C.; Sloan, J.; Tsang, S.C.; Green, M.L.H.; J. Catal. 180 (1998) 85 *(Section 4.3.2)*
123 Clarke, S.H.; Dicks, A.L.; Pointon, K.; Smith, T.A.; Swann, A.; Catal. Today, 38 (1997) 411 *(Section 2.3.2)*
124 Clausen, B.S.; Gråbæk, L.; Steffensen, G.; Hansen, P.L.; Topsøe, H.; Catal. Lett. 20 (1993) 23 *(Section 1.5.2)*
125 Cockerham, R.G.; Percival, G.; Trans. Inst. Gas Eng. 107 (1956–57) 390 *(Section 1.4.1)*
126 Cockerham, R.G.; Percival, G.; Yarwood, T.A.; Inst. Gas Eng. J. 5 (1965) 109 *(Sections 1.5.1, 3.1)*
127 Corella, J.; Toledo, J.M.; Molina, G.; Ind. Eng. Chem. Res., 46 (2007) 6831 *(Section 1.4.3)*
128 Cortright, R.D.; Davda, R.R.; Dumesic, J.A.; Nature, 418 (2002) 964 *(Section 1.4.1)*
129 Coute, N.; Ortego Jr., J.D.; Richardson, J.T.; Twigg, M.V.; Appl. Catal., B. 19 (1998) 175 *(Section 5.6)*
130 Crabtree, R.H.; Chem. Rev. 95 (1995) 987 *(Section 1.1.2)*

131 Cromarty, B.J.; Beedle, S.C.; Ammonia Plant. Saf. 33 (1992), 63 *(Section 3.2.5)*
132 Czernik, S.; French, R.; Feik, C.; Chornet, E.; Ind. Eng. Chem. Res. 41 (2002) 4209 *(Section 1.4.3)*
133 Czernik, S.; Bridgwater, A.V.; Energy Fuels, 18 (2004) 590 *(Section 1.4.3)*
134 Dagle, R.A.; Platon, A.; Palo, D.R.; Datye, A.K.; Vohs, J.M.; Wang, Y.; Appl. Catal., A 342 (2008) 63 *(Section 1.5.2)*
135 Dahl, S.; Logadottir, A.; Egeberg, R.C.; Larsen, J.H.; Chorkendorff, I.; Törnqvist, E.; Nørskov, J.K.; Phys. Rev. Lett. 83 (1999) 1814 *(Section 6.2)*
136 Daszkowski, T.; Eigenberger, G.; Chem. Eng. Sci. 47, (1992) 2245 *(Section 3.3.1)*
137 Daubert, T.E.; Danner, R.P.; "Physical and Thermodynamic Properties of Pure Chemicals". DIPPR Project 801 *(Section 1.2.1)*
138 Dauenhauer, P.J.; Salge, J.R.; Schmidt, L.D.; J. Catal. 244 (2006) 238 *(Section 1.4.1)*
139 Davies, H.S.; Humphries, K.J.; Hebden, D.; Percy, D.A.; Inst. Gas Eng. J. 7 (1967) 708 *(Sections 2.4.1, 3.1)*
140 Davies, H.S.; Tempelman, J.J.; Wrag, D.; Instn. Gas Engr. Commun. (1975) 979 *(Section 5.3.3)*
141 Davy, H.; Phil. Trans. Roy. Soc. London 107 (1817) 77 *(Preface)*
142 De Deken, J.C.; Devos, E.F.; Froment, G.F.; ACS Symp. Ser. 196 (1982) 181 *(Section 3.3)*
143 De Groote, A.M.; Froment, G.F.; Rev. Chem. Eng., 11 (2) (1995) 145 *(Section 3.3)*
144 Deam, R.J.; Leather, J.; deJong, K.P.; Hale, J.G.; Proc. 9th World Petr. Congress, Tokyo (1975) 5, 397 *(Section 2.6.1)*
145 deJong, K.P.; Catal. Today, 29 (1996) 171 *(Section 1.1.2)*
146 deJong, K.P.; Geus, J.W.; Catal. Rev.-Sci. Eng. 42 (2000), 129 *(Section 5.2.1)*
147 Deluga, G.A.; Salge, J.R.; Schmidt, L.D.; Verykios, X.E.; Science 303 (2004) 993 *(Section 1.4.1)*
148 Denbigh, K.G.; Chem. Eng. Sci. 6 (1956) 1 *(Section 2.1.3)*
149 Dent, F.J.; Cobb, J.W.; J. Chem. Soc. 2 (1929) 1903 *(Section 5.2.2)*
150 Dent, F.J.; Moignard, L.A.; Eastwood, A.H.; Blackburn, W.H.; Hebden, D.; Trans. Inst. Gas Eng. (1945–46) p. 602 *(Sections 3.1, 5.2.2)*
151 deWild, P.J.; Verkaak, M.J.F.M.; Catal. Today 60 (2000) 3 *(Section 1.4.1)*
152 Dibbern, H.C.; Olesen, P.; Rostrup-Nielsen, J.R.; Tøttrup, P.B.; Udengaard, N.R.; Hydrocarb. Process. 65 (1) (1986) 31 *(Sections 5.2.3, 5.5)*
153 Dissanayake, D.; Rosynek, M.P.; Lunsford, J.H.; J. Phys. Chem. 97 (1993) 3644 *(Section 1.3.3)*
154 Diver, R.B.; Chemtech 17 (1987) 606 *(Section 2.7)*

155 Diver, R.B.; Fish, J.D.; Levitan, R.; Levy, M.; Meirovitch, E.; Rosin, H.; Paripatyadar, S.A.; Richardson, J.T.; Sol. Energy 48 (1992) 21 *(Section 2.7)*

156 Diver, R.B.; Miller, J.E.; Allendorf, M.D.; Siegel, N.P.; Hogan, R.E.; J. Solar Energy Eng. Trans ASME, 130 (2008) 0410011 *(Section 2.7)*

157 Dixon, A.G.; Cresswell, D.L.; AIChE J, 25 (1979) 66 3 *(Section 3.3.2)*

158 Dixon, A.G.; Ind. Eng. Chem. Res. 36 (1997) 3053 *(Section 3.3.2)*

159 Dixon, A.G.; Nijemeisland, M.; Ind. Eng. Chem. Res. 40 (2001) 5246 *(Sections 3.3, 3.3.2)*

160 Dixon, A.G.; Taskin, M.E.; Nijemeisland, M.; Stitt, E.H.; Chem. Eng. Sci., 63 (2008) 2219 *(Sections 3.3.2, 3.3.3)*

161 DOE; "Hydrogen from Natural Gas and Coal", Hydrogen Coordination Group, DOE (June 2003) *(Section 2.2.1)*

162 Dowden, D.A.; Schnell, C.R.; Walker, G.T.; 4th ICC, Moscow (1968) prep. 62 *(Section 6.1)*

163 Dowden, D.A.; in "Chemisorption and Catalysis" (Hepple, E.P.; ed.) Inst. of Petroleum, London (1971) 1 *(Section 6.1)*

164 Dreyer, B.J.; Lee, I.C.; Krummenmacher, J.J.; Schmidt, L.D.; Appl. Catal., A. 307 (2006) 184 *(Section 1.3.3)*

165 Dry, M.; in "Catalysis Science and Technology" (Boudart, M.; Anderson, J.R.; eds.) 1, Springer-Verlag, Berlin (1985) 159 *(Section 2.6.5)*

166 Dudukovic, M.P.; Chem. Eng. Sci. 65 (2010) 3 *(Sections 3.3, 3.3.8)*

167 Duprez, D.; Mendez, M., Dalmon, J.A.; Appl. Catal. 21 (1986) 1 *(Section 4.2.1)*

168 Dybkjær, I.; Bohlbro, H.; Aldrige, C.L.; Riley, K.L.; Ammonia Plant Saf. 21 (1979) 145 *(Section 1.5.2)*

169 Dybkjær, I.; in "Ammonia" (Nielsen, A.; Dybkjær, I.; eds.), Springer-Verlag, Berlin (1995) 199 *(Sections 1.1.1, 2.5)*

170 Dybkjær, I.; Bøgild Hansen, J.; Stud. Surf. Sci. Catal. 107 (1997) 99 *(Sections 2.6.1, 2.6.4)*

171 Dybkjær, I.; Winter Madsen, S.E.L.; Hydrocarb. Eng., 3 (1) (1998) 56 *(Section 1.2.4)*

172 Dybkjær, I.; Christensen, T.S.; Stud. Surf. Sci. Catal. 136 (2001) 435 *(Sections 1.3.4, 2.6.5)*

173 Edgar, T.F.; Himmelblau, D.M.; "Optimisation of Chemical Processes", McGraw-Hill, New York (1988) *(Section 3.2.4)*

174 Edmonds, T.; McCarroll, J.J.; Pitkethly, R.C.; J. Vac. Sci. Technol. 8 (1971) 68 *(Section 4.2.1)*

175 Edwards, M.A.; Whittle, D.M.; Rhodes, C.; Ward, A.M.; Rohan, D.; Shannon, M.D.; Hutchings, G.J.; Kiely, C.J.; Phys. Chem. Chem. Phys. 4 (2002) 3902 *(Section 1.5.2)*

176 Edwards, T.J.; Newman, J.; Prausnitz, J.M.; AIChE J. 21 (1975) 248 *(Section 2.1.1)*

References

177 Edwards, T.J.; Maurer, G.; Newman, J.; Prausnitz, J.M.; AIChE J. 24 (1978), 966 *(Section 2.1.1)*
178 Efstathiou, A.M.; Kladi, A.; Tsipouriari, V.A.; Verykios, X.E.; J. Catal. 158 (1996) 64 *(Section 5.3.2)*
179 Egebjerg, R.C.; Ullmann, S.; Alstrup, I.; Mullins, C.B.; Chorkendorff, I.; Surf. Sci. 497 (2002) 183 *(Section 6.2)*
180 Eisenberg, B.; Fiato, R.A.; Maudlin, C.H.; Say, G.R.; Soled, S.L.; Stud. Surf. Sci. Catal. 119 (1998) 943 *(Section 1.3.3)*
181 Elnashaie, S.S.E.H.; Adris, A.M.; Soliman, M.A.; Al-Ubaid, A.S.; Can. J. Chem. Eng. 70 (1992) 786 *(Sections 3.3, 3.4.1, 3.4.3)*
182 Emonts, B.; Hansen, J.B.; Lægsgaard Jørgensen, S.; Höhlein, B.; Peters, R.; J. Power Sources 71 (1998) 288 *(Section 2.3.1)*
183 Emonts, B.; Hansen, J.B.; Grube, T.; Höhlein, B.; Peters, R.; Schmidt, H.; Stolten, D.; Tschauder, A.; J. Power Sources 106 (2002) 333 *(Section 1.4.1)*
184 Enger, B.C.; Lødeng, R.; Holmen, A.; Appl. Catal., A 346 (2008) 1 *(Section 1.3.3)*
185 Epstein, M.; Spiewak, I.; Proc. Int. Symp. Solar Conc. Technol. Moscow (1994) 958 *(Section 2.7)*
186 Erley, W.; Wagner, H.; J. Catal. 53 (1978) 287 *(Section 5.5)*
187 Ernst, W.S.; Venables, S.C.; Christensen, P.S.; Berthelsen, A.C.; Hydrocarb. Process. 79 (2000) 100C *(Section 1.3.2)*
188 Farrauto, R.J.; Lui, Y.; Ruettinger, W.; Ilinich, Q.; Shore, L.; Giroux, T.; Catal. Rev.-Sci. Eng. 49 (2007) 141 *(Sections 1.2.6, 2.3.1, 3.3.8)*
189 Fedders, H.; Harth, R.; Höhlein, B.; Nucl. Eng. Des. 34 (1975) 119 *(Section 2.7)*
190 Felcht, U.H.; in "Chemical Engineering: Visions of the World" (Darton, A.C; Prince, R.G.H.; Wood, D.G.; eds), Elsevier, Amsterdam, 2003, 41 *(Section 1.1.1)*
191 Feng, D.; Wang, Y.; Wang, D.; Wang, J.; Chem. Eng. J. 146 (2009) 477 *(Section 1.4.1)*
192 Fischer, F.; Tropsch, H.; Brennst. Chem. 3 (1928) 39 *(Section 3.1)*
193 Fleisch, T.H.; Sills, R.; Briscoe, M.; Freide, J.F.; in "Fundamentals of Gas to Liquids", Petroleum Economist (2003) 39 *(Section 2.6.5)*
194 Fleisch, T.H.; Sills, R.A.; Stud. Surf. Sci. Catal. 147 (2004) 31 *(Section 2.6.1)*
195 Fontell, E.; Kivisaari, T.; Christiansen, N.; Bøgild Hansen, J.; Pålsson, J.; J. Power Sources 131 (2004) 49 *(Section 2.3.2)*
196 Fox III, J.M.; Catal. Rev.-Sci. Eng. 35 (1991) 169 *(Section 1.1.2)*
197 Fredenslund, A.; Gmehling, J.; Rasmussen, P.; "Vapor-Liquid Equilibrium using UNIFAC", Elsevier, Amsterdam (1977) *(Section 2.1.1)*
198 Frennet, A.; Lienard, G.; Surf. Sci. 18 (1969) 80 *(Section 6.1)*
199 Froment, G.F.; Bischoff, K.B.; "Chemical Reactor Design and Analysis", John Wiley & Sons, New York (1979) *(Sections 3.3, 3.3.1, 3.3.2, 3.3.4, 3.4, 3.4.1, 3.4.2, 3.5.1)*

200 Froment, G.F.; Rev. Chem. Eng. 6 (1990) 293 *(Section 5.3.4)*
201 Froment, G.F.; J. Mol. Catal. A: Chem. 163 (2000) 147 *(Section 3.3)*
202 Frost, L.; Hartvigsen, J.; Elangovan, S.; Proc. Annual AIChE Meeting (2007) 8 *(Section 4.3.3)*
203 Gard, N.R.; Nitrogen 39 (1966) 25 *(Section 3.1)*
204 George, R.; Casanova, A.; Veyo, S.E.; Proc. Fuel Cell Seminar (2002) Palm Springs, 977 *(Section 2.3.2)*
205 Gierlich, H.H.; Fremery, M.; Skov, A.; Rostrup-Nielsen, J.R.; Stud. Surf. Sci. Catal. 81 (1980) 459 *(Sections 5.3.3, 5.3.5)*
206 Goetsch, D.A.; Schmidt, L.D.; Science, 271 (1996) 1560 *(Section 1.1.2)*
207 Gokhale, A.A.; Dumesic, J.A.; Mavrikakis, M.; J. Am. Chem. Soc. 130 (2008) 1402 *(Section 1.5.2)*
208 Gøl, J.N.; Dybkjær, I.; HTI Quarterly (Summer 1995) 27 *(Section 2.2.3)*
209 Göppert, U.; Maurer, G.; Fluid Phase Equilib. 41, (1988) 153 *(Section 2.1.1)*
210 Goring, R.L.; DeRosset, A.J.; J. Catal. 3 (1964) 341 *(Sections 5.4.3, 5.4.4)*
211 Grabke, H.; J. Mater. Corros. 49 (1998) 317 *(Section 5.2.2)*
212 Graboski, M.S.; Daubert, T.E.; Ind. Eng. Chem. Proc. Des. Dev. 18 (1979) 300 *(Section 2.1.1)*
213 Grenoble, D.C.; J. Catal. 51 (1978) 203 *(Section 4.3.1)*
214 Grenoble, D.C.; Estadt, M.M.; Ollis, D.F.; J.Catal. 67 (1981) 90 *(Section 1.5.2)*
215 Griffith, R.H.; Marsh, J.D.F.; "Contact Catalysis", Oxford Univ. Press, London (1957) *(Preface)*
216 Grunwaldt, J.-D.; Molenbroek, A.M.; Topsøe, N.-Y.; Topsøe, H.; Clausen, B.S.; J. Catal. 194 (2000) 452 *(Section 1.5.2)*
217 Hague, I.-U.; "Mono and Bi-metallic Catalysts for Steam Reforming", Phd Thesis, Univ. New South Wales, (1990) *(Section 5.3.2)*
218 Hammer, B.; Nørskov, J.K.; Adv. Catal. 45 (2000) 71 *(Section 6.3)*
219 Harms, H.; Höhlein, B.; Skov, A.; Chem. Ing. Tech. 52 (1980) 504 *(Section 2.6.6)*
220 Haryanto, A.; Fernando S., Murali N., Adhikari S., Energy Fuel 19 (2005) 2098 *(Section 1.4.1)*
221 Haugaard, J.; Livbjerg, H.; Chem. Eng. Sci. 53 (1998) 2941 *(Section 3.4.2)*
222 Häussinger, P.; Lohmüller, R.; Watson, A.M.; in "Ullmann's Encyclopedia of Industrial Chemistry", VHC, Weinheim, 1989, Vol A13, 395 *(Section 2.2.2)*
223 Haynes, B.S.; (private communication) (2009) *(Section 3.3.8)*
224 Heinemann, H.; in "Catalysis, Science and Technology" (Anderson, J.R.; Boudart, M.; eds.), vol. 1, chapt. 1, Springer, Berlin (1981) *(Sections 3.1, 5.3.2)*

225 Heinrich, E.; Dahmen, N.; Dinjus, E.; Biofuels, Bioprod. Biorefin. 3 (2009) 28 *(Section 1.4.3)*
226 Helveg, S.; López-Cartes, C.; Sehested, J.; Hansen, P.L.; Clausen, B.S.; Rostrup-Nielsen, J.R.; Abild-Pedersen, F.; Nørskov, J.K.; Nature 427 (2004) 426 *(Sections 5.2.1, 6.2)*
227 Helveg, S.; (unpublished results) (2006) *(Sections 5.2.1, 5.3.3)*
228 Heyen, G.; Maréchal, E.; Kalitventzeff, B.; Comput. Chem. Eng. 20, (1996), S539 *(Section 3.2.4)*
229 Hickman, D.A.; Schmidt, L.D.; Science 259 (1993) 343 *(Section 1.3.3)*
230 Higman, C.; Burgt, M.v.d.; "Gasification", GPP, Elsevier, Amsterdam (2003) *(Sections 1.4.2, 1.4.3)*
231 Hiller, H.; Gaswärme 8 (1959) 151 *(Section 1.3.3)*
232 Hinrichsen, K.O.; Kochloefl, K.; Muhler, M.; in "Handbook of Heterogeneous Catalysis" 2nd ed. (Ertl, G.; Knözinger, H.; Schüth, F.; Withkamp, J.; eds), Wiley-VCH, Weinheim (2008) Vol. 6, 2906 *(Section 1.5.2)*
233 Hoek, A.; Kersten, L.B.J.M.; Stud. Surf. Sci. Catal. 147 (2004) 25 *(Section 2.6.5)*
234 Hogan Jr., R.E.; Skocypec, R.D.; Diver, R.B.; Fish, J.D.; Garrait, M.; Richardson, J.T.; Chem. Eng. Sci., 45 (1990) 2751 *(Section 2.7)*
235 Højlund Nielsen, P.E.; Hansen, J.B.; J. Mol. Catal. 17 (1982) 183 *(Section 1.5.2)*
236 Højlund Nielsen, P.E.; Pedersen, K.; Rostrup-Nielsen, J.R.; Top. Catal. 2 (1995) 207 *(Section 2.6.6)*
237 Holladay, J.D.; Hu, J.; King, D.L.; Wang, Y.; Catal. Today 139 (2009) 244 *(Sections 2.2.1, 4.3.3)*
238 Holmen, A.; Catal.Today 142 (2009) 2 *(Section 1.1.2)*
239 Holm-Larsen, H.; Asia Nitrogen Conf., Bali (1994) *(Section 2.6.4)*
240 Holm-Larsen, H.; Stud. Surf. Sci. Catal. 136 (2001) 441 *(Section 2.6.2)*
241 Horiuchi, T.; Sakuma, K.; Fukui, T.; Kubo, Y.; Osaki, T.; Mori, T.; Appl. Catal. A 144 (1996) 111 *(Section 5.3.2)*
242 Horn, R.; Williams, K.A.; Degenstein, N.J.; Bitsch-Larsen, A.; Dalle Nogare, D.; Tupy, S.A.; Schmidt, L.D.; J. Catal., USA 249 (2007) 380 *(Section 1.3.3)*
243 Hottel, H.C.; Sarofim, A.F.; "Radiative Transfer", McGraw-Hill, New York (1967) *(Section 3.3.6)*
244 Hottel, H.C.; Ind. Eng. Chem. Fundam. 22 (1983) 153 *(Section 3.2.5)*
245 Hou, P.; Meeker, D.; Wise, H.; J. Catal. 80 (1983) 280 *(Section 1.5.2)*
246 Huang, H.; Young, N.; Williams, B.P.; Taylor, S.H.; Hutchings, G.J.; Catal. Lett. 110 (2006) 243 *(Section 1.5.1)*
247 Huber, G.W.; Shabaker, J.W.; Dumesic, J.A.; Science 300 (2003) 2075 *(Sections 1.4.1, 5.3.2)*
248 Huber, G.W.; Dumesic, J.A.; Catal. Today 111 (2006) 119 *(Section 1.4.1)*

249 Huber, G.W.; Shabaker, J.W.; Evans, S.T.; Dumesic, J.A.; Appl. Catal., B. 62 (2006) 226 *(Section 1.4.1)*
250 Hyman, M.H.; Hydrocarb. Process. 47 (1968) 131 *(Sections 3.3, 3.4.3)*
251 IEA; "Prospects for Hydrogen and Fuel Cells", IEA, Paris (2005) *(Sections 2.2.1, 2.2.2)*
252 Ioannides, T.; Verykios, X.E.; Catal. Lett. 47 (1997) 183 *(Section 3.3.8)*
253 Irving, P.M.; Pickles, J.; Abstr. 2006 Fuel Cell Seminar, Honolulu, HI (2006) 242 *(Section 2.3.1)*
254 Jackson, S.D.; Thomson, S.J.; Webb, G.; J. Catal. 70 (1981) 249 *(Section 5.3.3)*
255 Jacobsen, C.J.H.; Dahl, S.; Bahn, S.; Clausen, B.S.; Topsøe, H.; Logadottir, A.; Nørskov, J.K.; J. Am. Chem. Soc. 123 (2001) 8404 *(Section 2.5)*
256 Jacobsen, C.J.H.; Nielsen S.E., Ammonia Tech. Man. (2002) 212 *(Section 2.5)*
257 Jaeger, B.; Stud. Surf. Sci. Catal. 81 (1994) 419 *(Section 2.6.5)*
258 Jiang, C.J.; Trimm, D.L.; Wainwright, M.S.; Cant, N.W.; Appl. Catal., A 93 (1993) 245 *(Section 1.4.1)*
259 Jockel, H.; Triebskorn, B.E.; Hydrocarb. Process. 58 (9) (1979) 197 *(Section 2.4.1)*
260 Joensen, F.; Rostrup-Nielsen, J.R.; J. Power Sources, 105 (2002) 195 *(Section 1.4.1)*
261 Joensen, F.; Abstr. ACS Spring Meeting 2006, paper 181c, 527 *(Section 2.6.4)*
262 Johnsen, K.; Grace, J.R.; Elnashaie, S.S.E.H.; Kolbeinsen, L.; Eriksen, D.; Ind. Eng. Chem. Res. 45 (2006) 4133 *(Section 1.2.6)*
263 Johnsson, J.; Chem. Eng. News 85 (37) (2007) 10 *(Section 1.1.1)*
264 Jones, G.; Jakobsen, J.G.; Shim, S.S.; Kleis, J.; Anderson, M.P.; Rossmeisl, J.; Abild-Pedersen, F.; Bligaard, T.; Helveg, S.; Hinnemann, B.; Rostrup-Nielsen, J.R.; Chorkendorff, I.; Sehested, J.; Nørskov, J.K.; J. Catal. 259 (2008) 147 *(Sections 3.5.2, 4.3.1, 6.2, 6.4)*
265 Kane, L.; Romanow, S.; Hydrocarb. Process. 83 (2) (2004) 25 *(Section 2.2.2)*
266 Kappen, P.; Grunwaldt, J.-D.; Hammershøj, B.S.; Tröger, L.; Clausen, B.S.; J. Catal. 198 (2001) 56 *(Section 1.5.2)*
267 Karim, G.A.; Metwally, M.; Adv. Hydrogen Energy Syst. 2 (1979) 937 *(Section 4.3.3)*
268 Katzer, J.R.; Chem. Eng. Prog. 104 (2008) 515 *(Section 1.4.2)*
269 Kawazuishi, K.; Prausnitz, J.M.; Ind. Eng. Chem. Res. 26 (1987) 1482 *(Section 2.1.1)*
270 Kemball, C.; Discuss. Faraday Soc. 41 (1966) 190 *(Section 6.1)*
271 Khomenko, A.A.; Apel'baum, L.O.; Shub, F.S.; Snagovskii, Y.S.; Temkin, M.I.; Kinet. Catal. 12 (1971) 367 *(Section 3.5.2)*

272 Kikuchi, E.; Tanaka, S.; Yamazaki, Y.; Morita, Y.; Bull Jpn. Pet. Inst. 16 (2) (1974) 95 *(Sections 4.3.1, 6.4)*
273 Kikuchi, E.; Cattech. 1 (1997) 67 *(Section 2.2.4)*
274 Kjær, J.; "Measurement and Calculation of Temperature and Conversion Profiles in Fixed-Bed Catalytic Reactors", Haldor Topsøe, A/S, Vedbæk, DK (1976). Reprint of original thesis from 1958 *(Sections 3.3, 3.3.2)*
275 Kjær, J.; "Computer Methods in Catalytic reactor Calculations", Vedbæk, DK (1972) *(Section 3.3)*
276 Kjær, J.; "Computer Methods in Gas Phase Thermodynamics", Topsøe, Vedbæk, DK (1972) *(Section 1.2.1)*
277 Knudsen, J.; Nilekar, A.U.; Vang, R.T.; Schnadt, J.; Kunkes, E.L.; Dumesic, J.A.; Mavrikakis, M.; Besenbacher, F.; J. Am. Chem. Soc. 129 (2007) 6485 *(Section 1.5.2)*
278 Koch, A.J.H.M.; deBokx, P.K.; Boellard, E.; Klop, W.; Geus, J.W.; J. Catal. 96 (1985) 468 *(Section 5.2.1)*
279 Koerts, T.; Deelen, M.J.A.G.; van Santen, R.A.; J. Catal. 138 (1992) 101 *(Section 1.1.2)*
280 Koh, A.C.W.; Leong, W.K.; Chen, L.; Ang, T.P.; Lin, J.; Johnson, B.F.G.; Khimyak, T.; Catal. Commun. 9 (2008) 170 *(Section 1.4.1)*
281 Kohl, A.L.; Nielsen, R.B.; "Gas Purification", Gulf Publ. Co., Houston (1997) *(Section 1.5.3)*
282 Kolb, G.; "Fuel Processing for Fuel Cells", Wiley-VHC, Weinheim (2008) *(Sections 2.3.1, 3.3.8)*
283 Kopfle, J.; Hunter, R.; Iron Making and Steel Making 35 (2008) 254 *(Section 2.4.3)*
284 Kriebel, M.; in Ullmann's "Encyclopedia of Industrial Chemistry", VHC, Weinheim (1989) Vol A12, p. 249 *(Section 1.5.3)*
285 Kugeler, K.; Niessen, H.F.; Röth-Kamat, M.; Böcker, D.; Rüter, B.; Theis, K.A.; Nucl. Eng. Des. 34 (1975) 65 *(Section 2.7)*
286 Kundu, A.; Shul, Y.G.; Kim, D.H.; Adv. Fuel Cells 1 (2007) 419 *(Section 1.4.1)*
287 Kuo, J.C.W.; Kresge, C.T.; Palermo, R.E.; Catal. Today 4 (1989) 463 *(Section 1.1.2)*
288 Kvamsdal, H.M.; Svendsen, H.F.; Hertzberg, T.; Olsvik, O.; Chem. Eng. Sci, 54 (1999) 2697 *(Section 3.3.2)*
289 Lange, J.P.; Cattech 5 (2) (2001) 85 *(Section 1.1.1)*
290 Larsen, J.H.; Chorkendorff, I.; Surf. Sci. Rep. 35 (1999) 163 *(Section 6.2)*
291 Laursen, J.K.; Hydrocarb. Eng. 12 (8) (2007) 47 *(Section 1.5.3)*
292 Lee, J.S.; Oyama, S.T.; Catal. Rev.-Sci. Eng. 30 (1988) 249 *(Section 1.1.2)*
293 Lei, N.; Zhou, C.; Hu, G.; Chen, C.; J. Nat. Gas Chem. 18 (2009) 222 *(Section 5.2.2)*
294 Lerou, J.J.; Froment, G.F.; Chem. Eng. Sci. 32 (1977) 853 *(Section 3.3.1)*

295   Lerou, J.J.; Tonkovich, A.L.; Silva, L.; Perry, S.; McDonald, J.; Chem. Eng. Sci. 65 (2010) 380 *(Section 3.3.8)*
296   Li, C.-H; Finlayson, B.A.; Chem. Eng. Sci., 32 (1977) 1055 *(Sections 3.3.2, 3.3.3)*
297   Li, Z.; Lui, G.-J.; Ni, W.-D.; Proc. CSEE, 28 (8) (2008) 1 *(Section 2.6.4)*
298   Liguras, D.K.; Kondarides, D.I.; Verykios, X.E.; Appl. Catal., B 43 (2003) 345 *(Section 1.4.1)*
299   Lin, S.S.-Y.; Kim, D.H.; Ha, S.Y.; Appl. Cataly., A 355 (2009) 69 *(Sections 1.4.1, 4.3.2)*
300   Linden, H.R.; 7th World Petroleum Congress, Mexico City, 5 (1967) 139–156 *(Section 2.4.1)*
301   Linnhoff, B.; Flower, J.R.; AIChE J. 24 (1978) 633 *(Section 2.1.3)*
302   Llorca, J.; Homs, N.; Sales, J.; de la Piscina, P.R.; J. Catal. 209 (2002) 306 *(Section 1.4.1)*
303   Lloyd, L.; Ridler, D.E.; Twigg, M.V.; in "Catalyst Handbook", 2nd ed (Twigg, M.V.; ed.) Mansa Publishing, London (1989) 283 *(Section 1.5.2)*
304   Lobo, L.S.; Trimm, D.L.; J. Catal. 29 (1973) 15 *(Sections 5.2.1, 6.1)*
305   Lobo, W.E.; Evans, J.E.; Trans. of the AIChE, 35, (1939) 743 *(Section 3.3.6)*
306   Lovegrove, K.; Luzzi, A.; Soldiani, I.; Kreetz, H.; Solar Energy 76 (2004) 331 *(Section 2.7)*
307   Lunsford, J.H.; Catal.Today 63 (2000) 165 *(Section 1.1.2)*
308   Machiels, C.J.; Ruderson, R.B.; J. Catal. 58 (1979) 268 *(Section 3.5.3)*
309   Madron, F.; "Process plant performance. Measurement and data processing for optimisation and retrofits", Ellis Horwood (1992) *(Section 3.2.4)*
310   Mariani, N.J.; Mocciaro, C.; Martínez, O.M.; Barreto, G.F.; Ind. Eng. Chem. Res. 48 (2009) 1172 *(Section 3.4)*
311   Martin, G.A.; J. Catal. 60 (1979) 345 *(Section 6.1)*
312   Mayo, H.C.; Finneran, J.A.; Oil Gas J., 66 (12) (1968) 78 *(Section 3.1)*
313   Mazanec, T.; Petroleum Quarterly (Autumn 2003) 149 *(Section 3.3.8)*
314   McCarty, J.G.; Wise, H.; J. Catal. 52 (1979) 406 *(Sections 5.3.3, 5.5, 6.1)*
315   Metz, B.; Davidson, O.; de Coninck, H.; Loos, M.; Meyer, L.; IPCC Special Report on Carbon Dioxide Capture and Storage, Intergovernmental Panel on Climate Change, Cambridge University Press, Cambridge (2005) *(Section 1.4.2)*
316   Michelsen, M.L.; Mollerup, J.M.; "Thermodynamic Models: Fundamentals & Computational Aspects", Tie-Line Publications, Holte, DK (2004) *(Sections 1.2.1, 2.1.1)*

317 Miller, C.F.; Carolan, M.F.; Chen, C.M.; Minford, E.; Waldron, W.E.; Stephan, J.J.; ACS Meeting, Book of Abstracts (2007) *(Section 1.3.4)*
318 Mills, G.A.; Rostrup-Nielsen, J.R.; Catal. Today 22 (1994) 335 *(Sections 2.6.4, 2.7)*
319 Ming, Q.; Irving, P.M.; Harrison, J.W.; Proc. 2004 Fuel Cell Seminar, S. Antonio, TX (2004) *(Section 3.3.8)*
320 Mittasch, A.; "Geschichte der Ammoniaksynthese", Verlag Chemie, Weinheim (1951) p. 113 *(Section 2.5)*
321 Mogensen, M.; Lybye, D.; Kammer, K.; Bonanos, N.; Proc. Electrochem. Soc. PV (2005–07) 1068 *(Section 4.3.2)*
322 Moseley, F.; Stephens, R.W.; Stewart, K.D.; Wood, J.; J. Catal. 24 (1972) 18 *(Section 5.3.3)*
323 Mosqueda, B.; Toyir, J.; Kaddouri, A.; Gélin, P.; Appl. Catal., B 88 (2009) 361 *(Section 4.3.2)*
324 Müller, G.; Bender, E.; Maurer, G.; Ber. Bunsen. Phys. Chem. Chem. Phys. 92 (1988) 148 *(Section 2.1.1)*
325 Murty, C.V.S.; Murthy, M.V.K.; Ind. Eng. Chem. Res. 27 (1988) 1832 *(Section 3.3.6)*
326 Neelis, M.L.; van der Kooi, H.J.; Geerlings, J.J.C.; Int. J. Hydrogen Energy 29 (2004) 538 *(Section 2.2.1)*
327 Neumann, B.; Jacob, K.; Z. Elektrochem. 30 (1924) 557 *(Section 3.1)*
328 Nielsen, A.; in "An Investigation on Promoted Iron Catalysts for the Synthesis of Ammonia", 3rd ed., Gjelleup, Copenhagen (1968) *(Section 2.1.2)*
329 Nielsen, A.; Hansen, J.B.; Houken, J.; Gam, E.A.; Plant/Operations Design 1 (1982) 186 *(Section 1.5.2)*
330 Nielsen, P.; Christiansen, L.J.; Proc. 4th International Symposium on "Computational Technologies for Fluid/Thermal/Chemical Systems with Industrial Applications", Vancouver, Canada (2002) *(Section 3.3.6)*
331 Nitrogen 17 (1962) 35 *(Section 1.3.2)*
332 Nitrogen Syngas 305 (2010) 20 *(Section 2.6.5)*
333 Nørskov, J.K.; Bligaard, T.; Logadottir, A.; Bahn, S.; Hansen, L.B.; Bollinger, M.; Bengaard, H.S.; Hammer, B.; Sljivancanin, Z.; Mavrikakis, M.; Xu, Y.; Dahl, S.; Jacobsen, C.J.H.; J. Catal. 209 (2002) 275 *(Section 6.3)*
334 Nørskov, J.K.; Bligaard, T.; Hvolbæk, B.; Abild-Pedersen, F.; Chorkendorff, I.; Christensen, C.H.; Chem. Soc. Rev. 37 (2008) 2163 *(Section 6.3)*
335 Nummedal, L.; Røsjorde, A.; Johannessen, E.; Kjelstrup, S.; Chemical Engineering and Processing: Process Intensification, 44 (2005) 429 *(Sections 2.1.3, 3.3.6)*
336 Oudar, J; Catal. Rev.-Sci. Eng. 22 (1980) 171 *(Section 4.2.1)*

337 Ovesen, C.V.; Clausen, B.S.; Hammershøi, B.S.; Steffensen, G.; Askgaard, T.; Chorkendorff, I.; Nørskov, J.K.; Rasmussen, P.B.; Stolze, P.; Taylor, P.; J. Catal. 158 (1996) 180 *(Section 1.5.2)*
338 Pacheco, M.A.; Marshall, C.L.; Energy Fuels, 11 (1997) 2 *(Section 2.6.1)*
339 Paetsch, L.; Patel, P.S.; Maru, H.C.; Baker, B.S.; Abstr. 1986 Fuel Cell Seminar, Tuscon, AZ (1986) p. 143 *(Section 2.3.2)*
340 Park, G.-G.; Seo, D.J.; Park, S.-H.; Yoon, Y.-G.; Kim, C.-S.; Yoon, W.-L.; Chem. Eng. J. 101 (2004) 87 *(Section 1.4.1)*
341 Patil, C.S.; Annaland, M.v.S.; Kuipers, J.A.M.; Chem. Eng. Sci. 62 (2007) 2989 *(Sections 1.3.3, 1.3.4)*
342 Pedernera, M.N.; Pina, J.; Borio, D.O.; Bucalá, V.; Chem. Eng. J. 94 (2003) 29 *(Section 3.3)*
343 Peppley, B.A.; Amphlett, J.C.; Kearns, L.M.; Mann, R.F.; Appl. Catal., A 179 (1999) 21 *(Section 1.4.1)*
344 Perderau, M.; Acad. Sci. Ser., 267 (1968) 1107 *(Section 4.2.1)*
345 Periana, R.A.; Taube, D.J.; Gamble, S.; Taube, H.; Satoh, T.; Fuji, H.; Science 259 (1998) 560 *(Section 1.1.2)*
346 Phillips, T.R.; Mulhall, J.; Turner, G.E.; J. Catal. 15 (1969) 233 *(Section 3.5.3)*
347 Phillips, T.R.; Yarwood, T.A.; Mulhall, J.; Turner, G.E.; J. Catal. 17 (1979) 28 *(Section 3.5.3)*
348 Pichler, H.; Hector, A.; "Synthesegas" in Ullmann (Foerst, W.; ed.) vol. 16, p. 616, Urbane Schwarzenberg, München-Berlin (1965) *(Section 3.1)*
349 Pitt, R.; World Refining (2001) (Jan/Feb) 6 *(Section 2.2.1)*
350 Piwitz, M.; Larsen, J.S.; Christensen, T.S.; Proc. 1996 Fuel Cell Seminar, Orlando, FL (1996) 780 *(Section 5.3.3)*
351 Plácido, J.; Loureiro, L.V.; Comput. Chem. Eng. 22 (1998) S1035 *(Section 3.2.4)*
352 Plehiers, P.M.; Froment, G.F.; Chem. Eng. Technol., 12 (1989) 20 *(Section 3.3.6)*
353 Poling, B.E.; Prausnitz, J.M.; O'Connell, J.P.; "The Properties of Gases and Liquids", 5th ed., McGraw-Hill, New York (2000) *(Section 2.1.1)*
354 Ponec, V.; in "The Chemical Physics of Solid Surfaces and Heterogeneous Catalysis" (King, D.A.; Woodruf, D.P.; eds), Elsevier, Amsterdam, 1982, Chapt. 8 *(Section 6.1)*
355 Prettre, M.; Eichner, Ch.; Perrin, M.; Trans. Faraday Soc. 43 (1946) 335 *(Section 1.3.3)*
356 Qin, D.; Lapszewicz, J.; Catal. Today 21 (1994) 551 *(Section 6.4)*
357 Råberg, L.B.; Jensen, M.B.; Olsbye, U.; Daniel, C.; Haag, S.; Mirodatos, C.; Sjåstad, A.O.; J. Catal. 249 (2008) 250 *(Section 3.5.4)*
358 Rao, M.V.R; Plehiers, P.M.; Froment, G.F.; Chem. Eng. Sci. 43 (1988) 1223 *(Section 3.3.6)*

References

359 Rass-Hansen, J.; Christensen, C.H.; Sehested, J.; Helveg, S.; Rostrup-Nielsen, J.R.; Dahl, S.; Green Chem. 9 (2007) 1016 *(Section 1.4.1)*

360 Ratan, S.; Baade, W.; Wolfson, D.; Hydrocarb. Eng. 10 (7) (2005) 37 *(Section 2.2.1)*

361 Rethwisch, D.G.; Dumesic, J.A.; Appl. Catal. 21 (1986) 97 *(Section 1.5.2)*

362 Rhine, J.M.; Tucker, R.J.; "Modelling of Gas-Fired Furnaces and Boilers", British Gas in Association with McGraw-Hill, London (1991) *(Section 3.3.6)*

363 Richardson, J.T.; Paripatyadar, S.A.; Shen, J.C.; AIChE J. 34 (1988) 743 *(Section 2.7)*

364 Richardson, J.T.; Lei, M.; Turk, B.; Forster, K.; Twigg, M.V.; Appl. Catal., A 110 (1992) 217 *(Section 4.1.2)*

365 Richardson, J.T.; Twigg, M.V.; Stud. Surf. Sci. Catal. 91 (1995) 345 *(Section 3.2.1)*

366 Richardson, J.T.; Turk, B.; Twigg, M.V.; Appl. Catal., A 148 (1996) 97 *(Sections 4.1.2, 6.1)*

367 Richardson, J.T.; Hung, J.-K.; Zhao, J.; Stud. Surf. Sci. Catal. 136 (2001) 203 *(Section 5.3.2)*

368 Richardson, J.T.; Garrait, M.; Hung, J.-K.; Appl .Catal., A 255 (2003) 69 *(Section 2.7)*

369 Richter, B.; Schmidt, L.D.; Science, 305 (2004) 340 *(Section 1.4.1)*

370 Ridler, D.E.; Twigg, M.V.; in "Catalyst Handbook" (Twigg, M.V.; ed.), chapt. 5, Mansa Publishing, London (1989) *(Section 5.3.2)*

371 Rose, F.; Stahl Eisen, 95 (1975) 1012 *(Section 2.4.3)*

372 Rosenqvist, T.; J. Iron Steel Ind., London 176 (1954) 37 *(Sections 4.2.1, 5.4.1)*

373 Ross, J.R.H.; Steel, M.C.F.; J. Chem. Soc. Faraday Trans., 1, 69 (1973) 10 *(Section 6.1)*

374 Ross, L.L.; in "Pyrolysis: Theory and Industrial Practice" (Albright, L.F.; Crynes, B.L.; Corcocan, W.; eds.), Academic Press, New York (1983) 360 *(Sections 4.3.3, 5.3.2, 5.3.5)*

375 Rossini, F.D.; "Selected Values of Physical and Thermodynamic Properties of Hydrocarbons and Related Compounds", Carnegie Press, Washington (1953) *(Section 1.2.1)*

376 Rostrup-Nielsen, J.R.; J. Catal. 11 (1968) 220 *(Sections 4.2.1, 5.4.1, 5.4.2)*

377 Rostrup-Nielsen, J.R.; J. Catal. 21 (1971) 171 *(Sections 5.4.2, 5.4.4)*

378 Rostrup-Nielsen, J.R.; J. Catal. 27 (1972) 343 *(Sections 5.2.1, 5.2.2, 6.1)*

379 Rostrup-Nielsen, J.R.; J. Catal. 31 (1973) 173 *(Sections 3.5.2, 3.5.3, 4.3.1, 5.3.2, 5.4.5, 6.1, 6.2, 6.4)*

380 Rostrup-Nielsen, J.R.; J. Catal. 33 (1974) 184 *(Sections 5.2.1, 5.3.2, 6.1)*

381 Rostrup-Nielsen, J.R.; "Steam Reforming Catalysts", Danish Technical Press, Copenhagen (1975) *(Sections 1.2.1, 3.1, 3.3, 3.5.2, 3.5.3, 4.1.1, 4.2.1, 4.2.2, 4.2.3, 5.2.1, 5.2.2, 5.3.1, 5.3.2, 5.3.5, 5.4.2, 5.4.3, 5.4.4, 6.1)*

382 Rostrup-Nielsen, J.R.; Wrisberg, J.; Proc. 1st Conf. Natural Gas Utilisation, Dublin (1976) 5–23 *(Section 5.2.3)*

383 Rostrup-Nielsen, J.R.; Trimm, D.L.; J. Catal., 48 (1977) 155 *(Sections 5.2.1, 6.1, 6.2)*

384 Rostrup-Nielsen, J.R.; Chem. Eng. Prog. 73 (9) (1977) 87 *(Sections 1.2.5, 5.2.1, 5.3.1, 6.1)*

385 Rostrup-Nielsen, J.R.; Pedersen, K.; J. Catal. 59 (1979) 395 *(Section 5.5)*

386 Rostrup-Nielsen, J.R.; Tøttrup, P.B.; in Symp. "Science of Catalysis and its Applications in Industry", FPDIL, Sindri, India (1979) 379 *(Sections 3.5.3, 5.3.1, 5.3.3)*

387 Rostrup-Nielsen, J.R.; "Progress in Catalyst Deactivation" (Figueiredo, J.L.; ed.) Martin Nijhof, The Hague (1982) 209 *(Sections 5.4.2, 5.4.3)*

388 Rostrup-Nielsen, J.R.; in "Progress in Catalyst Deactivation" (Figueiredo, J.L.; ed), Martinus Nijhof, The Hague (1982) 127 *(Section 5.3.5)*

389 Rostrup-Nielsen, J.R.; in "Catalysis, Science and Technology" (Anderson, J. R.; Boudart, M.; eds.), Vol. 5, Chapter 1, Springer-Verlag, Berlin (1984) *(Sections Preface, 1.2.1, 1.2.4, 1.2.5, 2.4.3, 3.1, 3.2.1, 3.2.2, 3.3, 3.4.2, 3.5.1, 3.5.2, 3.5.3, 4.1.1, 4.1.2, 4.2.1, 4.2.2, 4.2.3, 4.3.1, 5.2.1, 5.2.3, 5.2.4, 5.3.1, 5.3.2, 5.3.4, 5.3.5, 5.4.2, 5.4.3, 5.4.4, 5.4.5, 5.5, 5.6, 6.2)*

390 Rostrup-Nielsen, J.R.; J. Catal. 85 (1984) 31 *(Sections 3.5.1, 5.2.1, 5.2.3, 5.3.2, 5.4.5, 5.5, 6.2)*

391 Rostrup-Nielsen, J.R.; Højlund Nielsen, P.E.; in "Catalyst Poisoning and Deactivation" (Wise, H.; Oudar, J.; eds.), Marcel Dekker, New York (1985) Chapter 7 *(Sections 1.5.1, 1.5.2, 5.6)*

392 Rostrup-Nielsen, J.R.; Skov, A.; Christiansen, L.J.; Appl. Catal. 22 (1986) 71 *(Sections 1.4.1, 5.3.3)*

393 Rostrup-Nielsen, J.R.; Christiansen, L.J.; Bak Hansen, J.-H.; Appl. Catal. 43 (1988) 287 *(Sections 3.3.3, 3.5.1)*

394 Rostrup-Nielsen, J.R.; Stud. Surf. Sci. Catal., 36 (1988) 73 *(Section 1.2.5)*

395 Rostrup-Nielsen, J.R.; Dybkjær, I.; Christiansen, L.J.; in NATO ASI "Chemical Reactor Technology for Environmentally Safe Reactors and Products", (de Lasa, H.J.; ed.), Kluwer Academic Publ., Dortrecht (1992) 249 *(Sections 2.5, 2.6.2, 3.3, 3.3.5)*

396 Rostrup-Nielsen, J.R.; Bak Hansen, J.-H.; J. Catal. 144 (1993) 38 *(Sections 2.4.2, 3.5.4, 5.2.1, 5.2.2., 5.2.3)*

397 Rostrup-Nielsen, J.R.; Catal. Today 18 (1993) 305 *(Sections 1.2.6, 2.6.4, 3.2.1)*

398 Rostrup-Nielsen, J.R.; Catal. Today 201 (1994) 257 *(Section 1.3.1)*
399 Rostrup-Nielsen, J.R.; Stud. Surf. Sci. Catal. 81 (1994) 25 *(Sections 1.2.2, 2.4.2)*
400 Rostrup-Nielsen, J.R.; Chem. Eng. Sci. 50 (1995) 4061 *(Section 1.1.1)*
401 Rostrup-Nielsen, J.R.; Christiansen, L.J.; Appl. Catal., A 126 (1995) 381 *(Sections 2.3.2, 4.3.1, 4.3.2)*
402 Rostrup-Nielsen, J.R.; Aasberg-Petersen, K.; Schoubye, P.S.; Stud. Surf. Sci. Catal. 107 (1997) 473 *(Sections 2.3.2, 2.7)*
403 Rostrup-Nielsen, J.R.; Bak Hansen, J.-H.; Aparicio, L.M.; J. Jpn. Pet. Inst. 40 (1997) 366 *(Sections 1.2.6, 2.4.2, 3.5.1, 4.2.1, 5.3.2)*
404 Rostrup-Nielsen, J.R.; Catal. Today, 37 (1997), 225 *(Section 1.1.1)*
405 Rostrup-Nielsen, J.R.; Dybkjær, I.; Christensen, T.S.; Stud. Surf. Sci. Catal. 113 (1998) 81 *(Sections 1.2.5, 5.3.2, 5.3.3, 5.3.4)*
406 Rostrup-Nielsen, J.R.; Proc. 15th World Petr. Congr., John Wiley & Sons, New York (1998) 767 *(Sections 2.5, 2.6.4)*
407 Rostrup-Nielsen, J.R.; Alstrup, I.; Catal. Today 53 (1999) 311 *(Sections 5.5, 6.1)*
408 Rostrup-Nielsen, J.R.; Catal. Today 63 (2000) 159 *(Sections 1.3.4, 1.3.5, 2.2.5)*
409 Rostrup-Nielsen, J.R.; Dybkjær, I.; Aasberg-Petersen, K.; Prepr.-Am. Chem. Socl, Div. Pet. chem.. 45 (2) (2000) 186 *(Sections 1.3.4, 2.6.5)*
410 Rostrup-Nielsen, J.R.; in "Combinatorial Catalysis and High Throughput Catalyst Design and Testing" (Derouane,E.G.; Lemos, F.; Corma, A.; Ribeiro, F.R.; eds), Kluwer Academic Publishers, Dortrecht (2000) 337 *(Sections 1.1.2, 1.3.3., 2.2.5)*
411 Rostrup-Nielsen, J.R.; Chem. Phys. Phys. Chem. 3 (2001) 283 *(Sections 1.4.1, 2.3.1, 2.3.2)*
412 Rostrup-Nielsen, J.R.; Stud. Surf. Sci. Catal. 139 (2001) 1 *(Section 5.2.1)*
413 Rostrup-Nielsen, J.R.; Catal. Today 71 (2002) 243 *(Sections 1.1.2, 1.3.2, 2.6.6, 3.1)*
414 Rostrup-Nielsen, J.R.; Rostrup-Nielsen, T.; Cattech 6 (4) (2002) 150 *(Sections 1.2.6, 1.4.1, 2.2.1, 2.2.2, 2.2.3, 2.2.4)*
415 Rostrup-Nielsen, J.R.; Sehested, J.; Nørskov, J.K.; Adv. Catal. 47 (2002) 65 *(Sections 1.2.1, 1.2.2., 1.2.5, 2.4.2, 2.6.6, 2.7, 3.3, 3.5.1, 3.5.2, 4.2.1, 4.2.3, 5.2.1, 5.2.2, 5.2.3, 5.3.1, 5.3.2, 5.3.3, 6.1, 6.2)*
416 Rostrup-Nielsen, J.R.; Appl. Catal., A 255 (2003) 3 *(Section 1.1.1)*
417 Rostrup-Nielsen, J.R.; Stud. Surf. Sci. Catal. 147 (2004) 121 *(Section 6.1)*
418 Rostrup-Nielsen, J.R.; Catal. Rev.-Sci. Eng. 46 (2004) 246 *(Sections 1.1.1, 1.4.2, 2.2.1, 2.3.2, 2.6.4)*
419 Rostrup-Nielsen, J.R.; Science 308 (2005) 1421 *(Sections 1.4.1, 1.4.3)*

420 Rostrup-Nielsen, J.R.; "Natural Gas: Fuel or Feedstock", Proc. NATO ASI "Sustainable Strategies for the Upgrading of Natural Gas: Fundamentals, Challenges and Opportunities, (Derouane, E.G.; Parmon, V.; Lemos, F.; Ribeiro, F.R.; eds), Springer, Dortrecht (2005) Chapt. 1, 3 *(Sections 1.1.1, 1.1.2, 2.3.2, 2.6.1, 2.6.5)*
421 Rostrup-Nielsen, J.R.; Hansen, J.B.; Helveg, S.; Christiansen, N.; Jannasch, A.-K.; Appl. Phys. A: Mater. Sci. Procss 85 (2006) 427 *(Section 5.4.5)*
422 Rostrup-Nielsen, J.R.; Nørskov, J.K.; Top. Catal. 40 (2006) 45 *(Section 6.3)*
423 Rostrup-Nielsen, J.R.; Pedersen, K.; Sehested, J.; Appl. Catal., A 330 (2007) 134 *(Sections 2.6.6, 4.2.3)*
424 Rostrup-Nielsen, J.R.; Stud. Surf. Sci. Catal. 167 (2007) 153 *(Section 5.4.5)*
425 Rostrup-Nielsen, J.R.; in "Handbook of Heterogeneous Catalysis", (Ertl, G.; Knözinger, H.; Schüth, F.; Weitkamp, J.; eds.) Weinheim (2008) Wiley-VCH Verlag, vol. 6, Chapt. 13.11, p. 2882 *(Sections 1.2.1, 1.2.4, 1.2.5, 3.1, 3.2.1, 4.2.1, 4.2.3, 5.2.1, 5.3.1, 5.3.2, 5.4.1, 6.1)*
426 Rostrup-Nielsen, J.R.; Winter-Madsen, S.; Prepr.-Am. Chem. Soc., Div. Pet. Chem. 53 (1) (2008) 82 *(Sections 1.2.4, 2.2.2, 2.2.3, 2.6.6)*
427 Rostrup-Nielsen, J.R.; Catal. Today 145 (2009) 72 *(Sections 1.2.4, 2.2.3, 2.6.5, 2.7, 5.3.2)*
428 Rostrup-Nielsen, T.; Joensen, F.; Madsen, J.; Højlund Nielsen, P.E.; Risø Int. Energy Conf., Roskilde, DK (2007) *(Section 2.6.4)*
429 Roy, S.; Pruden, B.B.; Grace, J.R.; Lim, C.J.; Chem. Eng. Sci. 54 (1999) 2095 *(Section 1.3.3)*
430 Ruan, L.; Besenbacher, F.; Steensgaard, I.; Lægsgaard, E.; Phys. Rev. Lett. 69 (1992) 3523 *(Sections 4.2.1, 5.4.2, 6.2)*
431 Saadi, S.; Hinnemann, B.; Helveg, S.; Appel, C.C.; Abild-Pedersen, F.; Nørskov, J.K.; Surf. Sci. 603 (2009) 162 *(Sections 5.2.2, 5.3.2)*
432 Sabatier, P.; Senderens, J-B.; CR Acad. Sci. 134 (1902) 514 *(Section 3.1)*
433 Sachtler, W.M.H.; Plank, P.v.d.; Surf. Sci. 18 (1969) 62. 596 *(Section 6.1)*
434 Salge, J.R.; Deluga, G.A.; Schmidt, L.D.; J. Catal. 235 (2005) 69 *(Section 1.3.3)*
435 Salge, J.R.; Dreyer, B.J.; Dauenhauer, P.J.; Schmidt, L.D.; Science 14 (2006) 801 *(Section 1.4.3)*
436 Salmi, T.; Wärnå, J.; Comput. Chem. Eng. 15 (1991) 715 *(Section 3.4.2)*
437 Salso, K.; (private information) *(Section 2.6.5)*
438 Sanfilippo, D.; Miracca, I.; Cornaro, U.; Mizia, F.; Malandrino, A.; Piccoli, V.; Rossini, S.; Stud. Surf. Sci. Catal. 147 (2004) 91 *(Section 1.2.6)*

References

351

439 Sanfilippo, D.; Miracca, I.; Catal. Today. 111 (2006) 133 *(Section 1.2.6)*
440 Satterfield, C.N.; "Heterogeneous Catalysis in Practice", McGraw-Hill, New York, 1980, 274 *(Sections 3,3,4, 3.4.1, 3.4.2, 5.3.5)*
441 Schaadt, A.; Arias, R.A.; Aicher, T.; Northorp, W.; Chem. Ing. Tech. 79 (2007) 1336 *(Section 1.4.1)*
442 Schildhauer, T.J.; Geissler, K.; Int. J. Hydrogen Energy 32 (2007) 1806 *(Section 1.4.1)*
443 Schmidt, L.D.; Stud. Surf. Sci. Catal. 130 (2006) 61 *(Section 1.3.3)*
444 Schmidt, L.D.; Prepr.-Am. Chem. Soc., Div. Pet. Chem. 52 (2) (2007) 82 *(Section 1.3.3)*
445 Schmieder, H.; Abeln, J.; Boukis, N.; Dinjus, E.; Kruse, A.; Kluth, M.; Petrich, G.; Sadri, E.; Schacht, M.; J. Supercrit. Fluids, 17 (2000) 145 *(Section 1.4.3)*
446 Schnell, C.R.; J. Chem. Soc. (B) (1970) 158 *(Section 6.1)*
447 Schulze, G.; Prausnitz, J.M.; Ind. Eng. Chem. Fundam. 20 (1981) 175 *(Section 2.1.1)*
448 Seglin, L.; (ed.), "Methanation of Synthesis Gas", Adv. Chem.Ser. 146, ACS, Washington, DC (1975) *(Section 2.6.6)*
449 Sehested, J.; Carlsson, A.; Janssens, T.V.W.; Hansen, P.L.; Datye, A.K.; J. Catal. 197 (2001) 200 *(Sections 4.2.1, 4.2.3)*
450 Sehested, J.; Jacobsen, C.J.H.; Rokni, S.; Rostrup-Nielsen, J.R.; J. Catal. 201 (2001) 206 *(Section 4.3.2)*
451 Sehested, J.; Seier Christensen, P.; Jacobsen, J.; Helveg, S.; Rostrup-Nielsen, J.R.; Abstracts 2005 ACS Meeting (2005) p. PETR-137 *(Section 5.2.3)*
452 Sehested, J.; Catal. Today 111 (2006) 103 *(Section 4.2.3)*
453 Sehested, J.; (unpublished results) (2006) *(Sections 4.2.3, 5.3.3)*
454 Selimovic, A.; Pålsson, J.; Sjunneson, L.; J. Power Sources 86 (2000) 442 *(Section 2.7)*
455 Seris, E.L.C.; Abramowitz, G.; Johnston, A.M.; Haynes, B.S.; Chem. Eng. Res. Des. 83 (2005) (6) 619 *(Sections 1.2.6, 2.3.1)*
456 Shabaker, J.W.; Huber, G.W.; Dumesic, J.A.; J. Catal. 222 (2004) 180 *(Sections 1.4.1, 5.3.2)*
457 Shakoor, A.; Henriksen, F.A.; Feddersen, M.; Ammonia Tech. Man. (2002) 279 *(Section 1.2.5)*
458 Sheldon, R.A.; J. Chem. Technol. Biotechnol. 68 (4) (1997) 381 *(Section 1.1.1)*
459 Sickafus, E.N.; Surf. Sci., 19 (1970) 181 *(Section 5.4.2)*
460 Simonsen, S.B.; Dahl, S.; Johnson, E.; Helveg, S.; J. Catal. 255 (2008) 1 *(Section 5.3.5)*
461 Sinbeck, D.R.; Dickenson, R.L.; Oliver, E.D.; "Coal Gasification Status", EPRI AP-3109, Synthetic Fuels Associates, Inc., Palo Alto, CA (1983) *(Section 1.4.2)*
462 Sinfelt, J.H.; Catal. Rev. 3 (1969) 175. 603 *(Sections 1.4.1, 6.1)*

463 Singh, C.P.P.; Saraf, D.N.; Ind. Eng. Chem. Proc. Des. Dev. 18 (1979), 1 *(Section 3.3.6)*
464 Singh, J.; Niessen, H.F.; Harth, R.; Fedders, H.; Reutler, H.; Panknin, W.; Müller, W.D.; Harms, H.; Nucl. Eng. Des. 78 (1984) 179 *(Section 3.3.5)*
465 Skjøth-Rasmussen, M.S.; Glarborg, P.; Østberg, M.; Johannesen, J.T.; Livbjerg, H.; Jensen, A.D.; Christensen, T.S.; Combust. Flame 136 (2004) 91 *(Section 1.3.2)*
466 Skjøth-Rasmussen, M.S.; Holm-Christensen, O.; Østberg, M.; Christensen, T.S.; Johannesen, J.T.; Jensen, A.D.; Glarborg, P.; Livbjerg, H.; Comput. Chem. Eng. 28 (2004) 2351 *(Section 1.3.2)*
467 Skov, A.; Pedersen, K.; Chen, C.; Coates, R.L.; Prepr. Symp.-Am. chem.. Soc., Div. Fuel Chem. 31 (1986) 137 *(Section 2.6.6)*
468 Smith, J.M.; van Ness, H.C.; Abbott, M.M.; "Chemical Engineering Thermodynamics", 6th edition, McGraw-Hill, Boston (2001) *(Sections 1.2.1, 2.1.2, 2.1.3)*
469 Snoeck, J.-W.; Froment, G.F.; Fowles, M.; Ind. Eng. Chem. Res. 41 (2002) 4253 *(Section 5.2.4)*
470 Soave, G.; Chem. Eng. Sci. 27 (1972) 1197 *(Section 2.1.1)*
471 Sogge, J.; Strøm, T.; Stud. Surf. Sci. Catal., 107 (1997) 561 *(Section 2.2.4)*
472 Solymosi, F.; Tolmacsov, P.; Kedves, K.; J. Catal. 216 (2003) 377 *(Section 3.5.4)*
473 Sorin, M.; Paris, J.; Comput. Chem. Eng. 23 (1999) 497 *(Section 2.1.3)*
474 Stagg, S.M.; Resasco, D.E.; Stud. Surf. Sci. Catal. 111 (1997) 543 *(Section 5.3.2)*
475 Stahl, H.; Rostrup-Nielsen, J.R.; Udengaard, N.R.; Abst. 1985 Fuel Cell Seminar, Tuscon, AZ (1985) 83 *(Section 3.3.5)*
476 Stefanidis, G.D.; Vlachos, D.G.; Kaisare, N.S.; Maestri, M.; AIChE J. 55 (2009) 180 *(Section 3.3.8)*
477 Steynberg, A.P.; Vogel, A.P.; Price, J.G.; Nel, H.G.; Proc. "Natural Gas Utilisation", AIChE Spring Nat. Meeting, Houston, TX (2001) 195 *(Section 2.6.5)*
478 Stigsson, L.; Int. Chem. Recovery Conf., Tampa, FL (1988) 2, 663 *(Section 1.4.3)*
479 Stitt, E.H.; Proc. NATO ASI "Sustainable Strategies for the Upgrading of Natural Gas: Fundamentals, Challenges and Opportunities", (Derouane, E.G.; Parmon, V.; Lemos, F.; Ribeiro, F.R.; eds), Springer, Dortrecht (2005) 185 *(Sections 2.6.2, 3.3.5)*
480 Strub, R.A.; Imarisio, G. "Hydrogen as an Energy Vector", Reidel, Dordrecht (1980) *(Section 2.2.1)*
481 Sundaram, K.M.; Froment, G.F.; Chem. Eng. Sci. 35 (1980) 364 *(Section 5.3.4)*
482 Sussman, M.V.; Energy, 5 (1980) 793 *(Section 2.1.3)*

483	Tanaka, Y.; Kikuchi, R.; Takeguchi, T.; Eguchi, K.; Appl. Catal., B 57 (2005) 211 *(Section 1.4.1)*
484	Taskin, M.E.; Dixon, A.G.; Nijemeisland, M.; Stitt, E.H.; Ind. Eng. Chem. Res. 47 (2008) 5966 *(Sections 3.3.1, 3.3.2, 3.4)*
485	Taylor, C.E.; Howard, B.H.; Myers, C.R.; Ind. Eng. Chem. Res. 46 (2007) 8906 *(Section 1.4.1)*
486	TerHar, I.W.; Vogel, J.R.; Proc. 6th World Petr. Congr. 4, Frankfurt (1969) 383 *(Section 1.3.1)*
487	Tessie du Motay, M.M.; Maréchal, M.; Bull. Soc. Chim. France 9 (1868) 334 *(Section 3.1)*
488	Tonkovich, A.Y.; Perry, S.; Wang, Y.; Qiu, D.; la Plante, T.; Rogers, W.A.; Chem. Eng. Sci. 59 (2004) 4819 *(Sections 1.2.6, 2.3.1, 3.3.8)*
489	Tomita, T.; Moriya, A.; Shinjo, T.; Kikuchi, K.; Sakamoto, T.; J. Jpn. Pet. Inst. 23 (1980) 69 *(Section 4.3.2)*
490	Töpfer, H.J.; Gas Wasserfach 117 (1926) 412 *(Sections 2.4.3, 5.5)*
491	Topp-Jørgensen, J.; Kiilerich Hansen, H.; Rostrup-Nielsen, J.R.; Jørn, E.; Paper at AIChE Fuels and Petrochem. Div., New Orleans (April 1986) *(Section 2.6.4)*
492	Topp-Jorgensen, J.; Stud. Surf. Sci. Catal. 36 (1988) 293 *(Sections 2.6.1, 2.6.4, 5.3.3)*
493	Topsøe, H.F.A.; Inst. Gas Eng. J, 6 (1966) 401 *(Sections 2.4.1, 3.1, 3.3)*
494	Topsøe, H.F.A.; Fossum Poulsen, H.; Nielsen, A.; Chem. Eng. Progr. 63 (1967) 67 *(Section 2.5)*
495	Topsøe, H.F.A.; Rostrup-Nielsen, J.R.; Scan. J. Metall. 8 (1979) 168 *(Section 2.4.3)*
496	Torbati, R.; Cimino, S.; Lisi, L.; Russo, G.; Catal. Lett. 127 (2009) 260 *(Section 5.4.5)*
497	Tøttrup, P.B.; Appl. Catal. 4 (1982) 377 *(Section 3.5.3)*
498	Trimm, D.L.; Catal. Rev.-Sci. Eng. 16 (1977) 155 *(Section 5.2.1)*
499	Trimm, D.L.; in "Pyrolysis: Theory and Industrial Practice" (Albright, L.F.; Crynes, B.L.; Corocan, W.H.; eds.), Academic Press, New York (1983) Chapt. 9, 203 *(Sections 5.3.4, 5.3.5)*
500	Trimm, D.L.; Catal. Today, 37 (1997) 233 *(Section 5.3.2)*
501	Trimm, D.L.; Önsan, Z.I.; Catal. Rev.-Sci. Eng. 43 (2001) 30 *(Section 1.4.1)*
502	Udengaard, N.R.; Christiansen, L.J.; Summers, W.A.; Rastler, D.M.; "Endurance Testing of a High-Efficiency Steam Reformer for Fuel Cell Power Plants", EPRI AP-6071, Final Report (October 1988) *(Section 3.3.5)*
503	Udengaard, N.R.; Bak Hansen, J.-H.; Hanson, D.C.; Stal, J.A.; Oil Gas J. 90 (10) (1992) 62 *(Section 5.5)*
504	Udengaard, N.R.; Olsen, A.; Wix-Nielsen, C.; Pittsburgh Coal Conf. (2006) *(Section 2.6.6)*
505	van der Lee, M.K.; van Dillen, A.N.; Geus, J.W.; deJong, K.P.; Bitter, J.H.; Carbon 44 (2006), 629 *(Section 5.2.1)*

506 van Grootel, P.W.; Hensen, E.J.M.; van Santen, R.A.; Surf. Sci. 603 (2009) 3275 *(Section 6.2)*
507 van Hardeveld, R.; Hartog, F.; Surf. Sci. 15 (1969) 180 *(Sections 6.1, 6.2)*
508 van Santen, R.A.; Neurock, M.; Shelly, S.G.; Chem. Rev. 110 (2010) 2005 *(Section 6.2)*
509 Vang, R.T.; Honkala, K.; Dahl, S.; Vestergaard, E.K.; Schnadt, J.; Lægsgaard, E.; Clausen, B.S.; Nørskov, J.K.; Besenbacher, F.; Nat. Mater. 4 (2005) 160 *(Section 6.2)*
510 Vannby, R.; Nielsen, C.S.; Kim, J.S.; Hydrocarb. Tech. Int. (1993) *(Sections 2.4.2, 3.2.2)*
511 Venkataraman, K.; Wanat, E.C.; Schmidt, L.D.; AIChE J. 49 (2003) 1277 *(Section 3.3.8)*
512 Villadsen, J.; Michelsen, M.L.; "Solution of Differential Equation Models by Polynomial Approximation", Prentice-Hall, Englewood Cliffs, NJ (1978) *(Sections 3.3 1, 3.4.1)*
513 Voecks, G.E.; in "Handbook of Fuel Cells" (Vielstich, E.; Lamm, A.; Gassteiger, H.A. eds), vol 3, 229, J. Wiley, Chichester (2003) *(Section 4.3.3)*
514 Vortmeyer, D.; Schuster, J.; Chem. Eng. Sci. 38 (1983) 1691 *(Section 3.3.1)*
515 Wagner, W.; Kruse, A.; "Properties of Water and Steam", The Industrial Standard IAPWS-IF9, Springer, Berlin (1998) *(Section 2.1.1)*
516 Wei, J.M.; Iglesia, E.; J. Catal. 224 (2004) 370 *(Sections 3.5.2, 3.5.4)*
517 Wei, J.M.; Iglesia, E.; J. Phys. Chem. B. 108 (2004) 4094 *(Sections 3.5.2, 3.5.4, 4.3.1, 6.2, 6.4)*
518 Weisz, P.B.; Goodwin, R.B.; J. Catal. 6 (1966) 227 *(Section 5.3.5)*
519 Wesenberg, M.H.; Svendsen, H.F.; Ind. Eng. Chem. Res. 46 (2007) 667 *(Section 3.3.5)*
520 Wijngaarden, R.J.; Westerterp, R.; Chem. Eng. Sci., 44 (1989) 1653 *(Section 3.3.3)*
521 Winter-Madsen, S.; Bak Hansen, J.-H.; Rostrup-Nielsen, J.R.; "Industrial Aspects of CO_2-reforming", AIChE Meeting, Houston (1997) *(Section 2.4.2)*
522 Winter-Madsen, S.; Olsson, H.; Hydrocarb. Eng. (July 2007), 37 *(Section 2.2.3)*
523 Wise, H.; McCarty, J.; Oudar, J.; in "Deactivation and Poisoning of Catalysts" (Wise, H.; Oudar, J.; eds.), Marcel Dekker, New York (1985) *(Sections 4.2.1, 5.4.1, 5.4.2)*
524 Wörner, A.; Tamme, R.; Catal. Today 46 (1998) 165 *(Section 2.7)*
525 Xu, J.; Froment, G.F.; AIChE J. 35 (1989) 88 *(Sections 3.3, 3.3.5, 3.3.6, 3.4, 3.4.1, 3.4.3, 3.5.1, 3.5.2)*
526 Xu, J.; Chen, L.; Tan, K.F.; Borgna, A.; Saeys, M.; J. Catal. 261 (2009) 158 *(Section 5.3.2)*

References

527 Xu, M.; Lunsford, J.H.; Goodman, D.W.; Bhattacharyya, A.; Appl. Catal., A 149 (1997) 289 *(Section 2.6.1)*
528 Yan, W.; Kontogeorgis, G.M.; Stenby, E. H.; Fluid Phase Equilib. 276 (2009) 75 *(Section 2.1.1)*
529 Yaripour, F.; Baghaei, F.; Schmidt, I.; Perregaard, J.; Catal. Commun. 6 (2005) 147 *(Sections 2.6.1, 5.3.3)*
530 Yaseneva, P.; Pavlova, S.; Sadykov, V.; Alinka, G.; Lykashevich, A.; Rogov, V.; Belockapkine, S.; Ross, J.R.H.; Catal. Today. 137 (2008) 23 *(Section 1.4.1)*
531 Yurchak, S.; Stud. Surf. Sci. Catal. 36 (1988) 251 *(Sections 2.6.1, 2.6.4, 5.3.3)*
532 Zdonik, S.B.; Green, E.J.; Hallee, L.P.; Oil Gas J. 65 (26) (1967) 96 *(Sections 1.2.5, 4.3.3)*
533 Zdonik, S.B.; in "Pyrolysis: Theory and Industrial Practice" (Albright, L.F.; Crynes, B.L.; Corcoran, W.H.; eds.), Academic Press, New York (1983) 377 *(Section 4.3.3)*
534 Zeng, Z.; Natesan, K.; Chem. Mater. 17 (2005) 3794 *(Section 5.2.2)*
535 Zhang, J.; Schneider, A.; Inden, G.; Corros. Sci. 50 (2008) 1020 *(Section 5.2.2)*
536 Zhang, P.; Rong, G.; Wang, Y.; Comput. Chem. Eng. 25, (2001), 941 *(Section 3.2.4)*
537 Zhang, Q.; He, D.; Zhu, Q.; J. Nat. Gas Chem. 12 (2003) 81 *(Section 1.1.2)*
538 Zhang, Z.; Verykios, X.E.; Catal.Lett. 38 (1996) 175 *(Section 3.5.4)*
539 Zhang, L.; Liu, J.; Li, W.; Guo, C.; Zhang, J.; J. Nat. Gas Chem. 18 (2009) 55 *(Section 1.4.1)*
540 Ziemecki, S.B.; Stud. Surf. Sci. Catal. 38 (1987) 625 *(Section 5.2.2)*
541 Zyskin, A.G.; Avetisov, A.K.; Kuchaev, V.L.; Shapatina, E.N.; Christiansen, L.J.; Kinet. Catal. 48(3), 2007, 337 *(Sections 3.4.3, 3.5.2)*
542 Brit. Pat. 1182829 (1967), (Haldor Topsøe (Rostrup-Nielsen, J.R.)) *(Section 5.3.2)*
543 DRP 51572 (1889) (Mond, L.; Langer, B.;) Chem. Zentralbl II, 32 (1890) *(Section 3.1)*
544 DRP 296866 (1912) (BASF (Mittasch, A.; Schneider, C.)) *(Section 3.1)*
545 EU Patent 0470 626 B1 (1990) (Hague, I.-U.; Trimm, D.L.) *(Section 5.3.2)*
546 US Pat. 1934836 (1927) (I.G. Farben (Wietzel, G.; Haller, W.; Hennicke, W.;)) *(Section 3.1)*
547 US Patent 3766278 (1973) (Haldor Topsøe (Andersen, K.J.; Rostrup-Nielsen, J.R.; Wrisberg, J.)) *(Section 4.3.3)*
548 US Pat. 3795485 (1974) (M.J.P. Bogart) *(Section 1.2.6)*
549 US. Pat. 5624964 (1997) (Mobil (Cimini, R.J.; Marler, D.O.; Shinnar, R.; Teitmann, G.J.)) *(Section 1.2.6)*

Author index

A
Aasberg-Petersen, K. [1], [2], [3], [4], [34], [41], [402], [409]
Abbott, M.M. [468]
Abeln, J. [445]
Abild-Pedersen, F. [5], [6], [7], [8], [226], [264], [334], [431]
Abramowitz, G. [455]
Adhikari, S. [9], [220]
Adjiman, C.S. [12]
Adris, A.M. [10], [181]
Agnew, J.B. [11]
Agrawal, P.K. [39]
Aguiar, P. [12]
Aicher, T. [441]
Akers, W.W. [13]
Alberton, A.L. [14]
Aldrige, C.L. [168]
Alinka, G. [530]
Allendorf, M.D. [156]
Alstrup, I. [15], [16], [17], [21], [47], [49], [179], [407]
Al-Ubaid, A.S. [181]
Alzarmora, L.E. [18]
Amphlett, J.C. [343]
Andersen, K.H. [19], [20]
Andersen, K.J. [76], [547]
Andersen, N.T. [21]
Andersen, S.L. [112]
Anderson, M.P. [264]
Andrew, S.P.S. [22]
Ang, T.P. [280]
Annaland, M.v.S. [341]
Aparicio, L.M. [23], [403]
Apel'baum, L.O. [58], [59], [271]
Appel, C.C. [431]
Arias, R.A. [441]
Aris, R. [24]
Armstrong, P.A. [25]
Arnstad, B. [26]

Ashcroft, A.T. [27]
Askgaard, T. [337]
Avetisov, A.K. [28], [29], [541]
Azar, C. [30]

B
Baade, W. [360]
Back, M.H. [31]
Back, R.A. [31]
Badmaev, S.D. [32]
Baghaei, F. [529]
Bahn, S. [255], [333]
Bak Hansen, J.-H. [2], [28], [111], [120], [393], [396], [403], [503], [521]
Baker, B.S. [339]
Baker, R.T.K. [33]
Bakkerud, P. [34]
Balandin, A.A. [35]
Balashova, S.A. [35]
Barber, M.A. [33]
Barcicki, J. [36]
Barreto, G.F. [310]
Bartholomew, C.H. [37], [38], [39]
Basile, A. [40]
Basini, L. [41]
Bauer, R. [42], [43]
Becker, M. [44]
Beckerle, J.D. [98]
Beebe Jr., T.P. [45]
Beeck, O. [46]
Beedle, S.C. [131]
Belockapkine, S. [530]
Belyaev, V.D. [32]
Bender, E. [324]
Bengaard, H.S. [47], [48], [333]
Bennett, D.L. [99]
Bernardo, C.A. [49]
Berthelsen, A.C. [187]
Besenbacher, F. [50], [51], [277], [430], [509]

Beutier, D. [52]
Beyer, F. [88]
Bharadwaj, S. [53]
Bhatta, K.S.M. [54]
Bhattacharyya, A. [527]
Bischoff, K.B. [199]
Bitsch-Larsen, A. [242]
Bittencourt, R.C. [14]
Bitter, J.H. [55], [505]
Blackburn, W.H. [150]
Blakely, D.W. [56]
Bligaard, T. [8], [57], [264], [333], [334]
Böcker, D. [285]
Bodrov, I.M. [58], [59]
Boellard, E. [278]
Bogart, M.J.P. [548]
Bøgild Hansen, J. [60], [61], [62], [63], [64], [65], [66], [96], [170], [195], [183], [182], [235], [329], [421]
Bohlbro, H. [68], [67], [69], [168]
Bollinger, M. [333]
Bonanos, N. [321]
Borgna, A. [526]
Borio, D.O. [342]
Borowiecki, T. [36], [70]
Bose, A.C. [71]
Bossen, B.S. [72]
Boudart, M. [73], [75], [74],
Boukis, N. [76], [445]
BP [78], [77]
Bradford, M.C.J. [79], [81], [80]
Brandon, N.P. [12]
Braunschweig, B. [82]
Bredesen, R. [26]
Breen, J.P. [83]
Bresler, S.A. [84]
Bridger, G.W. [85], [86], [87]
Bridgwater, A.V. [133]
Brightling, P. [88]
Briscoe, M. [193]
Broman, E. [89]
Bromberg, L. [90]
Brown, A. [91]
Brungs, A.J. [122]
Brüniche-Olsen, N. [118]
Bucalá, V. [342]

Burgt, M.v.d. [230]
Bussani, G. [92]
ByrneJr., P.J. [93]

C
Camp, D.P. [13]
Cant, N.W. [258]
Cao, C. [94]
Carlsson, A. [449]
Carolan, M.F. [95], [99], [317]
Carstensen, J.H. [20], [89] [96], [97], [118]
Casanova, A. [204]
Ceyer, S.T. [98]
Cheetham, A.K. [27]
Chen, C. [293]
Chen, C. [467]
Chen, C.M. [95], [99], [317]
Chen, L. [526], [280]
Chen, Z. [100]
Cheng, Z. [101]
Chew, K. [102]
Chiari, M. [92]
Chin, Y.-H. [103]
Chincen, G.C. [85]
Chorkendorff, I. [5], [47], [50], [104], [135], [179], [264], [290], [334], [337]
Chornet, E. [132]
Choudhary, V.R. [105]
Christensen, T.S. [2]
Christensen, C.H. [57], [106], [334], [359]
Christensen, P.S. see Seier Christensen, P.
Christensen, T.S. [107], [108], [109], [110], [111], [172], [350], [405], [465], [466]
Christiansen, L.J. [112], [113], [114], [115], [116], [117], [118], [119], [120], [330], [392], [393], [395], [401], [502], [541]
Christiansen, N. [195], [421]
Chun, C.M. [121]
Cimini, R.J. [549]
Cimino, S. [496]
Clarigde, J.B. [122]
Clarke, S.H. [123]

Clausen, B.S. [17], [48], [50], [124], [216], [226], [255], [266], [337], [509]
Coates, R.L. [467]
Cobb, J.W. [149]
Cockerham, R.G. [125], [126]
Cohn, D.R. [90]
Corella, J. [127]
Cornaro, U. [438]
Cortright, R.D. [128]
Coute, N. [129]
Crabtree, R.H. [130]
Cresswell, D.L. [157]
Cromarty, B.J. [131]
Czernik, S. [132], [133]

D
Dagle, R. [103]
Dagle, R.A. [134]
Dahl, S. [57], [135], [255], [333], [359], [460], [509]
Dahmen, N. [225]
Dalle Nogare, D. [242]
Dalmon, J.A. [167]
Daniel, C. [357]
Danner, R.P. [137]
Daszkowski, T. [136]
Datye, A.K. [134], [449]
Daubert, T.E [137], [212]
Dauenhauer, P.J. [138], [435]
Davda, R.R. [128]
Davidson, O. [315]
Davies, H.S. [139], [140]
Davy, H. [141]
De Deken, J.C. [142]
De Groote, A.M. [143]
de la Piscina, P.R. [302]
Deam, R.J. [144]
de Bokx, P.K. [278]
de Coninck, H. [315]
Deelen, M.J.A.G. [279]
Degenstein, N.J. [242]
de Jong, K.P. [144], [145], [146], [505]
Deluga, G.A. [147], [434]
Denbigh, K.G. [148]
Denis, A. [36]
Dent, F.J. [149], [150]
DeRosset, A.J. [210]

Devos, E.F. [142]
de Wild, P.J. [151]
Dibbern, H.C. [152]
Dickenson, R.L. [461]
Dicks, A.L. [123]
Dinjus, E. [76], [225], [445]
Dissanayake, D. [153]
Diver, R.B. [154], [155], [156], [234]
Dixon, A.G. [54], [157], [158], [159], [160], [484],
Djéga-Mariadassou, G. [75]
DOE [161]
Dohnalkova, A.C. [103]
Dowden, D.A. [162], [163]
Dreyer, B.J. [164], [435]
Dry, M. [165]
Dudukovic, M.P. [166]
Dumesic, J.A. [128], [207], [247], [248], [249], [277], [361], [456]
Duprez, D. [167]
Dybkjær, I. [2], [4], [3], [34], [109], [111], [168], [169], [170], [171], [172], [208], [395], [405], [409]

E
Eastwood, A.H. [150]
Edgar, T.F. [173]
Edmonds, T. [174]
Edwards, M.A. [175]
Edwards, T.J. [177], [176]
Efstathiou, A.M. [178]
Egeberg, R.C. [135], [179]
Egeblad, K. [106]
Eguchi, K. [483]
Eichner, Ch. [355]
Eigenberger, G. [136]
Eisenberg, B. [180]
Elangovan, S. [202]
Elnashaie, S.S.E.H. [100], [181], [262]
Emonts, B. [182], [183]
Engbæk, J. [5]
Enger, B.C. [184]
Epstein, M. [185]
Eriksen, D. [262]
Erley, W. [186]
Ernst, W.S. [187]
Estadt, M.M. [214]

Evans, J.E. [305]
Evans, S.T. [249]

F
Faravelli, T. [92]
Farnell, P. [88]
Farrauto, R.J. [37], [188]
Feates, F.S. [33]
Fedders, H. [189], [464]
Feddersen, M. [457]
Feik, C. [132]
Felcht, U.H. [190]
Feng, D. [191]
Fernando, S. [9], [220]
Fiato, R.A. [180]
Finlayson, B.A. [296]
Finneran, J.A. [312]
Fischer, F. [192]
Fish, J.D. [155], [234]
Fleisch, T.H. [193], [194]
Flower, J.R. [301]
Fontell, E. [195]
Fontes, C.E. [14]
Foord, J.S. [27]
Forster, K. [364]
Fossum Poulsen, H. [494]
Foster, C. [88]
Foster, E.P. [25], [99]
Fowles, M. [469]
Fox III, J.M. [196]
Fredenslund, A. [197]
Freide, J.F. [193]
Fremery, M. [205]
French, R. [132]
Frennet, A. [198]
Froment, G.F. [142], [143], [199], [200], [201], [294], [352], [358], [469], [481], [525],
Frost, L. [202]
Fuji, H. [345]
Fukui, T. [241]

G
Galla, U. [76]
Gallucci, F. [40]
Gam, E.A. [329]
Gamble, S. [345]

Gani, R. [82]
Gard, N.R. [203]
Garrait, M. [234], [368]
Geerlings, J.J.C. [326]
Geissler, K. [442]
Gélin, P. [323]
George, R. [204]
Geus, J.W. [146], [278], [505]
Gierlich, H.H. [205]
Gioventu, G. [92]
Giroux, T. [188]
Glarborg, P. [465], [466]
Gmehling, J. [197]
Goetsch, D.A. [206]
Gohr, R.J. [93]
Gokhale, A.A. [207]
Gøl, J.N. [34], [208]
Golebiowski, A. [70]
Goodman, D.W. [45], [527]
Goodwin, R.B. [518]
Göppert, U. [209]
Goring, R.L. [210]
Gråbæk, L. [124]
Grabke, H. [211]
Graboski, M.S. [212]
Grace, J.R. [10], [262], [429]
Greeley, J. [6], [8]
Green, E.J. [532]
Green, M.L.H. [27], [122]
Grenoble, D.C. [213], [214]
Grey, C.P. [27]
Griffith, R.H [215]
Grottoli, M.G. [92]
Grube, T. [183]
Grunwaldt, J.-D. [216], [266]
Grzegorzyk, W. [36]
Guarinoni, A. [41]
Gunnardson, H.H. [25]
Guo, C. [539]

H
Ha, S.Y. [299]
Haag, S. [357]
Hague, I.-U. [217], [545]
Hale, J.G. [144]
Hallee, L.P. [532]
Haller, W. [546]

Author Index

361

Hammer, B. [50], [218], [333]
Hammershøi, B.S. [97], [266], [337]
Hansen, J.B. see Bøgild Hansen, J.
Hansen, L. [109]
Hansen, L.B. [333]
Hansen, P.L. [124], [226], [449]
Hanson, D.C. [503]
Harms, H. [44], [219], [464]
Harris, P.S. [33]
Harrison, J.W. [319]
Harth, R. [189], [464]
Hartog, F. [507]
Hartvigsen, J. [202]
Haryanto, A. , [9], [220]
Haslam, R.T.[93]
Haugaard, J. [221]
Häussinger, P. [222]
Haynes, B.S. [223], [455]
He, D. [537]
Hebden, D. [139], [150]
Hector, A. [348]
Heinemann, H. [224]
Heinrich, E. [225]
Helveg, S. [7], [226], [227], [264], [359], [421], [431], [451], [460]
Hennicke, W. [546]
Henningsen, T. [76]
Henriksen, F.A. [457]
Hensen, E.J.M. [506]
Hershkowitz, F. [121]
Hertzberg, T. [288]
Heyen, G. [228]
Hickman, D.A. [229]
Higman, C. [230]
Hiller, H. [231]
Himmelblau, D.M. [173]
Hinnemann, B. [264], [431]
Hinrichsen, K.O. [232]
Hochgreb, S. [90]
Hoek, A. [233]
Hogan, R.E. [156], [234]
Höhlein, B. [182], [183], [189], [219]
Højlund Nielsen, P.E. [64], [65], [235], [236], [391], [428]
Holladay, J.D. [237]
Holm-Christensen, O. [466]
Holmen, A. [26], [184], [238]

Holmestad, R. [26]
Holm-Larsen, H. [239], [240]
Homs, N. [302]
Honkala, K. [509]
Horiuchi, T. [241]
Horn, R. [242]
Hottel, H.C. [244], [243]
Hou, P. [245]
Houken, J. [329]
Howard, B.H. [485]
Hu, G. [293]
Hu, J. [103], [237]
Huang, H. [246]
Huber, G.W. [247], [248], [249], [456]
Humphries, K.J. [139]
Hung, J.-K. [367], [368]
Hunter, R. [283]
Hutchings, G.J. [175], [246]
Hvolbæk, B. [334]
Hyman, M.H. [250]

I

IEA [251]
Iglesia, E. [516], [517]
Ilinich, Q. [188]
Imarisio, G. [480]
Inden, G. [535]
Ioannides, T. [252]
Ireland, J.O. [84]
Irving, P.M. [253], [319]

J

Jackson, S.D. [254]
Jacob, K. [327]
Jacobsen, C.J.H. [255], [256], [450]
Jacobsen, J. [451]
Jaeger, B. [257]
Jakobsen, J.G. [264]
Jannasch, A.-K. [421]
Janssens, T.V.W. [449]
Jarvan, J.E. [113]
Jensen, A.D. [465], [466]
Jensen, M.B. [357]
Jiang, C.J. [258]
Jockel, H. [259]
Joensen, F. [60], [260], [261], [428]
Johannesen, J.T. [465], [466]

Johannessen, E. [335]
Johnsen, K. [262]
Johnson, A.D. [98]
Johnson, B.F.G. [280]
Johnson, E. [460]
Johnsson, J. [263]
Johnston, A.M. [455]
Jones, G. [264]
Jorgensen, M.H. [69]
Jørn, E. [491]

K
Kaddouri, A. [323]
Kaisare, N.S. [476]
Kalitventzeff, B. [228]
Kammer, K. [321]
Kane, L. [265]
Kappen, P. [266]
Karim, G.A. [267]
Katzer, J.R. [39], [268]
Kawazuishi, K. [269]
Kay, B.D. [45]
Kearns, L.M. [343]
Kedves, K. [472]
Kemball, C. [270]
Kersten, L.B.J.M. [233]
Khimyak, T. [280]
Khomenko, A.A. [271]
Kiely, C.J. [175]
Kiilerich Hansen, H. [491]
Kikuchi, E. [272], [273]
Kikuchi, K. [489]
Kikuchi, R. [483]
Kim, C.-S. [340]
Kim, D.H. [286]
Kim, D.H. [299]
Kim, J.S. [510]
King, D.L. [237]
Kivisaari, T. [195]
Kjær, J. , [274], [275], [276]
Kjelstrup, S. [335]
Kladi, A. [178]
Kleis, J. [264]
Klette, H. [26]
Klop, W. [278]
Kluth, M. [445]
Knudsen, J. [277]

Koch, A.J.H.M. [278]
Kochloefl, K. [232]
Koerts, T. [279]
Koh, A.C.W. [280]
Kohl, A.L. [281]
Kolb, G. [282]
Kolbeinsen, L. [262]
Kondarides, D.I. [298]
Kontogeorgis, G.M. [528]
Kopfle, J. [283]
Kreetz, H. [306]
Kresge, C.T. [287]
Kriebel, M. [284]
Kruissink, E.C. [18]
Krummenmacher, J.J. [164]
Kruse, A. [445], [515]
Kubo, Y. [241]
Kuchaev, V.L. [28], [541]
Kugeler, K. [285]
Kuipers, J.A.M. [341]
Kundu, A. [286]
Kunkes, E.L. [277]
Kuo, J.C.W. [287]
Kvamsdal, H.M. [288]

L
Lægsgaard Jørgensen, S. [1], [182]
Lægsgaard, E. [430], [509]
Lange, J.P. [289]
Langer, B. [543]
laPlante, T. [488]
Lapszewicz, J. [356]
Larsen, J.H. [135], [290]
Larsen, J.S. [350]
Larsen, T. [76]
Larson, E. [30]
Laursen, J.K. [291]
Leather, J. [144]
Lee, I.C. [164]
Lee, J.S. [292]
Lee, M.B. [98]
Lei, M. [364]
Lei, N. [293]
Leong, W.K. [280]
Lercher, J.A. [55]
Lerou, J.J. [294], [295]
Levitan, R. [155]

Levy, M. [155]
Li, C.-H. [296]
Li, J. [101]
Li, W. [539]
Li, Z. [297]
Lienard, G. [198]
Liguras, D.K. [298]
Lim, C.J. [10], [429]
Lin, J. [280]
Lin, S.S.-Y. [299]
Linden, H.R. [300]
Lindgren, K. [30]
Linnhoff, B. [301]
Lisi, L. [496]
Liu, J. [539]
Livbjerg, H. [221], [465], [466]
Llorca, J. [302]
Lloyd, L. [303]
Lobo, L.S. [304]
Lobo, W.E. [305]
Lødeng, R. [184]
Logadottir, A. [135], [255], [333]
Lohmüller, R. [222]
Loos, M. [315]
López-Cartes, C. [226]
Loureiro, L.V. [351]
Lovegrove, K. [306]
Lui, G.-J. [297]
Lui, Y. [188]
Lunsford, J.H. [153], [307], [527]
Luzzi, A. [306]
Lybye, D. [321]
Lykashevich, A. [530]
Lytken, O. [5]

M

Machiels, C.J. [308]
Madron, F. [309]
Madsen, J. [428]
Maestri, M. [476]
Malandrino, A. [438]
Mann, R.F. [343]
Maréchal, E. [228]
Maréchal, M. [487]
Mariani, N.J. [310]
Marler, D.O. [549]
Marquez-Alvarez, C. [122]

Marsden, C.C. [106]
Marsh, J.D.F. [215]
Marshall, C.L. [338]
Martin, G.A. [311]
Martínez, O.M. [310]
Maru, H.C. [339]
Mathiesen, J. [57]
Maudlin, C.H. [180]
Maurer, G. [177], [209], [324]
Mavrikakis, M. [207], [277], [333]
Mayo, H.C. [312]
Mazanec, T. [313]
McCarroll, J.J. [174]
McCarty, J. [523]
McCarty, J.G. [314]
McDonald, J. [295]
Meeker, D. [245]
Meirovitch, E. [155]
Mendez, M. [167]
Metwally, M. [267]
Metz, B. [315]
Meyer, L. [315]
Michelsen, M.L. [316], [512]
Miller, C.F. [95], [317]
Miller, J.E. [156]
Mills, G.A. [318]
Minford, E. [95], [317]
Ming, Q. [319]
Miracca, I. [438], [439]
Mirodatos, C. [357]
Mittasch, A. [320], [544]
Mizia, F. [438]
Mocciaro, C. [310]
Mogensen, M. [321]
Moignard, L.A. [150]
Molenbroek, A.M. [48], [50], [216]
Molina, G. [127]
Möllersten, K [30]
Mollerup, J.M. [316]
Mond, L. [543]
Mori, T. [241]
Morita, Y. [272]
Moriya, A. [489]
Moseley, F. [322]
Moses, P.G. [8]
Mosqueda, B. [323]
Muhler, M. [232]

Mulhall, J. [346], [347]
Müller, G. [324]
Müller, W.D. [44], [464]
Mullins, C.B. [179]
Munter, T.R. [8]
Murali, N. [220]
Murrell, A.J. [27]
Murthy, M.V.K. [325]
Murty, C.V.S. [325]
Myers, C.R. [485]

N
Natesan, K. [534]
Nazimek, D. [36]
Neelis, M.L. [326]
Nel, H.G. [477]
Neumann, B. [327]
Neurock, M. [508]
Newman, J. [176], [177]
Ni, W.-D. [297]
Nielsen, A. [328], [329], [494]
Nielsen, C.S. see Stub Nielsen, C.
Nielsen, G. [5]
Nielsen, L.P. [48]
Nielsen, N.K. [19]
Nielsen, P. [330]
Nielsen, R.B. [281]
Nielsen, S.E. [256]
Niemantsverdriet, J.W. [104]
Niessen, H.F. [285], [464]
Nijemeisland, M. [159], [160], [484]
Nilekar, A.U. [277]
Nørskov, J.K. [5], [6], [7], [8], [47], [48], [50], [57], [135], [218], [226], [255], [264], [334], [333], [337], [415], [422], [431], [509]
Northorp, W. [441]
Nummedal, L. [335]

O
O'Brien, C. [90]
O'Connell, J.P. [353]
Olesen, P. [152]
Oliver, E.D. [461]
Ollis, D.F. [214]
Olsbye, U. [357]
Olsen, A. [504]

Olsen, C. [17]
Olsson, H. [522]
Olsvik, O. [288]
Önsan, Z.I. [501]
Ortego Jr., J.D. [129]
Osaki, T. [241]
Østberg, M. [41], [465], [466]
Oudar, J [336], [523]
Ovesen, C.V. [337]
Oyama, S.T. [292]

P
Pacheco, M.A. [338]
Paetsch, L. [339]
Palermo, R.E. [287]
Palo, D.R. [94], [134]
Pålsson, J. [63], [195], [454]
Panknin, W. [464]
Pannell, R.B. [38]
Paripatyadar, S.A. [155], [363]
Paris, J. [473]
Park, G.-G. [340]
Park, S.-H. [340]
Patel, P.S. [339]
Patil, C.S. [341]
Paturzo, L. [40]
Pavlova, S. [530]
Pedernera, M.N. [342]
Pedersen, K. [236], [385], [423], [467]
Pedersen, P.S. [96]
Peppley, B.A. [343]
Percival, G. [125], [126]
Percy, D.A. [139]
Perderau, M. [344]
Periana, R.A. [345]
Perregaard, J. [4], [529]
Perrin, M. [355]
Perry, S. [295], [488]
Peters, R. [182], [183]
Petrich, G. [445]
Phillips, T.R. [346], [347]
Piccoli, V. [438]
Pichler, H. [348]
Pickles, J. [253]
Pierucci, S. [92]
Pina, J. [342]
Pinto, J.C. [14]

Author Index

Pitkethly, R.C. [174]
Pitt, R. [349]
Piwitz, M. [350]
Plácido, J. [351]
Plank, P.v.d. [433]
Platon, A. [134]
Plehiers, P.M. [352], [358]
Pointon, K. [123]
Poling, B.E. [353]
Ponec, V. [354]
Prabhakar, B. [105]
Prausnitz, J.M. [176], [177], [269], [353], [447]
Prettre, M. [355]
Price, J.G. [477]
Primdahl, I.I. [107], [109], [111]
Pruden, B.B. [10], [429]

Q

Qin, D. [356]
Qiu, D. [488]

R

Råberg, L.B. [357]
Rabinovich, A. [90]
Rajput, A.M. [105]
Ramanarayanan, T.A. [121]
Rao, M.V.R [358]
Rasmussen, P. [197]
Rasmussen, P.B. [337]
Rass-Hansen, J. [106], [359]
Rastler, D.M. [502]
Ratan, S. [360]
Renon, H. [52]
Resasco, D.E. [474]
Rethwisch, D.G. [361]
Reutler, H. [464]
Rhine, J.M. [362]
Rhodes, C. [175]
Ricci, G. [92]
Richardson, J.T. [129], [155], [234], [363], [364], [365], [366], [367], [368]
Richter, B. [369]
Ridler, D.E. [303], [370]
Riley, K.L. [168]
Røen, S. [15]
Rogers, W.A. [488]

Rogov, V. [530]
Rohan, D. [175]
Rokni, S. [450]
Romanow, S. [265]
Rong, G. [536]
Rose, F. [371]
Rosenqvist, T. [372]
Rosin, H. [155]
Røsjorde, A. [335]
Ross, J.R.H. [18], [83], [373], [530]
Ross, L.L. [374]
Rossini, F.D. [375]
Rossini, S. [438]
Rossmeisl, J. [8], [264]
Rostrup-Nielsen, J.R. [2], [7], [15], [17], [21], [28], [47], [48], [49], [64], [108], [152], [205], [226], [236], [260], [264], [318], [359], [376], [377], [378], [379], [380], [381], [382], [383], [384], [385], [386], [387], [388], [389], [390], [391], [392], [393], [394], [395], [396], [397], [398], [399], [400], [401], [402], [403], [404], [405], [406], [407], [408], [409], [410], [411], [412], [413], [414], [415], [416], [417], [418], [419], [420], [421], [422], [423], [424], [425], [426], [427], [450], [451], [475], [491], [495], [521], [542], [547]
Rostrup-Nielsen, T. [414], [428]
Rosynek, M.P. [153]
Röth-Kamat, M. [285]
Roy, S. [429]
Ruan, L. [430]
Ruderson, R.B. [308]
Ruettinger, W. [188]
Russo, G. [496]
Rüter, B. [285]

S

Saadi, S. [431]
Sabatier, P. [432]
Sacthler, W.M.H. [433]
Sadri, E. [445]
Sadykov, V. [530]
Saeys, M. [526]
Sakamoto, T. [489]
Sakuma, K. [241]

Sales, J. [302]
Salge, J.R. [138], [147], [434], [435]
Salmi, T. [436]
Salso, K. [437]
Sanfilippo, D. [438] [439]
Saraf, D.N. [463]
Sarofim, A.F. [243]
Satoh, T. [345]
Satterfield, C.N. [440]
Say, G.R. [180]
Schaadt, A. [441]
Schacht, M. [445]
Schildhauer, T.J. [442]
Schinski, W.L. [99]
Schlünder, E. [43]
Schmidt, H. [183]
Schmidt, I. [529]
Schmidt, L.D. [53], [138], [147], [164], [206], [229], [242], [369], [434], [435], [444], [443], [511]
Schmieder, H. [445]
Schnadt, J. [277], [509]
Schneider, A. [535]
Schneider, C. [544]
Schnell, C.R. [162], [446]
Schoubye, P. [402]
Schrøder, M. [118]
Schulze, G. [447]
Schuster, J. [514]
Schwaab, M. [14]
Seglin, L. [448]
Sehested, J. [7], [48], [57], [226], [264], [359], [415], [423], [449], [450], [451] [452], [453]
Seier Christensen, P. [2], [111], [120], [187], [451]
Selimovic, A. [454]
Senderens, J-B. [432]
Seo, D.J. [340]
Seris, E.L.C. [455]
Seshan, K. [55]
Shabaker, J.W. [247], [249], [456]
Shakoor, A. [457]
Shannon, M.D. [175]
Shapatina, E.N. [28], [541]
Sheldon, R.A. [458]
Shelly, S.G. [508]

Shen, J.C. [363]
Shim, S.S. [264]
Shinjo, T. [489]
Shinnar, R. [549]
Shore, L. [188]
Shub, F.S. [271]
Shul, Y.G. [286]
Siegel, N.P. [156]
Sills, R.A. [193], [194]
Silva, L. [295]
Sickafus, E.N. [459]
Simonsen, S.B. [460]
Sinfelt, J.H. [462]
Singh, C.P.P. [463]
Singh, J. [464]
Sjåstad, A.O. [357]
Sjunneson, L. [454]
Skjøth-Rasmussen, M.S. [465], [466]
Skocypec, R.D. [234]
Skov, A. [205], [219], [392], [467]
Skúlason, E. [8]
Sljivancanin, Z. [333]
Sloan, J. [122]
Slovokhotova, T.A. [35]
Smith, J.M. [468]
Smith, T.A. [123]
Smits, R.H.H. [17]
Snagovskii, Y.S. [271]
Snoeck, J.-W. [469]
Soave, G. [470]
Sobyanin, V.A. [32]
Sogge, J. [471]
Soldiani, I. [306]
Soled, S.L. [180]
Soliman, M.A. [181]
Solymosi, F. [472]
Somerjai, G.A. [56]
Sorin, M. [473]
Spiewak, I. [185]
Stagg, S.M. [474]
Stahl, H. [475]
Stal, J.A. [503]
Stasinska, B. [70]
Steel, M.C.F. [373]
Steensgaard, I. [50], [430]
Stefanidis, G.D. [476]
Steffensen, G. [124], [337]

Stenby, F.H [528]
Stephan, J.J. [317]
Stephens, R.W. [322]
Steppan, J.J. [95]
Stewart, K.D. [322]
Steynberg, A.P. [477]
Stigsson, L. [478]
Stitt, E.H. [160], [479], [484]
Stolten, D. [183]
Stolze, P. [337]
Strøm, T. [471]
Strub, R.A. [480]
Stub Nielsen, C. [1], [2], [3], [4], [120] [510],
Studt, F. [8]
Sinbeck, D.R. [461]
Summers, W.A. [502]
Sundaram, K.M. [481]
Sussman, M.V. [482]
Svendsen, H.F. [288], [519]
Swann, A. [123]

T
Taarning, E. [106]
Takeguchi, T. [483]
Tamme, R. [524]
Tan, K.F. [526]
Tanaka, S. [272]
Tanaka, Y. [483]
Taskin, M.E. [160], [484]
Taube, D.J. [345]
Taube, H. [345]
Tavares, M.T. [16]
Taylor, C.E. [485]
Taylor, D.M. [99]
Taylor, P. [337]
Taylor, S.H. [246]
Teitmann, G.J. [549]
Temkin, M.I. [58], [271]
Tempelman, J.J. [140]
TerHar, I.W. [486]
Tessie du Motay, M.M. [487]
Theis, K.A. [285]
Thomson, S.J. [254]
Toledo, J.M. [127]
Tolmacsov, P. [472]
Tomita, T. [489]

Tonkovich, A.Y.L. [94], [295], [488]
Töpfer, H.J. [490]
Topp-Jorgensen, J. [491], [492]
Topsøe, F. [21]
Topsøe, H. [124], [216], [255]
Topsøe, H.F.A. [493], [494], [495]
Topsøe, N.-Y. [216]
Torbati, R. [496]
Törnqvist, E. [135]
Tøttrup, P.B. [152], [386], [497]
Toyir, J. [323]
Triebskorn, B.E. [259]
Trimm, D.L. [258], [304], [383], [498], [499], [500], [501], [545]
Tröger, L. [266]
Tropsch, H. [192]
Tsang, S.C. [122]
Tschauder, A. [183]
Tsipouriari, V.A. [178]
Tucho, W.M. [26]
Tucker, R.J. [362]
Tupy, S.A. [242]
Turk, B. [364], [366]
Turner, G.E. [346], [347]
Twigg, M.V. [129], [303], [364], [365], [366], [370]

U
Udengaard, N.R. [152], [475], [502], [503], [504]
Ullmann, S. [47], [179]

V
van der Kooi, H.J. [326]
van der Lee, M.K. [505]
van Dillen, A.N. [505]
van Grootel, P.W. [506]
van Hardeveld, R. [507]
van Ness, H.C. [468]
van Reijnen, L.L. [18]
van Santen, R.A. [279], [506], [508]
Vang, R.T. [51], [277], [509]
Vannby, R. [510]
Vannice, M.A. [79], [80], [81]
Venables, S.C. [187]
Venkataraman, K. [511]
Venvik, H. [26]

Verkaak, M.J.F.M. [151]
Vernon, P.D.F. [27]
Verykios, X.E. [147], [178], [252], [298], [538]
Vestergaard, E.K. [509]
Veyo, S.E. [204]
Villadsen, J. [512]
Vlachos, D.G. [476]
Voecks, G.E. [513]
Vogel, A.P. [477]
Vogel, J.R. [486]
Vohs, J.M. [134]
Volkova, G.G. [32]
Vortmeyer, D. [514]

W
Wagner, H. [186]
Wagner, W. [515]
Wainwright, M.S. [258]
Waite, R.J. [33]
Waldron, W.E. [95], [317]
Walker, G.T. [162]
Walmsley, J.C. [26]
Wanat, E.C. [511]
Wang, D. [191]
Wang, J. [191]
Wang, Y. [94], [103], [134], [191], [237], [488]
Wang, Y. [536]
Ward, A.M. [175]
Wärnå, J. [436]
Watson, A.M. [222]
Webb, G. [254]
Wei, J.M. [516], [517]
Weisz, P.B. [518]
Wesenberg, M.H. [519]
Westerterp, R. [520]
Whittle, D.M. [175]
Wietzel, G. [546]
Wijngaarden, R.J. [520]
Williams, B.P. [246]
Williams, K.A. [242]
Winter-Madsen, S. [2], [171], [426], [521], [522]
Wise, H. [245], [314], [523]
Wix-Nielsen, C. [504]
Wolfson, D. [360]

Wood, J. [322]
Wörner, A. [524]
Wrag, D. [140]
Wrisberg, J. [382], [547]
Wu, Q. [101]
Wyrwas, W. [86]

X
Xu, J. [526]
Xu, J. [525]
Xu, M. [527]
Xu, Y. [333]

Y
Yamazaki, Y. [272]
Yan, W. [528]
Yang, Q.Y. [98]
Yaripour, F. [529]
Yarwood, T.A. [126], [347]
Yaseneva, P. [530]
YatesJr., J.T. [45]
Yoon, W.-L. [340]
Yoon, Y.-G. [340]
York, A.P.E. [122]
Young, N. [246]
Yurchak, S. [531]

Z
Zdonik, S.B. [532], [533]
Zeng, Z. [534]
Zhang, J. [535]
Zhang, J. [539]
Zhang, L. [539]
Zhang, P. [536]
Zhang, Q. [537]
Zhang, Z. [538]
Zhao, J. [367]
Zhou, C. [293]
Zhu, Q. [101], [537]
Ziemecki, S.B. [540]
Zyskin, A.G. [28], [541]

Subject index

acetic acid, 29, 106, 117, 118, 126, 145
acetylene, 13, 42, 55, 230, 237, 238
acidic supports, 228
activation energy, 204, 207, 210, 211, 218, 229, 238, 272, 285, 299, 306, 307
active sites, 304
ADAM-EVA, 135, 179
adsorption energy, 302, 306, 307, 308
affinity for carbon, 247, 248, 289
Ag. See silver
ageing, 274, 275
air, 12, 36, 38, 42, 43, 48, 49, 53, 63, 70, 82, 83, 87, 93, 96, 98, 102, 103, 105, 111, 114, 116, 130, 139, 140, 150, 152, 178, 183, 259, 274, 275, 284, 293
Air Products, 48, 49, 129
air separation, 48
Al, 121, 218
alcohols, 15, 29, 51, 54, 97, 106, 119, 145
alkali, 34, 209, 213, 214, 228, 232, 260, 261, 262, 272, 293, 295, 296, 297, 299
alumina, 67, 69, 213, 214, 218
  β-, 214, 261
  γ-, 213, 216
  η-, 216
ammonia, 4, 29
ammonia decomposition, 139, 297
ammonia plant, 36, 113, 114, 115, 116, 144, 173, 280, 281
ammonia production, 6, 7, 8, 112
ammonia synthesis, 3, 8, 80, 113, 114, 115, 126, 134, 144
Andritz, 133
anode, 96, 97, 101, 102, 104, 179
antracite, 56, 57
approach to equilibrium, 23, 31, 46, 78, 189, 198, 199, 325, 326

aromatics, 15, 36, 40, 85, 118, 212, 230, 234, 259, 260, 264, 267, 268
arsenic, 293
ash, 56, 58, 59, 60, 63
associated gas, 5
ATR, 14, 39, 40, 41, 42, 43, 47, 48, 49, 50, 54, 109, 114, 119, 120, 121, 122, 123, 128, 129, 130, 132, 161, 179, 246
Au. See gold
automotive, 6, 53, 70, 95, 97
autothermal reforming. See ATR
axial temperature profile, 211, 292
B. See boron
B_5 sites, 297, 298, 302
back diffusion, 219
BASF, 41, 143
Benfield, 70
benzene, 13, 106, 211, 212, 237, 267
BGC, 105, 106
BGL, 60
BGL gasifier, 60
Bi. See bismuth
bimetallic catalysts, 296
binding energy, 310
biodiesel, 6, 54, 117, 118
biogas, 105, 119
biomass, 3, 6, 15, 39, 44, 53, 55, 57, 63, 64, 123, 133, 134, 135, 136
Biot, 170
bismuth, 263
bituminous carbon, 56
bituminous coal, 57
black liquor, 63, 64
boron, 263
Boudouard reaction, 241, 292, See CO decomposition
breakthrough curves, 281
British Gas, 106, 144
Brønsted–Evans–Polyani, 308
brown coal, 56

369

burner, 33, 39, 40, 42, 43, 131, 146, 149, 151, 152, 153, 180, 181, 183, 184, 185, 186, 187
C adsorption, 310
C1 chemistry, 10
calcium, 293
calcium aluminate, 230, 261
calcium oxide, 143
capture, 88
carbide, 68, 69, 202, 228, 229
carbon activity, 254
carbon adsorption, 308
carbon dioxide, 3, 14, 97, 111, 143, 212
carbon formation, 17, 28, 34, 36, 54, 106, 109, 112, 131, 135, 136, 139, 145, 146, 161, 162, 168, 212, 214, 219, 233, 236, 237, 238, 239, 240, 245, 246, 247, 248, 249, 252, 253, 254, 255, 256, 257, 258, 260, 262, 263, 264, 273, 274, 288, 289, 290, 291, 296, 302, 303, 304
carbon limit, 107, 112, 145, 245, 248, 249, 251, 252, 253, 255, 256, 269, 290
carbon limit diagram, 108, 250
carbon limit temperature, 172, 245, 246, 251, 252, 253
carbon monoxide, 3, 97, 221, 235, 238, 241, 245, 247, 292, 308
carbon profile, 273
carbon reactions, 25, 26, 234
carbon whisker, 235, 263, 274
Carbona, 63
Carnot, 84, 99, 100, 101
catalysed hardware, 99, 102, 148, 189
catalyst activation, 217
catalyst activity, 36, 64, 147, 148, 149, 152, 172, 187, 228
catalyst particle models, 197
catalyst shape. *See* particle shape
catalytic burner, 104
catalytic partial oxidation, 39, 43, *See* CPO
C-C bonds, 15
CCS, 10, 38, 55, 62, 63, 71, 86, 87, 95
Ce, 262, 263
ceramic membrane, 49
cerium oxide, 274

CFD, 42, 153, 159, 161, 162, 163, 168, 170, 173, 183, 187, 190, 191, 193
C-H bond, 10, 11
CH_4 decomposition, 25, 172, 244, 288
CH_4 profile, 197
chemical equilibrium constant, 324
chemical recuperation, 34, 94, 131, 138, 139, 140
chemisorption, 65, 66, 69, 220, 221, 277, 282, 284, 299
  CH_4, 299
  H_2S, 222, 223, 276, 278, 279, 285
  hydrogen, 221, 222, 223
  hydrogen sulpbide steam, 279
  hydrogen sulphide, 222
  oxygen, 280
  sulphur, 223
  water, 228, 260, 261, 262, 279, 296
Chemrec, 64
Chevron, 129
c-hexane, 267
chlorine, 56, 67, 293
chromia, 213
CH_x intermediates, 308
Claus process, 71, 123
cleaning and WGS, 123
Co. *See* cobalt
CO decomposition, 25, 256
CO/CO_2, 123
CO_2 acceptor, 37
CO_2 emission, 10, 133
CO_2 footprint, 4, 6
CO_2 production, 58, 134
CO_2 reforming. *See* reforming CO_2
CO_2 removal, 76, 124
CO_2 sequestration. *See* CCS
CO_2-free power, 87
coal, 3, 5, 10, 13, 39, 53, 55, 56, 58, 59, 60, 61, 62, 63, 64, 67, 68, 69, 76, 86, 105, 123, 125, 127, 134, 137, 296
coal gas, 29
coal gasification, 55, 60, 76
coal gasifier, 58, 60, 64, 68, 127
cobalt, 128, 130, 237, 276, 310
coke, 8, 104, 190, 230, 231, 232, 233, 234, 260, 262, 270, 271, 272, 273, 274, 275, 293

coking, 105, 272, 273
combustion, 6, 34, 38, 41, 42, 55, 58, 61, 62, 74, 81, 82, 83, 84, 93, 99, 100, 111, 114, 115, 116, 139, 140, 146, 150, 152, 161, 178, 183, 184, 185, 190, 326
combustion engines, 6
compression, 95
concentration gradient, 203
consecutive reactions, 12
conversion per pass, 9, 11, 12, 14, 127, 130, 134
CoO, 217
copper, 26, 38, 51, 53, 54, 65, 66, 67, 68, 69, 70, 121, 217, 223, 242, 263, 276, 293, 297, 308, 309, 310
co-production, 85, 86, 126, 127, 136, 231
corrosion, 56, 131, 245, 246, 247, 292
COS, 56, 58, 60, 65, 67
CPA equation, 75
CPO, 13, 39, 40, 43, 44, 45, 46, 47, 48, 50, 51, 64, 86, 95, 96, 228, 287
cracking, 40, 230, 231, 272, 275
creep, 152
critical steam-to-carbon ratio, 255, 256, 258, 259, 269
crystal size, 216, 225, 251, 252
CSTR, 203, 229
Cu. *See* copper
cyclic process, 38
cyclohexane, 106, 237
d-bands, 306
deactivation, 8, 173, 190, 228, 267, 268, 275, 285, 293, 303
dealkylation
steam, 228
decane, 267
decoking, 56, 275
decomposition of carbon monoxide, 242
decomposition of methane, 239, 242, 243, 245, 253, 255, 303
dehydrogenation, 40, 271, 297, 300
density functional theory. *See* DFT
desulphurisation, 35, 66, 85, 114, 211, 275, 293
deuterium exchange, 295, 299

DFT, 112, 239, 300, 301, 302, 303, 304, 306, 307, 308, 309
diesel, 13, 30, 35, 85, 97, 99, 105, 118, 128, 130, 133, 211, 230, 261, 268, 287
diffusion coefficient, 192, 195, 196, 282
diffusion of nickel, 218
diffusion restrictions, 204, 282
dimethyl, 267
dimethyl carbonate, 117, 118
dimethylether. *See* DME
direct conversion, 10, 11, 13, 45, 96
direct reduction, 28, 29, 79, 110, 111, 172, 291
direct reduction of iron ore, 85, 110, 111
dispersion, 224, 227, 305, 323
axial, 161, 202
radial, 164
DME, 6, 29, 51, 53, 55, 79, 80, 118, 125, 126, 127, 136, 137
Dow, 60
d-states, 306
duty, 16, 36, 75, 83, 90, 93, 107, 116, 122, 175, 185
E Gas, 60
economy of scale, 6, 50
E-factor, 10
effective diffusion coefficients, 196
effective rate, 164, 200, 201
effectiveness factor, 147, 164, 177, 191, 193, 194, 197, 198, 200, 201, 205, 248
effectiveness factors, 198
efficiency, 4, 8, 9, 13, 59, 62, 63, 83, 84, 86, 88, 90, 91, 93, 99, 100, 101, 103, 104, 105, 112, 113, 115, 116, 117, 122, 130, 133, 134, 139, 140, 150, 183, 325
E-gasifier, 61
ejector, 269
electrolytes, 78
electron microscopy, 220, 221, 225, 238, 240, 274, 296
electronic factor, 307
encapsulation of catalyst, 234
energy balance, 81

energy consumption, 5, 8, 10, 88, 90, 91, 111, 112, 127, 144
energy diagram. See DFT
enriched air, 41
ensemble, 263, 297, 303
ensemble control, 256, 263, 297
enthalpy, 16, 17, 18, 19, 32, 79, 80, 81, 83, 92, 115, 116, 323
entropy, 18, 19, 82, 83, 100, 278, 309, 325
equilibrated gas, 108, 248, 252, 254, 256, 289, 290, 291
equilibrium constant, 17, 18, 20, 24, 25, 46, 78, 199, 200, 214, 216, 217, 234, 241, 242, 243, 254, 262, 263, 284, 304
equilibrium quotient, 246, 325
Ergun, 176
ethane, 12, 13, 208, 210, 211, 296
ethanol, 6, 15, 36, 44, 51, 53, 259
ethylene, 4, 13, 35, 53, 54, 55, 70, 230, 237, 259, 272
exchange studies, 209, 300
exergy, 82, 83, 84, 90, 91, 92, 100, 101, 115, 116, 135, 139, 181, 182
exit temperature, 34, 42, 48, 59, 83, 88, 95, 102, 106, 119, 120, 130, 181
Faraday equation, 99
Fe. See iron
film diffusion, 45
firing, 151, 153, 173, 181, 183, 186, 188
first principle. See DFT
Fischer–Tropsch, 137
Fischer–Tropsch synthesis, 7, 16, 29, 79, 127, 128, 130, 133, 134, 268
flame temperature, 178
flue gas, 34, 38, 62, 82, 83, 84, 87, 93, 109, 111, 115, 150, 151, 152, 159, 178, 179, 180, 181, 183, 184, 185, 186, 187
formaldehyde, 12, 117, 118
Foster Wheeler, 63
free energy, 16, 17, 18, 19, 25, 79, 81, 82, 83, 99, 216, 308, 309
FT diesel, 126, 129
FT synthesis. See Fischer–Tropsch synthesis

fuel cell efficiencies, 100
fuel cells, 4, 6, 7, 48, 49, 53, 79, 86, 95, 96, 97, 99, 100, 101, 118, 134, 287
fuel processing, 96, 97, 98, 104
fugacity coefficient, 17, 325
furnace, 33, 34, 94, 110, 111, 112, 116, 122, 146, 149, 150, 151, 153, 157, 167, 181, 182, 183, 184, 185, 186, 187, 188, 197, 228
furnace model, 183, 186
gas film, 190, 192, 202, 273
gas oil, 211, 231
gas turbine, 55, 127, 139, 140
gasification, 6, 55, 56, 57, 58, 59, 60, 61, 62, 63, 64, 69, 70, 76, 86, 97, 123, 124, 134, 137, 138, 255, 260, 274
gasoline, 13, 85, 117, 118, 124, 125, 127, 269
GE gasifier. See Texaco
Gibbs energy. See Gibbs free energy
Gibbs free energy, 82, 323
glucose, 54
glycerol, 55
gold, 157, 246, 263, 302, 303, 308, 310
graphene islands, 301, 304
graphite, 25, 241, 242, 244, 245, 248, 250, 252, 254, 323
Group VIII metals, 54, 135, 136, 209, 227, 228, 275, 309, 310
GTL, 4, 118, 128, 129, 130, 131, 132, 133, 138, 179
gum formation, 233, 234, 270, 292
H/C ratio, 55, 57, 58, 119, 290
H_2/CO, 125
H_2/CO ratio, 26, 27, 58, 67, 108, 119, 125, 130
H_2S, 56, 58, 60, 65, 66, 67, 69, 70, 71, 76, 77, 78, 85, 115, 124, 134, 222, 223, 272, 276, 277, 278, 279, 281, 282, 284, 286, 287, 290, 291, 302
H_2S wash, 123
HCN, 58, 60, 65, 76, 78
HDS, 35, 96, See hydrodesulphurisation
heat flux, 34, 35, 36, 93, 149, 151, 152, 158, 166, 167, 168, 171, 173, 174,

175, 181, 182, 184, 185, 188, 189,
190, 201, 219, 228
heat flux profile, 35, 151, 168, 182
heat flux profiles, 151
heat of reaction, 18, 31, 32, 45, 51, 103,
135, 138, 139, 146, 153, 181, 201,
323
heat pipe, 138
heat recovery, 14, 34, 61, 85, 96, 115,
133, 135, 178
heat transfer, 163, 173
heat transfer coefficients, 157, 172,
174, 175, 178
heating value, 56
heavy oil, 39, 61, 86
Henry's law, 77, 78
heptane, 55, 106, 210, 212, 236, 237,
257
Hg. *See* mercury
higher hydrocarbons, 13, 15, 16, 20, 35,
36, 42, 104, 107, 131, 201, 210,
211, 212, 219, 233, 234, 253, 255,
257, 260, 262, 264, 269, 270
high-temperature shift, 67, 68
hot bands, 147
hot tubes, 34, 147, 233
HRTEM, 238, 265, 302, 328
HTER, 94, 132, 328
HTS, 115
hydration, 214
hydrocracking, 85, 128, 129, 130, 260,
264, 265, 267
hydrodesulphurisation, 66, 85, 115
hydrogen, 4, 29, 126
hydrogen capacity, 221
hydrogen costs, 88
hydrogen economy, 4, 7, 86
hydrogen plant, 24, 32, 83, 89, 90, 91,
92, 93, 94, 95, 97, 112, 143, 179,
180, 197, 198, 266
hydrogen production, 91
hydrogen sulphide. *See* H2S
hydrogenolysis, 54, 210, 296, 297, 303
hydrotreating, 3, 85, 88, *See*
hydrodesulphurisation
ICI, 105, 106, 143, 144, 295
IGCC, 55, 62, 63, 126, 127, 136
indirect conversion, 10, 13

induction period, 239, 247, 257, 263
induction time, 239, 257, *See* induction
period
inlet temperature, 36, 68, 88, 135, 139,
178, 180
internal recycle reactor, 203, 206
intrinsic rate, 164, 165, 199, 200
investments, 9, 14, 55, 62, 86, 118,
130, 134
Ir. *See* iridium
iridium, 209, 240, 276, 305, 310
iron, 68, 69, 70, 85, 110, 111, 112, 223,
237, 276, 310
iron ore, 29, 110
isobars, 221, 277, 278
isotherm, 200, 277, 282, 283
ITM technology, 49
jet fuel, 230
Kellogg, 144
Kelvin equation, 244
kerosene, 30, 35, 261
kinetic severity factor, 230, 231, 232,
271, 273
kinetic severity function, 230, 273
kinetics, 199
    CO_2 reforming, 212
    ethane reforming, 208, 210
    intrinsic, 204
    methane reforming, 204, 207, 209,
    210
    micro, 205, 300
    n-heptane reforming, 210
    two-step, 208
Knudsen diffusion, 195, 196
Koppers Totzek atmospheric gasifier,
61
KSF. *See* kinetic severity factor
K_w, 204, *See* chemisorption, water
La, 262, 263
laboratory reactors, 161, 201
Langmuir, 204, 277
LEED, 223
Leva, 176
LHV, 4, 8, 30, 62, 80, 81, 82, 83, 84,
88, 90, 91, 100, 105
lignite, 56, 57, 60
liquid hydrocarbons, 44, 76, 105, 119,
212, 261

LNG, 5, 118, 128
logistic fuels, 86, 230
low-temperature shift. *See* LTS
LPG, 6, 118, 143
LTS, 67, 68, 69, 114, 115, 293
Lurgi, 43, 60, 105, 106
Lurgi–Sasol, 60
M, 29
M module, 119, 120, 135
magnesia, 34, 208, 209, 210, 212, 214, 215, 216, 218, 236, 261, 262, 297
magnesium alumina spinel, 43, 204, 206, 207, 220, 226, 240
mari. *See* most abundant reaction intermediate
Martin–Hou equation, 80
mass velocity, 150, 165, 170, 175, 203
Maxted model, 285, 288
MCFC, 97, 99, 101, 102, 103
MEA, 70
membrane reforming, 94
MeOH, 126, 137
mercury, 56
metal dusting, 131, 245, 246, 247, 256, 292
methanation, 37, 51, 55, 97, 105, 114, 134, 135, 136, 137, 138, 139, 143, 265, 269, 292, 296
methane, 55, 57, 101
methane chemisorption, 300
methane decomposition, 239, 247, 252, 253, 255, 304
methanisation, 115
methanol, 4, 29, 51, 53, 55, 127, 137
methanol plant, 7, 118, 119, 121, 122, 123, 127
methanol synthesis, 10, 13, 68, 80, 118, 123, 124, 126, 127, 128, 130, 133, 136, 293
$MgAl_2O_4$. *See* magnesium alumina spinel
MgO. *See* magnesia
micro-reactor, 146, 149, 159, 161, 162, 202
Midrex, 110, 111, 291
millisecond reactor, 44
molybdenum carbide, 228
molybdenum sulphide, 69

monoliths, 44, 148
monotube pilot, 107, 144, 172, 173, 232, 292
most abundant reaction intermediate, 207, 255
MTBE, 117, 118
MTG, 118, 124, 137
N_2, 43, 60, 76, 77, 78, 81, 114, 117, 272, 296, 297, 302, 306
nanotube, 234, 235, 238
naphtha, 3, 35, 52, 57, 66, 75, 92, 105, 106, 134, 143, 144, 211, 212, 230, 231, 258, 266, 267, 269, 272, 281
natural gas, 3, 4, 5, 6, 9, 10, 12, 13, 14, 30, 52, 61, 62, 65, 66, 76, 78, 86, 87, 89, 90, 92, 97, 98, 99, 100, 101, 104, 105, 106, 107, 110, 111, 112, 113, 114, 116, 117, 118, 119, 120, 123, 128, 130, 134, 135, 138, 139, 143, 144, 152, 178, 219
n-butane, 210, 211, 237, 240, 262
Nernst equation, 101
Newton–Raphson method, 19
NH_3, 58
NH_3 plant, 259
NH_3 synthesis, 115
n-hexane, 106, 237
Ni, 240, 276, 305, 308, 310
Ni particle size, 240
Ni,Au, 263
Ni,Cu, 242
Ni,Sn, 263
Ni/Mg, 217
$Ni/MgAl_2O_4$, 220, 226
Ni/MgO, 212, 261, 264, 297
$NiAl_2O_4$, 217, 218
nickel carbide, 235
nickel catalyst, 35, 202, 218, 219, 225, 247, 263, 272, 274, 293
nickel crystal, 225, 235, 241, 252
nickel crystallite size. *See* nickel particle size
nickel magnesium oxide, 216
nickel oxide, 216, 218
nickel particle size, 220, 239, 243, 244, 251
nickel sulphide, 222

nickel surface area, 33, 219, 221, 222, 224, 297
nitrogen, 3, 48, 56, 85, 96, 114, 152
nitrogen adsorption, 298, 302
NO_2, 9
noble metals, 13, 70, 107, 108, 240, 243, 252, 256, 264, 287, 302, 304
non-metal catalysts, 36
NO_x, 62, 152, 183
nucleation, 238, 239, 245, 246, 257, 260, 263, 288, 290, 291, 300, 302, 303, 304
Nusselt number, 169, 190
O adsorption, 310
O.N.I.A., 105
octopus carbon, 241
oil, 5
oil sand, 55
olefins, 13, 29, 35, 40, 85, 106, 117, 231, 232, 234, 245, 259, 272, 295
orimulsion, 57
orthogonal collocation, 168, 194
Otto, 105
oxidation degree, 111
oxidative coupling, 13, 36, 259
oxy-fuel, 62
oxygen, 3, 13, 15, 36, 38, 40, 41, 42, 44, 46, 48, 50, 51, 54, 58, 59, 61, 62, 63, 81, 99, 103, 110, 111, 114, 119, 123, 130, 131, 207, 247, 259, 274, 279, 287, 299, 302, 309
oxygenolysis, 15
PAFC, 97
palladium, 53, 95, 209, 217, 228, 240, 243, 305, 308, 310
partial oxidation, 26, 39, 40, 43, 44, 48, 50, 67, 95, 97, 98, 99, 106, 130, 228, 271
particle diameter, 148, 163, 170, 173, 177, 191, 200, 223, 323
particle shape, 147, 159, 169, 173, 177, 191, 193
particle size, 147, 161, 174, 177, 223, 225, 238, 243, 244, 252, 304
Pd. See palladium
Peclet number, 165, 166, 169
PEMFC, 29, 97, 98, 101, 104
petcoke, 39, 55, 56, 123

phosphor, 293
platinum, 54, 97, 157, 209, 210, 217, 223, 228, 240, 262, 263, 276, 305, 310
poisoning, 64, 233, 268, 276, 281, 285, 288, 293
pore condensation, 211
pore diffusion, 281
pore volume distribution, 195, 196
porosity, 195, 282
potassium, 67, 231, 232, 239, 260, 263, 299, 300, 302, 303
potential energy diagram, 308
power law, 208
POX. See partial oxidation
Prandtl number, 170
pre-exponential factor, 261, 299, 300
preheat temperature, 35, 48
preheater. See inlet temperature
Prenflo, 60
Prenflo gasifier, 61
preparation methods, 224, 261
prereformer, 35, 36, 42, 66, 88, 91, 93, 104, 105, 106, 114, 122, 129, 131, 146, 167, 199, 201, 231, 251, 252, 254, 267, 280, 281
prereforming, 266, 269
pressure drop, 43, 146, 147, 148, 149, 153, 162, 164, 167, 168, 171, 173, 176, 177, 194, 233, 323
principle of actual gas, 254, 255, 258, 291
principle of equilibrated gas, 68, 247, 248, 250, 251, 252, 256, 257, 269, 289, 290, 291
process analysis, 80
production costs, 88
promotion, 68, 255, 260, 261, 262, 272, 297, 299
propane, 143
propylene, 4
PROX, 97, 98
PSA, 88, 89, 91, 92, 94, 96, 97, 126, 179
Pt. See platinum
Pt, Re, 263
p-xylene, 267

pyrolysis, 13, 36, 42, 64, 233, 234, 270, 272
pyrolytic carbon, 270
pyrolytic coke, 271
pyrometer, 157, 159, 187
quadratic programming method, 157
R/P ratio, 5
radial concentration gradients, 253
radial gradient, 253
radial temperature, 172
radial temperature gradient, 161
radiation, 34, 157, 158, 169, 178, 181, 183, 184, 185, 197, 327
radical abstraction, 11
radical reactions, 42
Raoult's law, 77
rate-determining step, 65, 209, 238, 300
reaction scheme, 309
reaction sequence, 208, 259, 260, 262, 263
reactor
   micro-channel, 190
   micro-scale, 190
reactor model
   heterogeneous, 160, 161, 162, 164, 179, 191, 202
   one-dimensional, 160, 161, 202
   two-dimensional, 156, 160, 161, 162, 163, 166, 168, 172, 179, 182, 193, 253
Recatro, 105, 106
Rectisol, 70, 124
recycle, 269
recycle ratio, 11, 135
reducing gas, 172, *See* direct reduction
refinery gas, 92
reformer
   adiabatic, 106, 146, 149, 161
   autothermal, 109, 122, 123, 131, 146, 149, 167, 214
   bayonet, 94, 178
   convective, 34, 51, 93, 94, 98, 116, 122, 131, 132, 138, 146, 147, 152, 153, 167, 178, 179, 180, 181, 189, 245
   convective heat exchange, 34, 36
   external, 102
   fired, 94, 101, 102, 117, 157, 183, 185, 188, 190
   heat exchange, 93, 98, 138, 139, 180
   heated, 153
   high flux, 252
   hydrogen, 149, 254
   methanol, 52
   micro-channel, 189
   monotube, 107, 291
   multichannel, 189, 190
   naphtha, 272
   plate type. *See* reformer:micro-channel
   primary, 115, *See* reformer:tubular
   secondary, 50, 114, 115, 116, 122
   steam, 44, 68, 75, 112, 115, 122, 146, 147, 148, 152, 153, 155, 159, 161, 182, 187, 197, 219, 225, 254, 259, 272, 281
   tubular, 24, 31, 32, 33, 34, 35, 36, 41, 47, 50, 52, 75, 78, 83, 84, 85, 89, 91, 94, 109, 114, 116, 119, 120, 121, 138, 143, 146, 147, 149, 150, 152, 153, 155, 158, 167, 170, 174, 178, 181, 182, 187, 197, 199, 201, 211, 213, 214, 218, 224, 231, 247, 257, 271, 280, 281
reformer furnace, 33
reformer tube, 94
reforming
   adiabatic. *See* prereformer
   air-blown, 87
   aqueous phase, 54
   autothermal, 26, 27
   butane, 262
   catalytic, 15, 85
   ceramic membrane, 49
   CO_2, 10, 28, 107, 109, 119, 139, 212, 228, 249, 260, 262
   convective, 94, 116, 133, 179
   convective steam, 180
   equilibrium, 95, 188
   ethane, 209
   external, 102
   gas heated. *See* reforming:convective

heat exchange, 139
high-pressure, 144
internal, 97, 99, 101, 102, 103, 105, 139, 229, 287
liquid phase, 54
membrane, 95
methane, 18, 20, 27, 94, 189, 190, 198, 247, 328
methanol, 52, 53
non-tubular, 36
primary, 116, 122
reaction, 15, 18
secondary, 119
solar, 138
stack heat integrated, 102
steam, 23, 27
stoichiometric, 28
sulphur passivated, 288, 290, 291, 292
thermal plasma, 230
tubular, 41, 50, 106, 109, 144, 150, 169, 211, 222
two-step, 41, 120, 121, 122
regeneration, 273, 274, 282, 283, 284
reheat scheme, 37
resid, 56, 57, 61
residence time, 13, 42, 44, 60, 230, 232, 234, 271, 273
resistance number, 267, 268
Reynolds number, 45, 170
Rh. *See* rhodium
Rheinbraun, 60
rhodium, 44, 209, 228, 240, 287, 305, 308, 310
Ru. *See* ruthenium
ruthenium, 223, 228, 229, 240, 264, 276, 305, 308, 310
Sasol, 129
SBA, 41
scale of operation, 49, 120
scale-up, 4, 153, 154, 159, 160, 171, 191
scaling model, 307, 308
Segas, 105
selectivity, 6, 8, 9, 11, 12, 14, 43, 46, 70
Selexol, 70, 124

separation, 9, 11, 43, 48, 49, 79, 91, 126, 130
sequestration, 87, 123, *See* CCS
shadowing effect, 271
shape factor, 177, 193, 326
Shell, 39, 60, 61, 127, 128
shift, 23, 91, 94, 126
shift reaction, 20, 23, 24, 26, 32, 43, 92, 130, 198, 205
Siemens, 60
silica, 213, 214
silver, 223, 227, 303, 308, 309, 310
sintering, 43, 68, 69, 135, 213, 220, 224, 225, 226, 268, 293
SiO_2, 242
SNG, 55, 59, 79, 126, 134, 135, 136, 144, 269, 296
SO_2. *See* sulphur dioxide
sodium, 138
SOFC, 49, 97, 99, 101, 102, 103, 104, 139, 140, 229, 287
solar energy, 138
soot formation, 39, 40, 42
soot precursor, 42, 43
sour shift, 69, 137
space velocity, 36, 146, 149, 150, 206
spalling, 273
SPARG, 108, 109, 246, 252, 292, 297, 303, 304
specific activity, 220, 225, 286, 297
specific rate, 210
spill-over, 262, 263, 295
SRK, 74, 75, 76
Standard Oil, 143
start-up, 53, 70, 95, 98, 147
steady-state activity, 247, 254, 255, 259
steady-state model, 155
steam, 131
steam adsorption, 263, *See* chemisorption, water
steam cracking, 4, 35, 231, 271, 272
steam export, 4, 92, 94, 131
steam generation, 75
steam refining. *See* refining
steam reforming of methane, 15, 17, 24, 32, 90, 200, 219, 227, 300, 301, 327
steaming, 214, 216, 274, 275, 284

steam-to-carbon ratio, 20, 27, 32, 42, 43, 53, 69, 88, 90, 106, 119, 121, 123, 130, 146, 175, 186, 234, 252, 257, 267, 274, 328
STEM, 279
step sites, 239, 297, 298, 299, 300, 302, 303, 304, 305, 306
sticking coefficient, 299, 300
STM, 223
stress corrosion, 67, 214, 261
substitute natural gas. *See* SNG
sulpbur coverage, 280
sulphur, 287, 302
sulphur capacity, 223
sulphur coverage, 277, 285, 289, 291, 292, 303, 304
sulphur dioxide, 9, 12, 65, 81, 284, 287
sulphur passivated, 107, 109, 112, 252, 263, 286, 288, 290, 291, 304
sulphur poisoning, 36, 145, 230, 233, 234, 241, 268, 275, 281, 286, 287, 303
superheated steam, 136
surface area, 221, 223, 224, 275
surface diffusion, 301
surface energy, 244
synfuel, 126
synfuel cycle, 137, 138
syngas composition, 29
synthesis
  once-through, 126
synthesis gas, 3, 4, 10, 13, 14, 16, 17, 18, 27, 28, 29, 35, 39, 41, 43, 44, 48, 49, 53, 55, 57, 58, 60, 62, 63, 64, 67, 68, 69, 70, 73, 74, 75, 76, 77, 79, 80, 81, 84, 106, 112, 114, 115, 116, 117, 118, 119, 122, 123, 124, 125, 127, 128, 130, 131, 134, 135, 138, 139, 145, 156, 161, 173, 226, 228, 231, 252
synthetic fuels, 3, 4, 6, 118, 127, 138
synthetic natural gas. *See* SNG
Tammann, 224, 225, 226
tar sand, 5, 55, 57, 86
TBR. *See* reformer:bayonet
TEM. *See* electron microscopy
Temkin identity, 204
Temkin isotherm, 277

temperature approach, 189, 198
temperature gradient, 152, 199, 201, 202, 203, 253, 301
temperature profile, 39, 136, 151, 153, 166, 170, 171, 173, 176, 188, 197, 230, 266, 267, 268, 273, 291
temperature window, 270
Texaco, 39, 60, 61
TGA, 201, 236, 237, 239, 240, 242, 248, 255, 257, 258, 261, 262, 264, 272, 274, 288, 289, 292, 296, 300
thermal conductivity, 163, 166, 169, 326
thermal cracking, 16, 36, 42, 190, 230, 259, 262, 270, 295
thermodynamics, 13, 14, 17, 26, 80, 81, 88, 110, 125, 145, 244, 247, 252, 276
Thiele modulus, 191, 200
thioether, 65
thiophene, 66
TIGAS, 118, 124, 126, 127, 137
TOF, 227, 228, 296, 297, 305, 310, *See* turn-over frequency
toluene, 228, 267
Topsøe, 41, 93, 94, 105, 106, 129, 143, 144, 154, 159, 178, 179, 236, 295, 300, 305, 311
tortuosity, 195
town gas, 105, 144
Toyo, 230
TPR, 266
TREMP, 135, 136, 138
trimethylbutane, 237
tube
  design, 152
  diameter, 36, 149, 158, 169, 174
  failure, 145, 152
  life, 157, 228
  wall, 120, 151, 152, 157, 158, 161, 164, 170, 174, 183, 197, 234, 235, 253, 254, 270, 271, 273
  wall temperature, 152, 153, 155, 157, 168, 171, 172, 173, 175, 181, 182, 189, 197, 219, 228, 233
tungsten, 228, 309
turn-over frequency, 220, 227
two-dimensional sulphide, 223, 279

two-step mechanism, 309
UNIQUAC model, 77, 80
vanadium, 57
view factor, 157, 158
void fraction, 147, 148, 163, 164, 177, 273
waste heat, 9, 33, 34, 56, 61, 92, 93, 101, 114, 115, 150, 151, 178
waste heat recovery, 61, 150
water. *See* chemisorption, water
water gas shift, 15, 26, 47, 67, 69, 70, 105, 134
wax, 128, 130
WGS. *See* water gas shift
whisker carbon, 108, 233, 234, 238, 241, 245, 246, 247, 250, 254, 260, 270, 273, 274, 290, 301, 302
whisker growth, 241, 255, 274, 304
Winkler, 60
Winkler gasifier, 60
WSA, 71, 329

X-ray, 216, 220
XRD, 68, 220, 221
yield, 8, 11, 12, 133, 230, 255
Z_{90}, 267
zinc oxide, 65, 66, 68, 69, 115, 121, 281, 293
zirconia, 231
Zn, 121
ZnO. *See* zinc oxide
ZrO_2, 262, 272
ZrO_2/K, 271, 272
α-Al_2O_3, 215, 217
β carbon, 265
γAl_2O_3, 217
ΔG_c, 245, 254, 289, 323